OXYGEN TRANSPORT TO TISSUE
Pharmacology, Mathematical Studies, and Neonatology

ADVANCES IN EXPERIMENTAL MEDICINE AND BIOLOGY

OXYGEN TRANSPORT TO TISSUE

Pharmacology, Mathematical Studies, and Neonatology

Edited by

Duane F. Bruley

Professor of Chemical Engineering
Clemson University
Clemson, South Carolina

and

Haim I. Bicher

Associate Professor of Pharmacology
University of Arkansas Medical Center
Little Rock, Arkansas

SPRINGER SCIENCE+BUSINESS MEDIA, LLC

Library of Congress Cataloging in Publication Data

International Symposium on Oxygen Transport to Tissue, Medical University of
South Carolina and Clemson University, 1973.
    Oxygen transport to tissue.

    (Advances in experimental medicine and biology, v. 37A-B)
    Editors' names in reverse order in v. 2.
    Sponsored by American Microcirculation Society and European Microcircu-
lation Society.
    Includes bibliographies.
    CONTENTS: [1] Instrumentation, methods, and physiology.—[2] Pharmacol-
ogy, mathematical studies, and neonatology.
    1. Tissue respiration—Congresses. 2. Biological transport—Congresses. 3.
Microcirculation—Congresses. I. Bicher, Haim I., ed. II. Bruley, Duane F., ed.
III. American Microcirculation Society. IV. European Microcirculation Society.
V. Title. VI. Series. [DNLM: 1. Biological transport—Congresses. 2. Oxygen—
Blood—Congresses. W1 AD599 v. 37 1973 / QV312 I613o 1973]
QP121.A1I56   1973   599'.01'24   73-13821

ISBN 978-1-4684-5091-0   ISBN 978-1-4684-5089-7 (eBook)
DOI 10.1007/978-1-4684-5089-7

The second half of the International Symposium on Oxygen
Transport to Tissue held in Charleston—Clemson, South Carolina,
April 22-28, 1973

© 1973 Springer Science+Business Media New York
Originally published by Plenum Press, New York in 1973
Softcover reprint of the hardcover 1st edition 1973

United Kingdom edition published by Plenum Press, London
A Division of Plenum Publishing Company, Ltd.
Davis House (4th Floor), 8 Scrubs Lane, Harlesden, London, NW10 6SE, England

INTERNATIONAL SYMPOSIUM ON
OXYGEN TRANSPORT TO TISSUE

HOSTS   MEDICAL UNIVERSITY OF SOUTH CAROLINA
Charleston, South Carolina

CLEMSON UNIVERSITY
Clemson, South Carolina

SPONSORS   AMERICAN MICROCIRCULATION SOCIETY

EUROPEAN MICROCIRCULATION SOCIETY

SUPPORT   MEDICAL UNIVERSITY OF SOUTH CAROLINA
CLEMSON UNIVERSITY
SCHERING CORPORATION
CIBA-GEIGY
INTERNATIONAL BIOPHYSICS CORPORATION
MEDISCIENCE TECHNOLOGY CORPORATION
BRISTOL LABORATORIES
BURROUGHS WELCOME
TRANSIDYNE GENERAL CORPORATION
SCIENTIFIC RESEARCH INSTRUMENTS
DAAD and DFG (GERMAN SCIENTIFIC AGENCIES)

INTERNATIONAL SOCIETY

OF

OXYGEN TRANSPORT TO TISSUE

PAST PRESIDENT - DR. MELVIN H. KNISELY, CHARLESTON, U.S.A.
PRESIDENT-ELECT - DR. GERHARD THEWS, MAINZ, GERMANY
SECRETARY - DR. HAIM I. BICHER, LITTLE ROCK, U.S.A.
TREASURER - DR. IAN A. SILVER, BRISTOL, ENGLAND

MEMBERS - INTERNATIONAL COMMITTEE

DR. HERBERT J. BERMAN, BOSTON, U.S.A.

DR. DUANE F. BRULEY, CLEMSON, U.S.A.

DR. BRITTON CHANCE, PHILADELPHIA, U.S.A.

DR. LELAND C. CLARK, JR., CINCINNATI, U.S.A.

DR. LARS-ERIK GELIN, GOTEBORG, SWEDEN

DR. JÜRGEN GROTE, MAINZ, GERMANY

DR. MANFRED KESSLER, DORTMUND, GERMANY

DR. DIETRICH W. LÜBBERS, DORTMUND, GERMANY

DR. DANIEL D. RENEAU, RUSTON, U.S.A.

DR. JOSE STRAUSS, MIAMI, U.S.A.

DR. WILLIAM J. WHALEN, CLEVELAND, U.S.A.

PREFACE

It can honestly be said that the scope and magnitude of this
meeting surpassed initial expectations with respect to the number
and quality of the papers presented. Our group has grown since
we last met in Dortmund in 1971. This is a good indication that
a spiraling of our interests has taken place with the effects of
the initial good work felt, not just in one corner of the globe,
but in all four. With such a start, it was only appropriate that
an international society was formed at the meeting to further
coordinate our mutual undertaking. Henceforth it shall be known
as the International Society of Oxygen Transport to Tissue.

A final note of acknowledgement should be made to those who
were in the supporting cast, not only in making the meeting in
Charleston and Clemson a success, but also in the compiling of
this book. Gratitude is due to Dr. Daniel H. Hunt for his efforts,
the end product of which you have in your hands. Considerable
service was rendered by Mr. Robert J. Adams, Mr. Buddy Bell and
Mr. Nathan Kaufman during the symposium itself. Much typing,
organizing and record keeping was done by our lovely secretaries,
Laura B. Grove, Muff Graham and Kaye Y. Zook.

Haim I. Bicher, MD, PhD

Duane F. Bruley, PhD

June, 1973

CONTENTS OF VOLUME 2   (VOLUME 37B)

Session III - THE PHARMACOLOGY OF IMPROVING
OXYGEN TRANSPORT TO TISSUE

Session IV - MATHEMATICAL STUDIES OF TISSUE OXYGENATION

PART B

Session V - OXYGEN TRANSPORT PROBLEMS IN NEONATOLOGY

Subsession: Placental Oxygen Transfer

Subsession: Oxygen Monitoring

CONTENTS OF VOLUME 1   (VOLUME 37A)

Session II - PHYSIOLOGY OF OXYGEN TRANSPORT TO TISSUE

Subsession: D P G

Subsession: Brain

Session III

THE PHARMACOLOGY OF IMPROVING OXYGEN TRANSPORT TO TISSUE

Chairmen: Dr. Lars-Erik Gelin, Dr. Konrad Messmer and

Dr. Mary P. Wiedeman

BLOOD CELL AGGREGATION AND PULMONARY EMBOLISM

John W. Irwin

Massachusetts Eye and Ear Infirmary; Massachusetts General

Hospital; Massachusetts Institute of Technology; Harvard University

INTRODUCTION

The lung is the organ of the mammal in which gas exchanges between inspired air and the blood take place. When an alveolus in a living lung is viewed under magnification of 200X, air and blood appear to be the main components, and the supporting stroma is rather skimpy. Blood flows through the pulmonary microcirculation, which consists of pulmonary arterioles, capillaries, and venules.

The movement of oxygen from inspired air to the hemoglobin with the erythrocyte involves the crossing of a number of barriers. There exists some conflict as to these barriers and their exact morphology, but potentially they are as follows: a layer of continuous epithelium or connective tissue between the alveolar air space and the capillary wall, endothelial cells of the capillary wall, basement membrane, plasma, and the red blood cell membrane. Information as to the amount and rate of oxygen diffusing through these barriers would be of interest, but exact data are not available. Krogh (1) showed that animal tissues are permeable to oxygen, and he measured the rate of diffusion in various tissues. He used two metal chambers, one 1.5 ml and the other about 50 ml capacity. The tissue tested separated the chambers. Blood filled both chambers. The blood in the larger chamber was saturated with oxygen at a time when the blood in the smaller chamber was oxygen-free. The time and amount of oxygen in the hemoglobin in the smaller chamber was measured to determine the diffusion constant and rate. This apparatus and modifications of it have been the tools generally used. More recently, microelectrodes have been used to measure oxygen tension. Duling and Berne (2) and Becher and Knisely (3) have done considerable work with various types of microelectrodes.

Blood contains a number of cellular elements: erythrocytes, white blood cells, and platelets. A number of proteins are in solution in plasma; these proteins under certain conditions can precipitate in the living animal. Aggregates of the cellular elements with or without protein precipitates could theoretically interfere with the pulmonary oxygenation of blood. An experimental model has been devised.

MATERIALS AND METHODS

The experimental animal was the rabbit. Methods of exposing the lung of living rabbits for microscopical observation have been described in detail by Irwin et al. (4).

Intravenous sodium pentobarbital was used in doses of 0.45 gram per kilogram of body weight. After tracheotomy, a cannula was tied into the trachea. A second cannula was inserted into the first, filling about half its lumen. The second cannula was attached by rubber tubing to an oxygen tank. Anesthesia and insufflation of oxygen suppressed respiratory movements.

A thoracotomy on the right between the 4th and 5th ribs exposed the middle lobe which was placed on a cushion of mammalian Ringer's solution, issuing from the tip of a quartz rod. Light from a 1000-watt tungsten lamp was transmitted down the rod to transilluminate the lung. Observations were made with various microscopes (48X to 900X).

For passive anaphylaxis, rabbits sensitized to bovine serum albumin were bled. Pools of serum from such blood were quantitatively analyzed for antibody. Then known quantities of antibody were injected intravenously into the prepared rabbit. The antibody injection was quickly followed by a calculated dose of specific antigen. During and after these injections, the pulmonary microcirculation was studied with microscopes. By varying the quantity of injected antibody, the severity of the reaction could be controlled.

In severe reactions immediately after cessation of the heart beat, the lungs were inflated with 10% formalin solution through the tracheal cannula. Samples of the lung were placed in paraffin blocks. Later, sections were made. These sections were stained by several methods: hematoxylin and

eosin, PAS and Weigert's, and phosphotungstic acid and hematoxylin.
These procedures have been described in detail by Walter et al. (5).

RESULTS

When the amount of antibody is very high (1000μg Ab-N per ml
serum), death occurs within a matter of one or two minutes after the
shocking dose of antigen. One does see aggregates of erythrocytes, white
emboli, and hyaline emboli in the vessels of the pulmonary microcirculation,
but the most obvious change is constriction of pulmonary arterioles and
venules to such a degree that the lung becomes pale.

When the amount of antibody is less (20 to 100μg Ab-N per ml serum),
the animal may survive an hour or may not even die. In these rabbits the
intravascular aggregates are predominant and include clumps of red cells,
aggregates of white blood cells and platelets, and hyaline emboli. The
aggregate that blocks the flow through some of the pulmonary arterioles,
capillaries, and venules appears to be the hyaline emboli. The white blood
cell and platelet emboli frequently pass through, and the red blood cell
aggregates tend to break up at the end arterioles.

In the histological sections emboli were found in arterioles and
venules of 200μ in diameter. These emboli were not attached to the wall.
Most of the emboli were hyaline but some had a ring of polymorphs about
them. Emboli frequently were found impacted in the pulmonary capillaries
and frequently distended these capillaries. The staining reactions of these
hyaline emboli, the deep staining with PTAH, the positive acid picro-
Mallory, and the retention of methyl violet with Weigert's suggest the
presence of fibrin. These stains, however, do not stain the whole embolus
uniformly so no doubt these emboli contain other substances.

DISCUSSION

It is tempting to think that the most efficient way for oxygen to enter
the hemoglobin of the red blood cell would be to have the elongated red
blood cell squeeze its way through the pulmonary capillary system. If some
of these capillaries are plugged by emboli, it would be logical to assume
that the efficiency of oxygenation of blood in lungs would decrease.
Perhaps this reasoning is correct in heavily sensitized rabbits which die
quickly on the injection of the specific antigen. In less sensitized rabbits
which survive the shocking dose, one might be a bit doubtful.

On occluding the trachea of an anesthetized rabbit with a hemostat, the cessation of its heart beat takes about ten minutes. A well-sensitized rabbit, however, can be dead (no heart beat) within a minute of the shocking dose. At least in anaphylaxis there would appear to be more than just the lack of oxygen.

SUMMARY

This model experiment in anaphylaxis suggests that emboli do have some effect on the oxygenation of pulmonary capillary blood. Data obtained, however, are not really quantitative. It would appear unlikely that the transport of oxygen from inspired air to hemoglobin in the erythrocyte will not be fully understood until oxygen measurements can be checked at each barrier in relation to time.

REFERENCES

(1) Krogh, A. The rate of diffusion of gases through animal tissues, with some remarks on the coefficient of invasion. J. Physiol. 52: 391 (1919).

(2) Duling, B.R., and Berne, R.M. Longitudinal gradients in periarterial oxygen tension in the hamster cheek pouch. Fed. Proc. 29: 320 (1970).

(3) Becher, H.I., and Knisely, M.H. Brain tissue re-oxygenation time, demonstrated with a new ultramicro oxygen electrode. J. Appl. Physiol. 28: 387 (1971).

(4) Irwin, J.W., Burrage, W.S., Aimar, C.E., and Chesnut, R.W. Microscopical observations of the pulmonary arterioles, capillaries, and venules of living guinea pigs and rabbits. Anat. Rec. 119: 391 (1954).

(5) Walter, J.B., Frank, J.A., and Irwin, J.W. Hyaline emboli in the microcirculation of rabbits during anaphylaxis. J. Exp. Path. 42: 603 (1961).

SLUDGED BLOOD, HUMAN DISEASE AND CHEMOTHERAPY

Edward H. Bloch

Department of Anatomy, The Medical School

Case Western Reserve University, Cleveland, Ohio

The fluidity of the blood can be reduced during the course of some human diseases and this change in the fluidity has been described under the generic name of "Sludged Blood" (e.g. Knisely and Bloch, 1942; Knisely et.al., 1947; Bloch, 1956). The potential consequences of such pathology for structures and functions have been demonstrated in experimental animals. It was found that sludged blood could produce thrombosis and infarction resulting in tissue degeneration or necrosis as well as producing symptoms (e.g. muscle pain)(e.g. Knisely et.al., 1945; Bloch,1953; Gelin, 1956; Fajers and Gelin, 1959; Stalker, 1967; Bicher and Beemer, 1967). Such pathology could be prevented or ameliorated by de-sludging the blood (e.g. Knisely et.al., 1945; Gelin, 1956; Zederfelt, 1965; Bicher, 1972). The need for specific chemo-therapy for de-sludging is essentially restricted to those diseases where no specific effective therapy against the etiological agent exists because when effective therapy is instituted it destroys the etiological agent and the blood is de-sludged. Thus the use of de-aggregation drugs is limited to acute processes such as trauma, burns, and incompatible blood transfusions and perhaps for such chronic diseases as 'degenerative' arthritis, alcoholism and during the course of some neoplasms. Non-specific therapy exists (e.g. low molecular dextrans) for de-sludging blood in such acute processes as trauma, and burns, so that the major therapeutic problem resides in the treatment of sludge in chronic diseases to reduce or prevent thrombo-embolism.

What is a drug to be effective against? An answer to this question can be derived at by having a clear conception of what "sludged blood" is. Sludged blood is the result of a complex of changes in the circulating blood, as observed principally in the microvascular system. The changes affect erythrocytes, leucocytes, platelets and plasma, usually by aggregating the cells and subsequently altering the colloidal properties of the plasma by decreasing its albumin content and increasing various globulin fractions (e.g. Bloch, 1956). Thus in the design of any effective de-sludging substance it is desirable to establish what component of the blood is involved. This may be difficult to establish in man as such low magnifications are used that the optical resolution is low, about 3 micra (2.7 micra) due to numerical apertures of 0.10 and 0.11. Thus erythrocyte aggregates are recognized most readily as they present the greatest contrast then anemia of a rather marked degree. It is more difficult to recognize "hyaline" aggregates and changes in plasma viscosity.

After establishing what component of a sludge is involved the interaction of the sludge with a microvascular system must be assessed as quantitatively as possible. This requires the measurement and identification of the vessel, and such intra-vascular components as the size, shape and plasticity of aggregates, the ratio of aggregates to free cells and the extent of their reduction in flow, and alterations in permeability of the vessel as well as the reactions of the vessel walls (see: Bloch, 1956, 1973 for a more complete description and illustrations of such measurements and how to record them). By applying such measurements to the course of a human disease it becomes possible to determine whether or not de-sludging might be indicated. On reaching such a decision a comparable disease in an animal is required not only for establishing the effectiveness of chemo-therapeutic agents but also to obtain insight in how the sludge affects such major organs as the liver, spleen, lung, brain and skeletal muscle. (It should be recalled that what is observed in the arterioles of the bulbar conjunctiva of man (probably the best site for assessing a sludge) is a statistical valid sample ⊛ of

---

⊛   A statistical valid sample can be secured by observing the blood flowing through the arterial system in any tissue or organ, with the exception of the pulmonary arterial system. Such a sample will be a statistically valid sample of all the arterial blood in respect to the presence or absence of cellular aggregates in vessels that are of similar dimensions and having similar flow velocities as those in the test site (e.g. the bulbar conjunctiva). Furthermore, the test site will depict accurately the visual composition of the blood flowing through the capillaries and venules in all those organs that are not involved in functions which involve the storage of blood like the liver, spleen and bone marrow, the primary site of disease, nor at those sites which are the seat of thrombo-embolism or sedimentation.

what is present in the circulation, but it does not necessarily
follow how the microvascular systems in specific organs respond to
a sludge).

In selecting an animal which is affected with a disease that
is comparable to that in man it is of course necessary to establish
whether or not the intravascular and vascular responses are
similar: They may not be. For example, the virus of human
anterior poliomyelitis may be transmitted to rhesus monkeys and
white mice and these animals will exhibit the signs and symptoms
of the disease in man but the intravascular reactions will be
dissimilar (Bloch, 1953). In man there are two phases in the
production of erythrocyte aggregation; one, during the acute phase
of the disease when nerve cell destruction occurs at which time
the blood is essentially a paste and the second, occurs during
muscle atrophy at which time showers of rather rigid red cell
aggregates produce intermittent embolization of arterioles.
However, neither in monkeys nor mice is there any significant
intravascular pathology throughout the course of the disease.
Another example of the difference in intravascular pathology
between a disease in man and experimental animals is exhibited by
malaria. All of the major malarias of man, vivax, malariae and
falciparum, produce extensive erythrocyte aggregation as well as
hyaline aggregates and anemia, as does knowlesi malaria in
monkeys (e.g. Knisely et.al., 1945; Bloch, 1956). In contrast
berghei malaria in mice (a malaria that is used extensively to
test antimalarial drugs) has minimal intravascular cellular
aggregation in spite of the fact that this malaria is just as
invariably fatal as knowlesi malaria is to monkeys (Bloch, 1973).
Furthermore, all the experimental malarias that have been examined
by the author have not produced an equivalent degree of erythrocyte
aggregation with a parasitemia that is as low as that in the human
disease (e.g. vivax malaria). These studies indicate that human
red cells are apparently more susceptible to aggregation than
those in experimental animals.

Another facet in the relationship of the circulation, disease
and an experimental model is illustrated by acute myocardial
infarction. This disease has been produced for example in rabbits
where tissue pathology and laboratory tests were similar to those
in man (see: Bicher, 1972 p. 30). However, the role of sludged
blood in inducing this disease in man is not as obvious as in the
experimental disease. Patients with acute myocardial infarction,
free of concurrent diseases, do not have intravascular pathology
for at least 24 to 72 hours following the onset of the disease
(Bloch, 1955). Thus it is somewhat dubious that acute myocardial
infarction in man is produced by sludged blood but this does not
imply that de-aggregating therapy may not be useful after onset
of infarction to reduce or prevent tissue damage by sludged blood.

Finally, de-aggregation may be as detrimental as the sludge if the surfaces of the de-aggregated cells are of such a nature as to be particularly prone to adhere to reticulo-endothelium. If they are, then with the improvement in linear velocity of flow more cells per unit time will be delivered to the phagocytes in the liver, spleen and bone marrow so that an extensive anemia and consequent reduction in blood volume may occur (Knisely et.al., 1945; Knisely, unpublished data).

From the above comments it is apparent that many factors need to be considered and numerous obstacles must be overcome in developing effective chemotherapy for de-aggregating sludged blood in chronic diseases. Of all the problems that exist the hope is that the physicochemical characteristics of the interfaces of the cells in the various sludges are similar enough so that it does not become necessary to produce specific therapeutic substances for each sludge.

Acknowledgements

The work on the malaria in mice was supported by
NIH Grant HE 09520.

References

Bloch, E.H. The in vivo microscopic intravascular and vascular
        reactions in acute poliomyelitis   Am. J. Med. Sciences 226:
        24-37, 1953.

Bloch, E.H. In vivo microscopic observations of the circulating
        blood in acute myocardial infarction. Am. J. Med. Sciences
        229: 280-293, 1955.

Bloch, E.H. Microscopic observations of the circulating blood in
        the bulbar conjunctiva in man in health and disease.
        Ergebnisse d. Anatomie u. Entwicklungsgeschichte 35: 1-98,
        1956.

Bloch, E.H. In vivo microscopic observations of the liver in mice
        infected with malaria. (Abstract of Motion Picture).
        Microvasc. Res., 1973. (In press)

Bicher, H.I. and Beemer, A.M. Induction of ischemic myocardial
        damage by red blood cell aggregation (sludge) in the rabbit.
        J. Atherosclerosis Res. 7: 409-414, 1967.

Bicher, H.I.   Blood cell aggregation in thrombotic processes.
     Charles C Thomas, Springfield, Illinois, 1972.

Fajers, C.M. and Gelin, L.E.   Kidney, liver and heart damages from
     trauma and from induced intravascular aggregation of blood
     cells.   Acta Pathol. Microbiol. Scand. 46: 97-104, 1959.

Gelin, L.E.   Studies in anemia of injury.   Acta Chirurgica Scand.,
     Supplement.210, 1956.

Knisely, M.H. and Bloch, E.H.   Microscopic observations of
     intravascular agglutination of red cells and consequent
     sludging of the blood in human diseases.   Anat. Rec. 82:
     426, 1942.

Knisely, M.H., Stratman-Thomas, W.K., Eliot, T.S. and Bloch, E.H.
     Knowlesi malaria in monkeys. I. Microscopic pathological
     circulatory physiology of rhesus monkeys during acute
     Plasmodium knowlesi malaria.   J. Nat'l. Malaria Soc. 4:
     285-300, 1945.

Knisely, M.H., Bloch, E.H., Eliot, T.S. and Warner, L.   Sludged
     blood.   Science 106: 431-440, 1947.

Stalker, A.L.   The microcirculatory effects of dextran.   J. Path.
     and Bact. 93: 191-212, 1967.

Zederfelt, B.   Rheological disturbances and their treatment in
     clinical surgery.   In: Part 4, Symposium on Bio-rheology,
     Pp. 397-408.   Ed. by Copley, A.L., Interscience Publishers,
     New York, 1965.

INTRAVASCULAR AGGREGATION OF BLOOD CELLS AND TISSUE METABOLIC DEFECTS

Lars-Erik Gelin

Department of Surgery I, University of Göteborg

Göteborg, Sweden

INTRODUCTION

Based on vital microscopy observations on the flow of blood, Knisely in the early 1940 stated the great importance of intra-vascular aggregation of blood cells for the pathology developing in different organs after shock, tissue injury and various diseases. The significance of aggregation of blood cells for exchange between blood and body cells as well as the exchange between body cells and blood for removal of CO_2 and metabolites is still under debate.

This presentation deals with cell distribution and oxygen transport when intravascular aggregation of blood cells is observed in the microcirculation.

FLOW DISTRIBUTION OF CELLS IN NARROW CAPILLARY TUBES

In 1931 Fåhraeus & Lindqvist made their classical experiment on flow of blood in narrow capillary tubes and found the hematocrit to decrease with decreasing diameter of the tube. This reduction in hematocrit was due to a more rapid flow rate of cells because of their orientation towards the faster axial stream. The diffe-rence in flow rate between cells and plasma indicated a separation which accelerated with increasing degree of aggregation of the cells. From this finding they assumed that aggregation was a favourable response in disease as it permits a more rapid trans-port of the oxygen carrying elements from heart to tissue.

The most striking microcirculatory phenomenon in the critically ill patient is, however, stagnation of aggregated cells in venules and sinusoids, a phenomenon which does not fit with this assumption. The straight capillary tube used by Fåhraeus only imitated the on-flow properties of blood but not the off-flow properties. Therefore another kind of a capillary model was elaborated for our experiments to study the effect of aggregation on the distribution of cell flow and plasma flow in a branching capillary system and in a postcapillary flow system. Semi-cylindrical channels, one longitudinal and another lateral, at right angels were made in two pieces of plexi-glass using an engraving technique in such a way that the length of the channels were exactly the same. When the two pieces were compressed by a screw-vice cylindrical cylindrical channels were formed. The diameters of the channels were varied between 500 mμ and 25 mμ. When blood was perfused through the capillary device a separation of cell flow and plasma

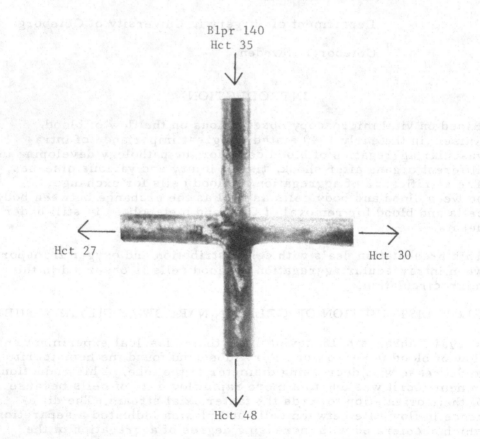

Fig. 1.   The streaming of blood through a branched capillary device with a diameter of 110 mμ.

flow became evident. A marked axial orientation of the cells in
the inlet channel appears in fig. 1. In the cross area some turbulence occurs. In the lateral channels there is plasma-skimming
and an axial orientation of the cells but the axis is dislocated
in the direction of initial stream. In the central outlet channel
there is no more an axial orientation of cells but a marked hemoconcentration. In horizontally placed outlet tubes connected to
the ends of the capillary device another phenomenon appears namely stasis of cells relative to plasma. Experiments with this
model show that there is a more rapid flow rate of cells on the
on-flow side but a slower flow rate of cells compared to plasma
on the off-flow side. When blood with marked aggregation of cells
flows through narrow tubes the cells travel even faster in the
axial stream on the on-flow side but much slower than plasma on
the off-flow side, which means stasis of cells in postcapillary
flow. This separation phenomenon and building up of stasis of
cells in postcapillary flow accelerated with decreasing diameter
of the tube, lowering of perfusion pressure, decreasing flow rate,
increasing aggregation tendency and increasing viscosity of blood
(Gelin 1961). In threatening gangrene and in the critically ill
patient this separation might result in hypostasis of cells which
can be demonstrated as long persisting pale pressure marks in a
red cyanotic skin, a condition we have named the "red shock"
syndrome.

OXYGEN CONSUMPTION IN SHOCK

Shock is defined as a low perfusion state either it is due to insufficient perfusion pressure, volume reduction, arteriolar hindrance,
decreased fluidity of the blood, heart failure or to primary asphyxia. The true internal environment is not reflected in the circulating blood unless there is a proper and adequate tissue perfusion.
This has long been recognized in hemorrhagic shock. In a surgical shock model we used 21 acutely splenectomised mongrel dogs
to provoke acute shock without blood loss by exteriorisation of
the small intestine (Bergentz et al. 1970). The intestines were
left outside of the abdominal cavity in wet towels for about two
hours. Blood pressure dropped during this 2 hours period from
an average of 140 mm Hg to about 70 mm Hg. After infusion of
Dextran 40 replacement of the gut and closure of the abdomen
blood pressure rose to an average of 110 mm Hg. In these animals both oxygen consumption and CO_2-elimination dropped markedly. In arterial blood there was only moderate changes in pH,
bicarbonate and PCO_2! There was a moderate increase of pyruvic acid and lactic acid. At the time of replacement of the gut
there was a sudden drop in pH, increase of PCO_2, pyruvic acid
and lactic acid, increase of oxygen consumption and CO_2-elimi-
nation. During the period of gut exteriorisation microcirculation

became markedly changed with severe aggregation of cells and venular stasis.

The important consequence of shock to the body as a whole is tissue anoxia through a lack of sufficient supply and an insufficient removal of metabolites both bringing on the gradual onset of death to body cells. Cellular metabolism looses its support because of inadequate exchange between tissue and blood. Blood during shock contains increased levels of glucose, lactate, pyrovate, amino acids, potassium and inorganic phosphate. Instead of an increased oxygen consumption brought about by an over-activity of the sympatho-adrenal system, a decreased oxygen consumption is most commonly recorded during shock, indicating not a decreased need of oxygen but a diminished availability of oxygen to the body as a whole. During shock there is often an even more significant decrease of CO_2-elimination. This change in the respiratory quotient indicates a decreased metabolic rate of the body cells, which however, does not fit with the increase of lactate, pyrovate and CO_2 in the blood. The poor tissue perfusion thus hides the metabolic defect from becoming apparent in the circulating blood. When an improved tissue perfusion was obtained there was a sudden increase both of oxygen consumption and of CO_2-elimination and a release of metabolites such as CO_2, lactic acid and pyrovic acid. Electrolytes released from the intracellular into the extracellular compartment are also hidden in shock (Haljamäe, 1970).

OXYGEN CONSUMPTION IN EXPERIMENTS WITH PROVOKED INTRAVASCULAR AGGREGATION

What significance did the aggregation and venular stasis have for the production of the observed metabolic disorders? To investigate the influence of rheologic disturbances on tissue metabolism and oxygen consumption experiments were performed in dogs on controlled respiration (Litwin et al. 1965). Pressures in aorta, pulmonary artery, left and right atrium were recorded and the cardiac output determined with the cardiogreen method. CO_2-elimination and oxygen consumption were estimated after sampling of respirated air in a Douglas sac. Arterial pH, PCO_2 and lactic acids were determined. The dogs were subjected to three experimental periods. After a control period of 2 hours the experimental dogs were given 1,5 g of Dextran 1000 and then followed for 3 hours. They were then given 3 g of Dextran 40 and followed for 2 hours. Following the infusion of Dextran 1000 intravascular aggregation and stasis built up in the microcirculation and viscosity increased, cardiac output decreased while aortic pressure increased. Right atrium pressure decreased indicating a poor venous return which explained the drop in cardiac output. The metabolic parameters showed that oxygen consumption as well as

carbon dioxide elimination decreased and that the respiratory quotient changed. There were, however, only moderate metabolic
changes in the arterial blood. After the infusion of Dextran 40
the carbon dioxide elimination suddenly increased as did oxygen
consumption and cardiac output and the microcirculation improved
with a release of stasis. A sudden acidosis appeared in arterial
blood with lowering of pH, increase of PCO_2 and lactic acid.
(Fig. 2) These experiments show that during a period of marked
aggregation with venular stasis both oxygen consumption and carbon dioxide elimination decreased without any marked metabolic
disturbances in the arterial blood. Hemodilution with Dextran 40
improved the impaired capillary flow, increased oxygen consumption and CO_2-elimination but induced acidosis in the arterial blood.
This effect of hemodilution showed that an accumulation of acid
material had occurred in the tissue during the period of aggregation and that the acid metabolites were not released to the blood
until aggregation and stasis was reversed.

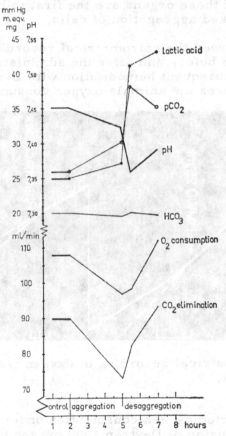

Fig. 2. Metabolic response after induced intravascular aggregation and disaggregation with Dextran 40.

Microscopic examination of organs from animals subjected to
either contusion injury, burn injury, cold injury or to the administration of high molecular weight dextran and sacrificed during
a period when they had a marked stasis of aggregated cells revealed the same type of morphologic damage namely central liver
necrosis, degeneration necrosis of tubular cells of the kidney and
anoxic damage of the anterior part of the pituitary gland. Animals
subjected to the same standardised injury but treated with Dextran
40 to counteract the flow disturbances which otherwise occurred
did not demonstrate these signs of organ damage (Fajers & Gelin
1959). Here it should be emphasized that there are three organ
systems in the body, namely the liver, the kidney and the anterior
part of the pituitary gland which are especially sensitive to damage from impaired flow properties of blood because they receive
their nutritional flow from a postcapillary flow system, i.e. the
portal flow to the liver, the postglomerular flow to the tubuli
ducts and the portal flow to the anterior part of the pituitary gland.
Therefore failure of these organs are the first to be seen as a
consequence of marked aggregation of cells.

Löfström 1959 followed the spirometrical recording of oxygen
utilisation in rabbits before and after the administration of
Dextran 1000 and subsequent hemodilution with Dextran 40. The
solid line in fig. 3 gives the animals oxygen consumption and the

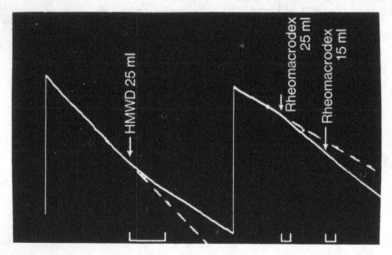

Fig. 3. Spirometrical recording of oxygen consumption in a
curarised rabbit.

dotted line the expected oxygen consumption under the two conditions. After the infusion of Dextran 1000 oxygen consumption
decreased while after infusion of Dextran 40 oxygen consumption
again increased. Similar experiments were performed by Hunt &

Zederfeldt 1968 on the local oxygen tension in healing wounds,
samples being obtained from an introduced microchamber in the
wound. Oxygen tension increased in the healing wound as with
time after infliction of the wound. If such animals were traumatized on day 25 after infliction of the wound oxygen tension in the
wound fluid decreased for the first 6 days and then increased
again. When treated with hemodilution with Dextran 40 this post-
traumatic drop of oxygen in the wound fluid was eliminated. This
indicates that hemodilution with Dextran 40 improved the oxygen
availability in the wound fluid despite a lowered hematocrit.

It is known that the microcirculatory insufficiency in shock varies
with different causes. In early phases of hemorrhagic shock ischemia from marked arteriolar constriction is characteristic while
venular stagnation of aggregated blood cells is more characteristic for shock due to tissue damage and sepsis. To investigate
the pattern of metabolic disturbance arising in renal tissue during
shock from acute hemorrhage and from contusion injury 33 rabbits
were used and divided into 3 groups, 10 for controls, 13 for hemorrhage and 10 for contusion injury (Gelin & Thomée 1973.
Kidney slices taken at different time intervals before and during
shock revealed that oxygen consumption was lower in kidney slices from animals in shock from contusion injury than in shock
from hemorrhage. (Fig.4) Lactic acid content was higher in kidney slices from animals in shock from contusion injury than in
shock from hemorrhage as was LDH production. This pattern of

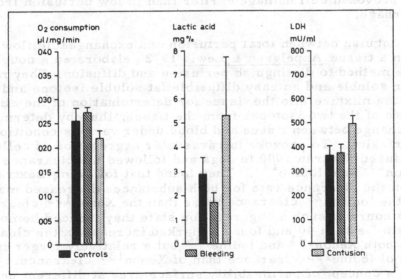

Fig. 4. Oxygen uptake, lactic acid and LDH production in kidney
slices from controls and rabbits in shock from hemorrhage and
from contusion injury.

metabolic disturbance in kidney slices matches the differences in microcirculatory flow pattern in the two shock models used. The acute hemorrhagic shock did not give rise to lactic acid accumulation or increased LDH production at this period of shock which the contusion injury did despite a higher perfusion pressure, which indicates that stagnation was more important than arteriolar constriction in producing cell damage. The metabolic activity in kidney slices was also followed for 2 hours after sacrifice. The control animals had a progressive decline in their kidney slices ability to use oxygen after death. Lactic acid increased progressively during the first 90 minutes after death but decreased thereafter while LDH-production increased during the first hour after death but decreased thereafter. Contusion injury was not followed by further increase in lactic acid or LDH production after death. Hemorrhagic shock was followed by a decrease of oxygen uptake and increase in lactic acid of kidney slices up to 60 minutes after death. These metabolic parameters indicate an increasing accumulation of metabolites and cell damage from non perfusion with time after death in the control animals. Two hours after death these changes were so severe that a large portion of the cells had lost their ability to metabolise. In the group dead after contusion injury there was severe metabolic defects already before death, hiding a further accumulation of metabolites from the non-perfusion period after death. In the group dead after hemorrhage there were signs of progressive damage from non perfusion after death. In relation to the microcirculatory disturbances observed it therefore seemed that aggregation of blood cells leading to venular stagnation provoked cell damage earlier than in low perfusion from hemorrhage.

To distinguish between total perfusion and exchangeable flow through a tissue Appelgren & Lewis 1972 elaborated a dougle isotope method to distinguish perfusion and diffusion. They mixed a water soluble and an easy diffusible fat soluble isotope and injected the mixture into the tissue for determination of the elimination rate of the two isotopes from the tissue, thereby determining the exchange between tissue and blood under various conditions of low perfusion. To provoke intravascular aggregation of cells they infused Dextran 1000 to dogs and followed the clearance rates of $Xenon^{133}$ and $Iodine^{125}$. They found that following Dextran 1000 infusion the clearance rate for both substances decreased markedly but the $Iodine^{125}$ clearance more than the $Xenon^{133}$ clearance. After 3 hours of such an aggregation state they induced hemodilution with Dextran 40 and found a marked increase in the clearance rate of both $Xenon^{133}$ and $Iodine^{125}$ but a relatively larger increase of $Iodine^{125}$ clearance than of $Xenon^{133}$ clearance. Using Renkin´s concept of permeability surface area at different perfusion rates they found in animals in shock and after infusion of Dextran 1000 a considerable decrease of the PS value. A subsequent hemodilution with Dextran 40 increased considerably the PS value both in case of shock and in case of aggregation from Dextran 1000. These data were interpreted as being consequences of changes in flow distribution.

In all discussions on disturbances of flow distribution in shock it must be kept in mind that such disturbances are the result of complex interactions between changes in perfusion pressure, vascular geometry, flow itself, viscosity of plasma, the sizes and the deformability of the blood corpuscles and their interaction with one and another and with the vascular endothelium. The similarity in oxygen utilization and CO_2-elimination in the shock models presented here make evident that in the presence of intravascular aggregation of cells, both after infusion of Dextran 1000 in hypervolemic animals and after contusion injury or other kinds of traumatic shock, the flow distribution of cells and plasma within the microcirculation is the dominant factor and more important than the volume of blood and the number of cells which are perfusing the tissue.

References:
K. L. Appelgren, & D. H. Lewis: Capillary Flow and Capillary Transport in Dog Skeletal Muscle in Hemorrhagic Shock. Europ. Surg. Res. 4:29, 1972.

M. H. Knisely & E. H. Bloch: Microscopic observations of intravascular agglutination of red cells and consequent sludging of the blood in human diseases. Anat. Rec. 82:426, 1942.

R. Fåhraeus & T. Lindquist: The viscosity of blood in narrow capillary tubes. Am. J. Physiol. 96:562, 1931.

L. - E. Gelin: A method for studies of aggregation of blood cells, erythrostasis and plasma skimming in branching capillary tubes. Biorheol. 1:119, 1963.

S. - E. Bergentz, A. Carlsten, L. - E. Gelin & J. Krebs: Hidden acidosis in experimental shock. Ann. Surg. 16:227, 1969.

Haljamäe, H.: "Hidden" cellular electrolyte responses to hemorrhagic shock and their significance. Rev. surg. 27:315, 1970.

M. S. Litwin, S. - E. Bergentz, A. Carlsten, L. - E. Gelin, C. - M. Rudenstam & B. Söderholm: Hidden acidosis following intravascular red blood cell aggregation in dogs. Effect of high and low viscosity dextran. Ann. Surg. 161:532, 1965.

B. Löfström: Intravascular aggregation and oxygen consumption. Acta Anaesth. Scand. 3:41, 1959.

L. - E. Gelin & M. Thomée: Oxygen consumption, lactic acid and LDH-production in kidney slices during shock. 7th Europ. Conf. Microcirc. Bibl. Anat. In Press.

ANTI-ADHESIVE DRUGS AND TISSUE OXYGENATION

H. I. Bicher, D. F. Bruley* and M. H. Knisely

Department of Anatomy, Medical University of South Carolina
Charleston, South Carolina, U.S.A.
*Department of Chemical Engineering, Clemson University
Clemson, South Carolina, U.S.A.

The intravascular agglutination of the formed elements of blood, which occurs under certain pathological conditions, results in peripheral circulation disturbances that may lead to blood flow stagnation and the plugging of small vessels with cell aggregates. [Bicher (1) and (2)] Of the blood cells involved, special attention has been given to the erythrocytes and platelets. These cells seem to possess the ability to start and accelerate the chain of events that leads to organic vessel occlusion.

The purpose of this paper is to assess the effect on oxygen transport to tissue of a group of drugs that share the common property of preventing both red cell and platelet aggregation, in vitro and in vivo, demonstrated with several objective methods devised or adapted to evaluate this action on a quantitative basis. Three different compounds showing these effects are introduced, and their possible action discussed, both at the platelet membrane function level and for the mechanism of their potential antithrombotic usefullness. These substances are called anti-adhesive drugs. [Bicher (2)]

An oxygen micro-electrode and an intra-arterial oxygen electrode were used to measure tissue and blood pO_2 in anaesthetized, curarized cats under positive pressure breathing. As a parameter for the ability of the circulation to oxygenate tissue, the 'reoxygenation time' (defined as the time required to reach the previous pO_2 level after a short period of anoxic anoxia) was determined on blood and cerebral cortex. The effect of inducing intravascular blood cell aggregation and reversing it with anti-adhesive drugs was thus evaluated.

MATERIALS

Adenosine diphosphate (ADP), Sigma Chemical Corporation, U.S.A. High molecular weight dextran, Pharmacia, Upsalla, Sweden. 2 Methyl 2 tert butyl-3-5-6 trioxotetrahydropyran (substance »86«) was synthetized according to Taub's method (7). 5-[1-(4-Chlorobenzoyl)-3 -indolylmethyl] tetrazole (BLR-743), Bristol Laboratories, Syracuse, U.S.A. 4-Isobutylcarboyl-5-carbobenzoxy-2, 3 dioxo-γ-lactone (substance »86-B«), supplied by Dr. M. Cais, Dr. W. Taub, and Dr. L. Vroman, Technion, Haifa, Israel. (Figure 1).

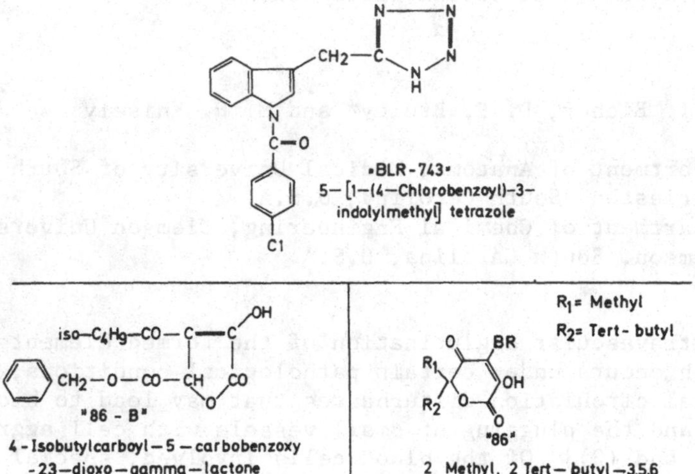

•BLR-743•
5-[1-(4-Chlorobenzoyl)-3-
indolylmethyl] tetrazole

"86-B"
4-Isobutylcarboyl-5-carbobenzoxy
-2,3-dioxo-gamma-lactone

R_1= Methyl
R_2= Tert-butyl

"86"
2 Methyl, 2 Tert-butyl-3,5,6
-trioxotetrahydropyran

Figure 1: Chemical structure of several anti-adhesive drugs.

METHODS

Oxygen Measurements: Tissue pO_2 was determined using our oxygen ultramicro electrode, as described elsewhere (8). Blood pO_2 was measured by placing our pO_2 catheter electrode (9), positioned in the aorta through the cannulated Femoral artery of the cats.

Determination of "re-oxygenation" time: When the respiratory gas administered to the animal is switched from air to 100% Nitrogen, pO_2 levels decrease in blood and brain. After giving N_2 for 60 seconds, the animal is again respired with air. The time necessary to reach pre-nitrogen O_2 levels is called "re-oxygenation" time.

Determination of Blood Cell Aggregation: Blood cell aggregation was determined using methodology previously described (2). The "rolling tube" test was used to measure platelet to glass adhesiveness. The "screen filtration pressure" (SFP) light aggrego-

meter and 'membrane capacitance' aggregometer were used to measure
platelet aggretation, and the erythrocytes sedimentation rate (ESR)
and direct microscopic observation of the bulbar conjunctiva were
used to determine degrees of red cell aggregation.

Operation:

Observations were made on forty cats. The animals were
anaesthetized with Nembutal (60 mg/Kg I.P.). Venous, arterial and
tracheal cannulations were performed. The head of the animal was
fixed in a David Kopf model 1730 stereotaxic instrument. The skin
over the skull was incised and the periosteum retracted. A trephine
hole was made in the skull over the frontal cortex and the dura
removed. The exposed surface of the brain was covered with liquid
paraffin. The electrode was fixed in a micromanipulator (David
Kopf, model 1760) and lowered on to the surface of the brain. Micro
movements of the electrode within brain tissue were achieved using
a David Kopf 1707B hydraulic microdrive. Measurements were made
only in the upper 3 mm of brain cortex.

In order to measure the brain 're-oxygenation time' the animal
was curarized using Flaxedil (1 mg/Kg I.V.) and maintained under
artificial respiration using a Harvard Apparatus (Dover, Massachusetts) Model 607 respiration pump. This procedure was necessary in
order to exclude respiratory compensation when nitrogen was administered instead of air. All gas mixtures were given to the animal
through the inspiratory inlet of the respiration pump. All experiments were performed in an electrically shielded cage.

Blood samples for the sedimentation rate and haematocrit
determinations were obtained from the cannulated artery.

Intravascular red cell aggregation was induced by the I.V.
injection of high molecular weight Dextran (1 g/Kg).

De-aggregation of the red cells was achieved by the I.V.
injection of substance »86«, 200 mg/Kg, »86B« and »BLR-743« 30 mg/Kg.

RESULTS

A. Prevention of Blood Cell Aggregation

Substance »86«, substance »86B« and »BLR-743« were active in
preventing platelet aggretation and adhesiveness in all tests performed in vitro. These results are summarized in Table I. Using
substance »86« as reference unit of activity, it can be concluded
that substance »86B« was five times more active than substance »86«
in preventing platelet aggregation in all tests, and »BLR-743« was

twice as active in preventing ADP induced platelet aggregation and ten times more active in preventing collagen induced platelet aggregation and platelet glass adhesiveness (Rolling Tube Test).

Table I. Relative potency of action of the anti-adhesive drugs-- in vitro experiments. Figures represent mean effective doses preventing blood cell aggregation in μg/ml.

DRUG	ESR	SFP	BORN (ADP)	(Collagen)	MCA	ROLLING TUBE
86	250	2,000	2,000	1,000	2,000	2,000
86B	25	400	400	-----	400	400
BLR-743	100	-----	1,000	100	1,000	100

EST	=	Erythrocyte sedimentation rate--modified Thorsen and Hint method.
SFP	=	Screen filtration pressure method.
BORN	=	Photoelectric method--platelet aggregation induced by ADP or collagen.
MCA	=	Membrane capacitance aggregometer--ADP induced platelet aggregation.
ROLLING TUBE	=	Platelet to glass adhesiveness.

In vivo, the mean i.v. effective doses preventing platelet aggregation and adhesiveness were 200 mg/Kg for substance »86« and 30 mg/Kg for substance »86B«. »BLR-743«prevented collagen-induced aggregation platelet adhesiveness and red cell aggregation at a dose of 30 mg/Kg, i.v.

B. Blood and Tissue pO₂--Re-oxygenation times

    1. Effect of intravascular red cell aggregation

    After intravascular red cell aggregation was induced by the I.V. administration of high molecular weight Dextran (1 g/Kg), sedimentation rates were maximally increased, from a normal of an average of 15 mm/hr up to 160 mm in but 15 min. Observations of the bulbar conjunctiva confirmed the presence of slow, stagnant flow, with red cell aggregates clearly visible both in the venular and arteriolar sides of the microcirculation.

    (a) Blood and tissue pO₂ levels. In most cases, blood from the Femoral artery and cerebral cortical pO₂ remained unchanged during the Dextran infusion. Six animals showed a transient decrease for 5-10 min, and then stabilized at the previous level.

(b) Blood re-oxygenation time. The blood re-oxygenation time was prolonged by an average of 80%. In some animals, re-oxygenation time almost doubled, while in others, it showed little change (Table II).

(c) Cerebral cortical re-oxygenation time. Cortical re-oxygenation time showed change after sludging. The actual time was prolonged an average of 100% (see Table III) and the secondary 'overshoot' phase disappeared.

2. Effect of anti-adhesive drugs

After the intravenous administration of anti-adhesive drugs, red cell aggregates disintegrated and the circulation became faster. The plasma spaces between red cell groups disappeared, especially as seen in the arterioles. Visible venular soft sludge remained in most cases. This effect lasted for 1 hr, after which time, our observations were discontinued. The most striking improvement occured during the first 15 min after injection.

The sedimentation of blood samples taken after high molecular weight Dextran administration, each read after 15 min of sedimenting,

Table II. Blood re-oxygenation times

Cat Number	Normal	Sludging	Anti-adhesive drug
1	245	450	280
2	360	350	300
3	325	375	330
4	240	325	270
5	250	330	280
6	225	250	210
7	250	275	245
8	250	275	---
9	200	225	175
10	225	275	225
11	245	280	245
12	260	320	270
Mean	256	311	257
S.D.	43.96	61.45	43.38
S.D.d		21.8	18.22
t test		-2.503	
P		$0.05 > P > 0.02$	$P > 0.5$

Each figure represents the mean of three determinations.

Table III. Tissue re-oxygenation times

Cat Number	Normal	Sludging	Anti-adhesive drug
1	90	135	92
2	110	150	115
3	70	145	85
4	165	345	100
5	50	125	75
6	100	150	110
7	80	125	75
8	75	135	80
9	50	95	60
10	80	160	93
11	105	180	80
12	82	135	90
Mean	88	157	88
S.D.	30.74	662.86	15.55
S.D.d		20.20	9.945
t test		-3.395	0.017
P		$P < 0.01$	$P > 0.5$

Each figure represents the mean of three determinations.

Table IV. Erythrocyte sedimentation rates

Cat Number	Normal	Sludging	Anti-adhesive drug
1	5	51	42
2	8	60	32
3	2	48	20
4	1	35	18
5	3	60	23
6	10	55	43
7	5	48	18
8	2	42	25
9	3	48	35
10	1	61	45
11	5	53	43
12	6	51	40
Mean	4	51	32
S.D.	7.80	7.66	10.66
S.D.d	---	2.36	3.18
t test	---	-19.849	-8.72
P	---	$P < 0.001$	$P < 0.001$

The sedimentation rates were read 15 min after setting up the tubes.

was prolonged; the rates of sedimenting were slowed down (see Table IV).

      (a)   Blood and tissue pO_2 levels. Both values remained unchanged after the intravenous injection of substance »86«.

      (b)   Blood re-oxygenation time. The Femoral artery blood re-oxygenation time returned to pre-sludging levels after the injection of substance »86«.

      (c)   Cerebral cortical re-oxygenation time. Cortical re-oxygenation time also was shortened, returning to pre-sludge values. In most experiments, the 'overshoot' phase was also restored.

DISCUSSION

The fact that all three compounds acted in the same way in preventing both red cell and platelet aggregation, indicated that this property of preventing blood cells from sticking to each other was more generalized and affected not only the blood platelets. O'Brien (10) described the ability of certain anti-malarials, local anaesthetics and imipramine derivatives to inhibit the ADP induced platelet aggregation. Constantine (11) reported that histamine possessed a similar effect and that this property was also shared by some antihistamines and anti-inflammatory compounds. O'Brien proposed the name of »anti-adhesive drugs« to include all chemicals sharing this type of activity.

Anti-adhesive drugs have been reported to induce platelet swelling also, probably by inhibiting a phosphoprotein kinase (12). They also affected the active uptake of serotonin by platelets, but not the passive histamine absorption (13). These results again suggest an interference with the functions of the platelet membrane. In previous publications,(1) and (2), we have reported a direct effect of drugs of this category on K^+ transport across the platelet membrane.

It is understandable that a poor capillary flow of red cells must impair tissue oxygen supply (14). Maintenance of blood volume and blood pressure does not guarantee adequate perfusion of the tissues with blood cells when the capillary flow is disturbed as described above. These pathological effects may be caused by several mechanisms, such as changes in peripheral blood flow, plugging of vessels and interference with red cell oxygen uptake or release.

The fact that cortical pO_2 levels were not changed or even slightly increased when intravascular red cell aggregation was induced should be attributed to the flow compensation mechanism of the brain, probably achieved in this case by vasodilation (15). Similar results during acute hypoglycemia were described by Silver (16). Obviously, the compensatory mechanisms responsible for maintaining an adequate cortical pO_2 were not overpowered by the degree of blood sludging we induced.

This reasoning also explains why the 're-oxygenation time' showed the effect of blood sludging on cortical oxygen levels. We can assume that during the re-oxygenation period, all the compensatory mechanisms were working at maximum capacity, or at least at maximum capacity for the given circumstances. Only cardiovascular or haematological regulatory mechanisms need be considered, as changes in respiratory mechanisms were excluded since the animal was curarized under artificial respiration.

If, with all compensatory mechanisms responding maximally, an additional load is placed on the system by the intravascular aggregation of the red cells, causing stagnant flow and plugging of some capillaries, then the 're-oxygenation time' should be accordingly prolonged.

Also, the re-oxygenation reaction seems to have two components, the quick pahse and the 'overshoot' (as described in the Results section). The effect of intravascular red cell aggregation seems to be mainly in damping the slow phase, which takes place then, mainly below the pre-nitrogen base line pO_2 levels.

If we consider, in view of the offered evidence, that the length of the re-oxygenation time and the presence of the 'overshoot' represents the ability of the blood circulation system to compensate for partial decreases in oxygen supply, then our methods provide a quantitative system to determine the deleterious effect that this pathological process may have for brain tissue survival.

The same effects that blood sludging has on the microcirculation throughout the body can also be applied for the lung circulation.

Direct microscopic observations of living lungs have shown that large cell masses, even though soft, plug the wide catch-trap pulmonary arteries for varying periods. When enormous numbers of masses plug pulmonary arterioles simultaneously, or within a brief period, death results immediately (17). It is significant that the lungs act continually as a sieve which catches, holds, and breaks up blood cell masses, thereby protecting all other parts of the body, including the heart wall itself, from embolization by masses too large and rigid to pass through lung capillaries.

De Bono & Gozetopoulos (18) have provided further objective
evidence of the crippling effect that intravascular red cell aggregation has on the pulmonary circulation. They demonstrated that
after polybrene-induced sludge, the plugging of pulmonary peripheral
vessels is so extensive that hypertension supervenes in the pulmonary
artery, with a concomitant fall in the systemic blood pressure.

These considerations explain the prolonged blood re-oxygenation
times that we recorded after blood sludging. Obviously, the pulmonary microcirculation became so impaired that the oxygenation process
was much delayed.

It is noteworthy that the tissue re-oxygenation times were
prolonged more than the blood re-oxygenation time after sludging.
These relults point to the fact that the tissue delay reflects the
delay, both in lung oxygen uptake by the erythrocytes, and its
subsequent release in the brain capillaries.

The possible use of anti-adhesive drugs in the treatment of
thrombosis had been implied (3) and (4). The present experiments
demonstrate that, by preventing intravascular red cell aggregation,
anti-adhesive drugs can counteract the impairment of the compensation ability of blood circulation to anoxia caused by the sludging,
thus improving tissue oxygenation.

SUMMARY

Both platelet and red cell aggregation have been related to
the etiology of thrombosis. This paper describes a new category
of drugs which prevent both types of aggregation and are devoid of
any other pharmacological activity. These substances are called
'anti-adhesive' drugs. A photoelectric method, a new 'membrane
capacitance' aggregometer, the screen filtration pressure method
and a new simple glass contact method were used to measure platelet
aggregation and adhesiveness. A sedimentation rate technique and
microcirculation observations determined red cell aggregation. The
anti-adhesive drugs counteracted blood cell aggregation in vitro
and in vivo. The effect of the anti-adhesive drugs on platelet
membrane mechanisms was determined using physiological and morphological techniques. An oxygen micro-electrode and intra-arterial O_2
catheter-electrode system were used to measure tissue and blood pO_2
in anaesthetized, curarized cats under positive pressure breathing.
As a parameter for the ability of the circulation to oxygenate tissue, the 'reoxygenation time' (defined as the time required to reach
the previous pO_2 level after a short period of anoxic anoxia) was
determined on blood and cerebral cortex. The re-oxygenation time
was significantly affected by these procedures. Sludging markedly
prolonged the re-oxygenation time, an effect counteracted by the use

of anti-adhesive drugs breaking up the red cell aggregates. The
improved oxygen transport to tissue achieved by the de-aggregating
effect of the tested drugs is discussed.

REFERENCES

1. Bicher, H. I. Blood Cell Aggregation in Thrombotic Processes,
    Charles C. Thomas, Springfield, Ill., 1972.

2. Bicher, H. I. Anti-adhesive Drugs in Thrombosis, Throm. et
    Diath. Haem., Suppl. XXXXII:197-214, 1970.

3. Bicher, H. I. A Current Concept of Anti-thrombotic Therapy
    Based on the Pharmacological Effects of Drugs Preventing
    Platelet and Red Cell Aggregation, Proc. of the Second
    International Symposium on Atherosclerosis, Springer-Verlag,
    New York, pp. 536-541, 1970.

4. Bicher, H. I. and A. Beemer. Prevention, by an Anti-adhesive
    Drug, of Thrombosis Caused by Blood Cell Aggregation,
    Angiology, 21:413-441, 1970.

5. Levi, J., H. I. Bicher and J. B. Rozenfeld. Platelet aggrega-
    tion in Hypertensive Platelets Determined with a Modified
    SFP (Screen Filtration Pressure) Method, J. of Medicine,
    1:132-141, 1970.

6. Bicher, H. I. 2-methyl-2-tert-butyl-ketolactone--an Anti-
    adhesive Drug Preventing Platelet and Red Cell Aggregation.
    Pharmacology, 4:152-162, 1970.

7. Taub, W. and M. Cais. The Synthesis of Ketolactones with Poten-
    tial Pharmacodynamic Properties, Bull. Res. Counc. Israel,
    114:18, 1962.

8. Bicher, H. I. and M. H. Knisely. Brain Tissue Reoxygenation
    Time, Demonstrated with a New Ultramicro Oxygen Electrode,
    J. of Appl. Physiol., 28:387-390, 1970.

9. Bicher, H. I. and J. W. Rubin. Clinical use of a new Intra-
    Arterial Catheter Electrode System, Oxygen Transport to
    Tissue, Plenum Press, New York, N. Y., in press.

10. O'Brien, J. R. Platelet Aggregation. Part. II. Some Results
    from a New Method of Study, J. Clin. Path., 15:452, 1962.

11. Constantine, J. W. Inhibition of ADP Induced Platelet Aggrega-
    tion by Histamine, Nature (Lond.) 207:91, 1965.

12.   Judah, J. D. Ciba Foundation Symposium on Enzymes and Drug
      Action, p. 339, Churchill, London, 1947.

13.   Markwardt, F., W. Barthel and E. Glusa. Changes in the Hista-
      mine and Serotonin Content of Blood Platelets due to the
      Effect of Local Anesthetics, Naunyn Schmiedeberg's Arch.
      Exp. Path. Pharm., 253:336, 1966.

14.   Knisely, M. H., D. D. Reneau and D. F. Bruley. The Develop-
      ment and use of Equations for Predicting the Limits on the
      Rates of Oxygen Supply to the Cells of Living Tissues and
      Organs, Angiology, 20:1-50, 1969.

15.   Opitz, E. and M. Schneider. Ueber Sauerstroffversorgung des
      Gehirns und den Mechanismus von Mangelwirkungen, Ergebn.
      Physiol., 46:126-260, 1950.

16.   Silver, I. The Measurement of Oxygen Tension in Tissue,
      Oxygen Measurements in Blood and Tissue and Their Signifi-
      cance, pp. 135-153, Boston: Little, Brown and Company,
      1968.

17.   Knisely, M. H. Intravascular Erythrocyte Aggregation (Blood
      Sludge), Handbook of Physiology, Circulation, Vol. III,
      Chap. 13, pp. 2249-2293, ed. Hamilton, W. F. & Dow, P.
      Baltimore: Williams and Wilkins, 1965.

18.   DeBono, A. and N. Gozetopoulos. Experimental Pulmonary Embolism
      due to Red Cell Aggregation, Thorax 19:244-250, 1964.

OXYGEN TRANSPORT AND TISSUE OXYGENATION DURING HEMODILUTION WITH DEXTRAN

K.Meßmer, L.Görnandt, F. Jesch, E. Sinagowitz,

L.Sunder-Plassmann and M.Kessler

Institut f. Surgical Research, University Munich

Max-Planck-Inst. Applied Physiology,Dortmund,W.Germany

The major problem in volume replacement therapy using erythrocyte free solutions is the maintenance of adequate oxygen supply to the tissues. Inspite of the wide spread use of dextran solutions the effects on oxygen transport and oxygen supply to the tissues are still not fully understood. The aim of this paper is therefore to present conclusive evidence about the effects of dextran on tissue oxygenation, discussing not only its effects on the macrohemodynamics and microcirculatory flow but discussing as well its influence on the respiratory function of the blood.

MATERIAL AND METHODS

This paper is dealing with experiments which have been performed during the last years with dextran of a mean molecular weight of 60.000 (Dextran 60[+]). The experimental protocols and techniques have been described in detail previously (see 6,8,9,12). The in vivo results have been obtained from splenectomized and anesthetized dogs, subjected to stepwise exchanges of arterial blood for equal amounts of dextran 60. Local tissue PO_2 was measured by

---

[+] Macrodex 6 %, Knoll AG, Ludwigshafen, Germany

platinum multiwire electrodes according to KESSLER (3). The oxygen affinity of hemoglobin has been studied by means of P_{50} determinations using the dissociation curve analyzer (DCA) designed by DUVELLEROY et al (1970) and manufactured by RADIOMETER, Copenhagen. Human blood was obtained from healthy volunteers.

1) Dextran and Respiratory Function of the Blood.

Apart from the work of KNORPP et al (1970) reports dealing with the effect of dextran on the oxygen affinity of hemoglobin are missing in literature. According to KNORPP et al dextran increases the oxygen affinity of hemoglobin by an unknown mechanism compromising therefore partly its favorable effects on the hemodynamics. To study the capability of dextran to alter the respiratory function of blood the oxygen dissociation curve has been established from the blood of five healthy volunteers before and after in vitro dilution with dextran 60.

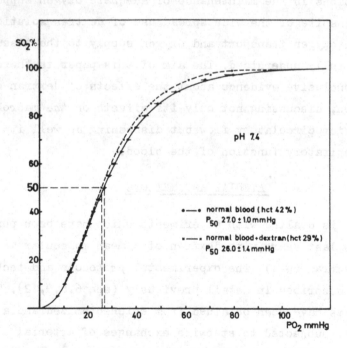

FIG. 1.   Dissociation curve after hemodilution with dextran 60. Controls = broken line (n = 5).

From fig.1 representing the mean dissociation curve it is evident
that neither the position nor the shape of the curve are significantly altered by dextran 60. At a mean hct of 29 % corresponding
to a dextran concentration of 2.5 g% no significant difference in
P_{50} was observed.

Concerning the in vivo effects of dextran, however, one has to
consider the changes in the distribution of the dextran fraction
as a result of capillary leakage of the smaller molecules. To
estimate the actual in vivo effects of dextran the hct of five
dogs has been decreased from a mean of 37.6 % to finally 10.4 % by
dextran 60. The dissociation curves were measured at different degrees of acute dilution and in two day intervals until day 9. Table
1 summarizes the changes in hct, P_{50} and the actual pH values of
the analysed blood. The P_{50} values were corrected to pH 7.4 according to SEVERINGHAUS.

Table 1

Effects of in vivo dilution with dextran 60 on P_{50}(n=5)

| | acute dilution | | | | days after dilution | | | |
					1	2	5	7	9
Hct %	37.6 ±1.3	28.3 1.0	19.7 0.6	10.4 0.3	12.1 0.9	12.9 1.2	18.5 1.0	24.3 0.8	27.3 2.1
P_{50} mmHg	26.3 ±0.4	27.8 1.07	27.4 0.6	28.4[a] 0.7	29.6[b] 0.5	30.0[b] 0.6	30.4[b] 1.3	29.7[b] 0.5	29.5[b] 0.9
pH	7.299 ±0.033	7.301 0.015	7.312 0.018	7.294 0.026	7.180 0.041	7.251 0.042	7.277 0.021	7.253 0.017	7.268 0.017

P_{50} corrected to 7.4 according to SEVERINGHAUS
a) p < 0.05, b) p < 0.005 as compared with initial values

These experiments have shown that the oxygen affinity of hemoglobin is not increased by high concentrations of dextran 60; the
resulting anemia was followed by a significant increase in P_{50}
within 24 hours; since the actual pH values were not changed significantly when compared with controls the increase in P50 reflects

true changes in oxygen affinity of hemoglobin. Thus dextran 60 does
neither increase the oxygen affinity of hemoglobin during acute
dilution nor does it interfere with the shift to the right of the
dissociation curve physiologically compensating for anemia. In
agreement with this VERSMOLD et al (1973) have reported normal 2,3
DPG levels following in vitro dilution of neonatal blood with dextran
60 as well as in puppies diluted in vivo with dextran 60.

2) Hemodynamic Compensation for the Decreased O_2 Capacity

The immediate compensation for the decrease in oxygen capacity due
to dilutional anemia consists in an increase in the total flow rate
(see 7).From previous findings as well as from literature there is
convincing evidence that the compensatory increase in cardiac output (CO) is closely related to the changes in whole blood viscosity,
rather than to hypoxic vasodilatation from the reduced oxygen capacity of the blood. Whereas the linear fall in hct is followed by an
exponential decrease in whole blood viscosity and an increase in
cardiac output, linear increments in hct above normal by means of
isovolemic hemoconcentration result in an exponential increase in
blood viscosity while cardiac output is declining immediately(8,12).
calculating from these data the systemic exygen transpert capacity
(O_2TC,product of cardiac output and oxygen capacity of the blood)
SUNDER-PLASSMANN et al (1971) obtained a bell shaped curve revealing
the   maximum at a hct of 30% thus confirming the theoretical assumption for the optimal oxygen transport by HINT (1968). Since the O_2TC
curve does not fall below the control values before the arterial hct
reaches hct-values of about 25 % one might assume that the oxygenation of the tissues should not be compromised despite of the significant reduction in oxygen content of the circulating blood. This
assumption,however,can only be substantiated by direct measurements
of nutritive capillary flow and local tissue oxygenation to exclude
perfusion of arterio venous shunts as cause of the increase in
cardiac output.

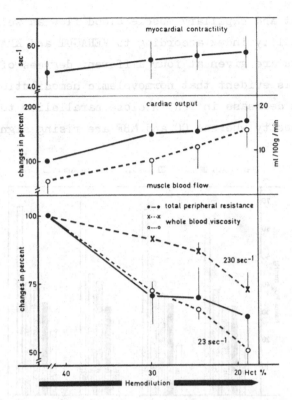

FIG. 2.   Hemodynamics and viscosity changes during normovolemic
hemodilution with dextran 60 (n = 3).

3) Local Tissue Oxygenation during Hemodilution

In a recent series of experiments in splenectomized dogs the state
of tissue oxygenation was studied in various organs; simultaneously capillary flow in skeletal muscle (MBF) as well as the macrohemodynamics and myocardial contractility were investigated (6,9).
Local tissue PO_2 was measured continuously on the surface of skeletal muscle, liver, kidney, small intestine and pancreas by a total
of 8 multiwire electrodes thus providing measurements from 47
individual tissue points (see 9). Fig. 2 illustrates the changes
in whole blood viscosity as measured at different shear rates with
the Brookfield microcone plate viscometer, the changes in TPR,

cardiac output and capillary muscle blood flow as well as myocardial contractility index according to VERAGUT and KRAYENBÜHL (1965).
All parameters are given at four different degrees of progressing
dilution. It is evident that normovolemic hemodilution is associated with a decrease in TPR in close parallelism to the decrease
in blood viscosity whereas CO and MBF are rising significantly.

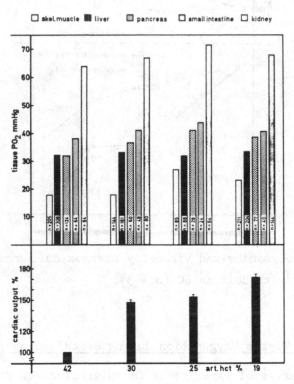

FIG. 3. Mean tissue PO_2 in various organs and cardiac output in
relation to the decrease in hct with normovolemic hemodilution in
3 animals. (n = mean computed from 3 subsequent measurements).

The myocardial contractility is not altered, which is in agreement
with findings of MURRAY et al (1969) and REPLOGLE (1972). One has,
however, to admit that the myocardial contractility index obtained
from analysis of the left ventricular pressure curves is to a certain degree dependant upon the actual preload. The latter is

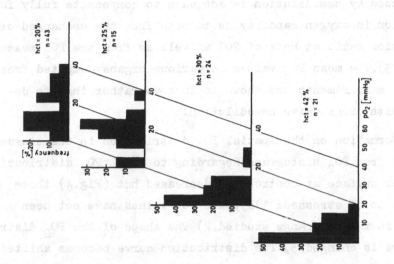

FIG. 5. PO$_2$ histogram from skeletal muscle during normovolemic hemodilution.

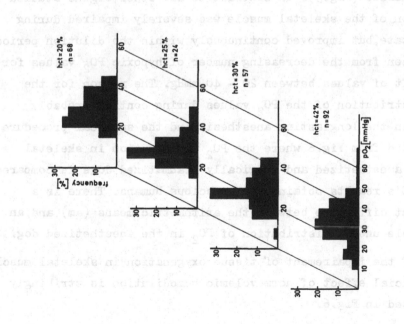

FIG. 4. PO$_2$ histogram from liver during normovolemic hemodilution.

slightly increased during hemodilution with dextran 60 as to see from a slight rise in RAP (9). Evidently the rise in cardiac output as induced by hemodilution is adequate to compensate fully for the reduction in oxygen capacity as to note from the unchanged oxygen extraction ratio at hcts of 20% as well as from the PO_2 measurements (Fig.3).The mean PO_2 values in various organs computed from 9500 single measurements are shown to improve rather than to deteriorate with progressive hemodilution.

Further information on the spatial PO_2 distribution in the tissues is provided from PO_2 histograms.Regarding to local PO_2 distribution on the liver surface at control and decreased hct (Fig.4) three points have to be stressed: 1)Hypoxic PO_2 values have not been registered in the hct range studied.2) The shape of the PO_2 distribution curve is changed.3) The distribution curve becomes shifted to the right towards higher PO_2 values.These changes of the PO_2 distribution during normovolemic hemodilution can only be interpreted as an improved and more even flow distribution on the microcirculatory level.This emphasis gains support from the findings in skeletal muscle (Fig.5). In contrast to all other organs studied the oxygenation of the skeletal muscle was severely impaired during control state,but improved continuously within the dilution period. This is seen from the decreasing number of hypoxic PO_2 values for the benefit of values between 20 - 40 mmHg. The reason for the uneven distribution of the PO_2 values during control probably consists in the longlasting anesthesia and the surgical procedure as suggested from Fig.6 where the PO_2 distribution in skeletal muscle of anesthetized and surgically traumatized dogs is compared with KUNZE's results obtained in conscious humans. There is a significant difference between the arithmetic means (am) and an indisputable uneven distribution of PO_2 in the anesthetized dog.

In view of the impairement of tissue oxygenation in skeletal muscle the beneficial effect of normovolemic hemodilution is strikingly demonstrated in Fig.6.

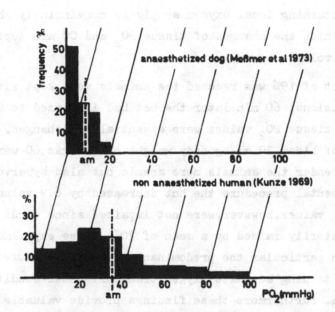

FIG. 6.   Comparison of PO_2 distribution in normal skeletal
muscle of conscious humans as obtained by micro-needle electrodes
and the PO_2 distribution in white skeletal muscle of dogs after
3 hours anesthesia and major surgery (see 9).   am = arithmetic
mean.

It has,however,to be stressed that his improvement in tissue nutrition can only be obtained by hemodilution if normovolemia as
well as cardiorespiratory compensation are given. If normovolemia
is not maintained during dilutional anemia total oxygen transport
and local tissue oxygenation are immediately diminished (8,9).

Hypoxia can thus be prevented during hemodilution if the total and
capillary flow are increased adequately. The decisive importance of
flow in maintaining local oxygen supply is convincingly shown in
Fig. 7, in which the changes of tissue PO_2 and CO are depicted
during hypervolemic hemodilution.

After the hct of 19% was reached the animals were kept without
further infusions; 60 min later the hct had increased to 21.4%
whereas the tissue PO_2 values were essentially unchanged. Without
withdrawal of blood,20 ml/kg body weight of dextran 60 were rapidly
infused to render the animals more anemic but also hypervolemic. By
this experimental procedure the hct decreased by 6.9 volume percent;
the mean PO_2 values,however,were not impaired since cardiac output
was compensatorily raised by a mean of 70%. These experiments are
stressing in particular the predominant role of the increased flow
rate in maintaining adequate oxygen transport under conditions of
acute anemia. Furthermore these findings provide valuable arguments
as far as the still open controversy on the mechanism of flow increase during hemodilution is concerned (8,10).

Since tissue hypoxia was not encountered in these experiments it
seems highly questionable that hypoxic vasodilatation should be regarded as a major cause for the rise in CO. It seems,however,justi-
fied to believe,that the reduction of the viscous resistance is the
predominant factor in decreasing the hearts afterload. Adequate
compensation is given in the hematocrit range of 45 - 20%; if hct
is allowed to fall below 20% hypoxia will take place also in normovolemic hemodilution. The oxygen supply will be endangered during
hemodilution as soon as hypovolemia and/or hypotension become involved.

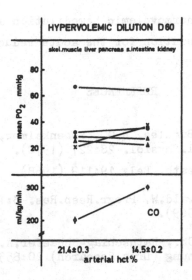

FIG. 7. Cardiac output and mean tissue PO$_2$ after hypervolemic hemodilution induced by rapid i.v. infusion of 20 ml/kg dextran 60 (n = 3).

Intentional hemodilution as therapeutic means for improvement of the microcirculation and to decrease the afterload can thereby only be performed if hypovolemic as well as hypotensive episodes can effectively be prevented.

SUMMARY

The respiratory function of blood is not affected by increasing concentrations of dextran 60 in serum neither in vitro nor in vivo. The progressing dilution of blood by dextran 60 in vivo is associated with an exponential decrease of whole blood viscosity and an increase in cardiac output and capillary flow in sceletal muscle. Due to the unaffected oxygen affinity of hemoglobin and the increase in nutritional flow the reduction in oxygen capacity is fully compensated during limited normovolemic hemodilution (hct 20%).This has been proved by the local PO$_2$ distribution in different tissues which was unimpaired even at hct values of 20%. Tissue hypoxia will not occur during limited normovolemic hemodilution as long as pulmonary and circulatory compensation for the acute reduction in O_2 capacity is provided.

REFERENCES

1. Duvelleroy, M.A.,Buckles, R.G., Rosenkaimer,S., Tung,C. and Laver,M.B. J. appl.Physiol. 28:227 (1970).

2. Hint,H. Acta Anaesth. Belg.19:119 (1968).

3. Kessler,M., Grunewald,W. Progr.Resp.Res. 3:147-152 (Karger, Basel/New York, 1969).

4. Knorpp,K., Bauer,Ch., Rathschlag-Schaefer,A.M., Bartels,H. Arzneimittelforschung (Drug Research) 20:853 (1970).

5. Kunze,K., Significance of oxygen pressure field measurements in human muscle with special remarks on PO_2 micro-neelde electrodes. In: Oxygen pressure recording in gases,fluids and tissues. Eds.: F.Kreuzer and H.Herzog, pp. 153 - 157, Basel/New York,Karger (1969).

6. Meßmer,K., Lewis,D.H., Sunder-Plassmann,L., Kloevekorn,W.P., Mendler,N. and Holper,K. Europ.Surg.Res. 4:55 (1972).

7. Meßmer,K., Schmid-Schönbein,H.(Ed.) Hemodilution: Theoretical basis and clinical application. Basel/New York,Karger (1972).

8. Meßmer,K., Sunder-Plassmann,L., Klovekorn,W.P., Holper,K. Adv. Microcirc. 4:1 - 70 (1972).

9. Meßmer,K., Sunder-Plassmann,L., Jesch,F., Görnandt,L., Sinago-witz,E. and Kessler,M. Res.exp.Med. 159:152 (1973).

10. Murray,J.F.,Escobar,E.,Rappaport,E. Amer.J.Physiol.216:638 (1969).

11. Replogle,R. Hemodynamic compensation of acute changes of the hemoglobin concentration,p.160. In: Meßmer,K.,Schmid-Schönbein, H.: Hemodilution theoretical basis and clinical application, Basel/New York, Karger (1972).

12. Sunder-Plassmann,L.,Klövekorn,W.P.,Holper,K.,Hase,U.,Meßmer,K. The physiological significance of acutely induced hemodilution. In:Ditzel/Lewis: Proc.6th Europ.Conf.Microcirculation,Aalborg 1970, p. 23 - 28, Basel/New York, Karger (1971).

13. Veragut,U.P.,Krayenbühl,H.P. Cardiologia 47:96 (1965).

14. Versmold,H.,Olbrich,G.,Hollmann,E.,Riegel,K. Klin.Paediat. 1973 (in press).

PLATELET AGGREGATES INDUCED BY RED BLOOD CELL INJURY

Mary P. Wiedeman, Ph.D.

Temple University School of Medicine

Thrombosis Specialized Center Grant #HL 14217

It has recently been demonstrated that a single-pulse ruby red laser beam of moderate strength produces enough heat to rupture red blood cells in the rapidly moving flow of an arterial vessel. The resulting clump of damaged red cells adheres to the vessel wall and is quickly covered with adhesive platelets. It was also demonstrated that the same strength laser beam did not cause a platelet aggregate if the vessel was perfused with saline or platelet rich plasma at the time the laser beam was delivered. (5) The fact that platelets did not adhere to the vessel wall when blood flow was restored suggested that the laser had not produced sufficient endothelial damage to initiate aggregation. McKenzie, Arfors, Hovig, and Matheson (3) have recently reported results of ultrastructure studies at sites of laser injury and state that there is minimal damage to endothelial cells.

It seemed important to pursue the studies concerning the role of the erythrocyte in platelet aggregation to supply additional evidence that rupture of red blood cells alone is an adequate stimulus to initiate platelet aggregation in vivo.

The investigation was conducted in the following manner. An unanesthetized bat was placed in a suitable holder with one wing extended across a large microscope slide. A branch of the major artery (a first order vessel) was exposed along its length by teasing away the covering epithelium. The vessel was covered with buffered saline and further protected by a cover slip which was supported by glass slivers. No disturbance in blood flow or vessel reactivity was noted. The major artery was cannulated with a glass cannula introduced at its most accessible distal portion so that selected solutions could be perfused retrograde to normal flow and completely replace blood in the denuded branch. On cessation of perfusion, normal blood flow would return immediately. Figure 1. Intravenous infusion was accomplished by cannulation of a vein in the tail membrane.

Before the retrograde perfusion of any solution, an injury was produced in the distal portion of the denuded arterial vessel by a single pulse laser beam of moderate strength (150 joules) as a control. Having established that a platelet aggregate would form as a result of the injury produced by a laser beam of this magnitude, the observed vessel was cleared of blood with the perfused solution and subjected to a laser beam. Normal flow was allowed to return.

It has been demonstrated that no aggregate formed when blood flow returned if the vessel was filled with saline or with PRP during the laser injury. It seemed expedient to test other substances. To overcome the criticism of using colorless fluids which might affect the absorption of heat produced by the ruby red laser beam, a solution of Congo Red was perfused with negative results. The next logical step was to separate whole blood into its two parts. Both plasma and washed red blood cells were perfused. No platelet aggregate formed after blood flow was restored following the perfusion of plasma. An aggregate did form, however, when washed red cells were perfused, similar to the aggregate in whole blood in the control conditions.

The next procedure involved hemolyzing the erythrocytes and the separate perfusion of the ghosts and the hemolysate. Restitution of normal flow after injury while perfusing red cell bodies did not result in an aggregate. A positive result was found when the hemolysate was perfused. As whole blood returned to the perfused vessel, a platelet aggregate formed at the site of the laser injury. This would suggest that the components of whole blood necessary for inducing platelet adhesiveness are contained in the contents of the erythrocyte. When the hemolysate was placed on a glass slide and hit with the laser beam, a coagulated mass was produced. This heat-coagulated material must act as the core of the platelet aggregate that was seen to form when blood flowed over the site that was subjected to the laser beam during the perfusion of the hemolysate.

This factual information can be interpreted to mean that vessel wall damage is not the sole means for the genesis of thrombus formation. It would seem that ruptured red blood cells can produce enough ADP to cause platelet adhesiveness and the subsequent aggregation of platelets. Damage to the vascular wall is not a prerequisite for thrombus formation.

Coupled with this, was a study to evaluate the protective action of acetylsalicylic acid against platelet aggregation.

There has been some difficulty in establishing where and how acetylsalicylic acid inhibits platelet aggregation. Many of the early investigations were in vitro, and the in vivo studies consisted of pre-treatment of animals with ASA followed by a test tube analysis of the anti-adhesive action. Other in vivo studies relied on indirect measurements such as changes in platelet counts or bleeding times. It was established through such studies that ASA inhibits collagen-induced platelet aggregation (4) and also that it inhibits the release of ADP from platelets, (6) the so-called release reaction.

It was also reported by Bygdeman (1) that preliminary experiments using an in vivo microscopic technique showed that ASA did not inhibit the formation of platelet aggregates at the site of a biolaser induced endothelial trauma. Evans et al (2) had shown that the addition of ASA to citrated platelet rich plasma did not supress platelet aggregation induced by ADP nor was there any inhibition of ADP-induced platelet aggregation in PRP prepared from the blood of rabbits pre-treated with ASA. This observation could be extended by investigating the effect of intravenous ASA on platelet aggregates produced by biolaser rupture of red blood cells which presumably has its effect by releasing ADP stored in the erythrocyte.

To establish a control baseline, an arterial vessel in the cecal mesentery of a rat was subjected to a laser beam of moderate strength, and the length of time that elapsed between the formation of the initial aggregate and the cessation of platelet activity was judged to occur when the initial mass no longer increased in size, indicating that platelets in the flowing blood were no longer being attracted to the site. Secondly, the termination of formation of platelet emboli was considered an end point.

Rats were infused intravenously with Bayer's Aspirin in the amount of 166 mgs/Kg of body weight, and platelet activity was determined during the first hour after aspirin, during the second hour, and during the third hour. There was no significant difference in platelet activity between the controls and aspirin treated rats up to three hours after intravenous infusion. Figure 2.

A film was shown which depicted platelet aggregates produced with laser injury to red blood cells and the responses in the various experimental conditions described such as perfusion with Congo Red, PRP, washed RBC's, Ghosts, and hemolysate and also the heat-coagulated mass produced from the hemolysate in vitro.

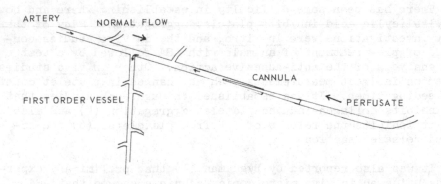

Fig. 1. A diagram of the arterial vessels in the
bat's wing which were used to study platelet aggregates resulting from biolaser injury. The laser
beam was focused on the first order vessel while
perfusing through the artery. Normal arterial
flow returned when perfusion ceased.

Duration of Platelet Activity Following Laser Injury

Before and After Intravenous Aspirin

	Animals	Trials	Average Duration of Activity
Controls	7	59	4 min. 19 sec.
0 - 1st hr.	5	24	4 min. 15 sec.
1st - 2nd hr.	7	40	4 min. 7 sec.
2nd - 3rd hr.	8	36	3 min. 53 sec.

Fig. 2. The average duration of platelet activity
shows no significant difference in time between
control values and variable periods after ASA
infusion in the rat.

Perfusate	Response
Whole Blood	Platelet aggregate
Washed RBC's	Platelet aggregate
Hemolysate	Platelet aggregate
Saline	No platelet aggregate
PRP	No platelet aggregate
Congo Red	No platelet aggregate
Plasma	No platelet aggregate
Ghosts	No platelet aggregate

Fig. 3. Platelet aggregates appeared in the first order arterial vessel when blood flow was allowed to return to normal after injury while perfusing with whole blood, washed red blood cells, and the hemolysate of red blood cells. Injury during perfusion of other substances did not produce a response.

Summary:

Rupture of erythrocytes with a biolaser results in platelet aggregation. Platelet aggregates did not form following restoration of blood after perfusing saline, PRP, Congo Red, Plasma, or Ghosts during laser injury but did form when the perfusate was whole blood, washed RBC's or hemolysate. Figure 3.

Aspirin did not prevent the formation of platelet aggregates, nor alter the duration of platelet activity resulting from biolaser injury.

REFERENCES

1. Bygdeman, S.: Effectiveness of different release inhibitors on collagen and adenosinediphosphate induced platelet aggregation. 6th Conference of European Society for Microcirculation, Aalborg 1970, 339-342, (Karger, Basel, 1971).

2. Evans, G., Packham, M.A., Nishizawa, E.E., Mustard, J.F., and Murphy, E.A.: The effect of acetylsalicylic acid on platelet function. J. Exp. Med. 128:877-894, 1968.

3. McKenzie, F.N., Arfors, K.-E., Hovig, T., and Matheson, N.A.: The mechanism of platelet aggregation at sites of laser-induced microvascular injury. 7th Conference of European Society for Microcirculation (Abstracts) p. 124, Aberdeen, 1972.

4. Weiss, H.J. and Aledort, L.M.: Impaired platelet connective-tissue reaction in man after aspirin ingestion. Lancet 2: 495-497, 1967.

5. Wiedeman, M.P. and Margulies, E.: Factors affecting production of platelet aggregates. 7th Conference of European Society for Microcirculation (Abstracts) p. 143, Aberdeen, 1972.

6. Zucker, M.B. and Peterson, J.: Inhibition of adenosine diphosphate induced secondary aggregation and other platelet functions by acetylsalicylic acid ingestion. Proc. Soc. Exp. Biol. Med. 127:547, 1968.

RECENT ADVANCES IN THE PREPARATION AND USE OF PERFLUORODECALIN

EMULSIONS FOR TISSUE PERFUSION

Leland C. Clark, Jr., Samuel Kaplan, Fernando Becattini and Virginia Obrock
Children's Hospital Research Foundation - University of Cincinnati College of Medicine

At the present time there are no perfusion media, aside from those containing red blood cells, which carry sufficient oxygen to maintain organs having a high QO_2. Resort is made to hypothermia, hyperbaria, and perhaps abnormally high flow rates. There is a need for better perfusion media and for artificial blood.

During the past eight years a variety of research projects have been initiated to exploit the unique properties of highly perfluorinated organic compounds (PFC) in biological systems. The ability of these compounds to dissolve large quantities of gases, oxygen and carbon dioxide being of particular interest to physiologists and the physical and chemical inertness of such compounds in industrial applications led to their experimental use in biology. Our earliest work involved the breathing of normobaric oxygen saturated PFC liquids by cats and mice with survival of the animals (1). Some mice also survived intravascular injection (1). Further proof of their potential value as a perfusion media was demonstrated by the intermittent perfusion of a beating heart with oxygen equilibrated PFC liquid and with Ringer's solution (2).

Sloviter (3) has perfused isolated rat brain with good results while Geyer (4,5) has completely replaced the blood of oxygen breathing rats with survival of at least some of these animals. Others have done research in liquid breathing (6,7,8), organ perfusion (9,10), whole body perfusion (11) and have sustained isolated cells in culture (12). Reference is made here to a symposium (13) which set the stage for future research. A more recent review (14) brings matters up-to-date.

Because PFC liquids are insoluble in water they must be used
as emulsions when used as artificial blood. Because they dissolve
40-60 volumes % oxygen a mixture of about 1/3 PFC and 2/3 saline
will dissolve about as much oxygen as normal whole blood. In order
to make satisfactory emulsions it is necessary to use a surface
active agent as an emulsifier. In addition, mechanical energy
must be imparted to break the PFC into particles. The size of the
particles depends mainly upon the type of PFC and emulsifier used,
the temperature, the kind of mechanical energy used and the concentration of PFC and emulsifier. <u>Under any given condition a
state is reached where further treatment with an ultrasonicator or
other mechanical emulsifier produces no further breakdown in particle size</u>. A plateau in optical density is reached at this time.

About two hours of sonication in a 400 ml batch in a flow-thru
cell at 10°C using a 400 watt ultrasonicator is required before
this stable condition exists.

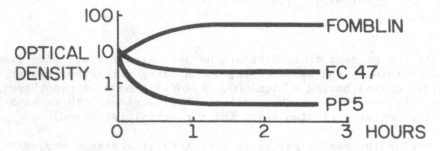

Somewhat less time is required with those based on shear such
as the Gaulin homogenizer. Generally speaking the higher the concentration of PFC the larger the particle while the larger the concentration of surfactant the smaller the particle. There are great
differences among PFC in their emulsifiability. Of course all
these methods produce particles having a distribution of sizes
clustered around a mean.

The stability of the final emulsion depends upon the type and concentration of PFC, the concentration and type of emulsifier, the temperature and other factors such as the presence of other surface active agents. Less is known about emulsion stability since it is common practice to make the emulsion the day, or the day, before it is to be used.

No one has yet found a non-toxic water soluble substance which can substitute for the erythrocyte. The case for having discrete particles, such as RBC or large PFC particles, in order to keep the microcirculation in good condition is crumbling rapidly so that perfusion of the body with a perfectly clear solution of the non-toxic water soluble oxygen solvent is at least a real possibility.

But at present we are only able to utilize emulsions for organ perfusion or as an artificial blood for whole animals.

There is nothing intrinsically unphysiological with the use of an emulsion intravascularly. Fat emulsions are widely used in Europe and are being used on a limited basis in the United States.

For the past several years we have been studying the physiological effects of various types of PFC emulsions in order to determine the best way to make them, to see their effects on oxygenation of the body tissues, and to learn their fate in the body. It is the purpose of this paper to share this knowledge, to tell of certain new and somewhat surprising results, and to glimpse the future especially in the way in which PFC emulsions may be able to serve as artificial blood in clinical medicine. Studies of the microcirculation as affected by PFC perfusion are just beginning. Work on brain microcirculation is being done by Rosenblum (15) and studies of muscle circulation are being conducted by Whalen.

When a PFC emulsion is given to an intact animal it gradually leaves the blood over several days. The major part of it is deposited in the liver where it is found, strangely enough, in the hepatocytes and not in the Kupfer cells.

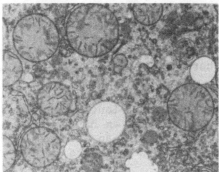
Mouse

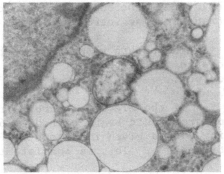
Dog

Some is deposited in the spleen and the remainder seems to be scattered around in very low concentrations in other parts of the body with perhaps more in the kidney than in muscle and other tissues.

All PFC emulsions tested to date behave in this general way although there are variations: for example, it appears that the finer the particle the less the tendency for it to be deposited in the liver.

Surprisingly some PFC stay in the liver for the lifetime of the animal and, in view of this, equally surprising is the fact that some leave in a matter of days.

Before we go into this I should like to summarize some of the physiological findings.

First, in the pentobarbitalized dog, but apparently not in other animals, there is an abrupt drop in arterial blood pressure in about 3/4 of the animals even when a tiny (0.1 ml) dose of PFC emulsion is given. The cause of this drop in blood pressure is not known. It is said that it is not found in urethane anesthesia in the dog. This drop in pressure is easily overcome by infusion of larger quantities, e.g., 200 ml, of PFC emulsion.

After the infusion of PFC emulsion the blood level of PFC usually reaches about 10% by volume of PFC as determined by chemical analysis. The blood level gradually falls until it is hardly detectable by the end of a week.

As soon as PFC emulsion is infused the mixed venous (pulmonary artery) oxygen pressure begins to climb and in the oxygen breathing dog reaches values as high as 300 torr. In about 2 to 3 hours the mixed venous oxygen tension returns to normal.

Analysis of blood containing PFC reveals that the oxygen capacity of this blood is equal to the sum of the oxygen capacity of the red cells, the plasma, and the PFC. PFC blood which has circulated in the body does not lose the ability to dissolve and transport gas intravascularly.

The return to normal of the mixed venous pO_2 is not due to a change in the blood or the PFC but rather to a change in the cardiac output.

If the electric current from a chronically implanted platinum polarographic brain cathode (16) is measured while an oxygen breathing cat is given infusions of PFC emulsion there is an increase in tissue pO_2. This is consistent with the findings reported with respect to blood.

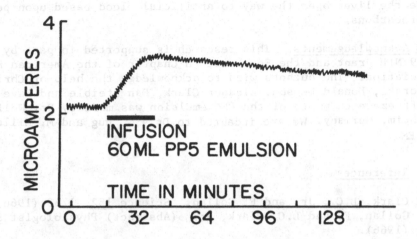

INFUSION
60 ML PP5 EMULSION

TIME IN MINUTES

Other means to evaluate the effects of artificial blood are those of partial or total replacement, of destroying the function of the RBC by low pressures of carbon monoxide, and of analyzing the performance of isolated organs such as the heart. Many such studies were postponed because of the finding that the commonly used PFC, such as FC47 remain in the liver for life. The finding of Ullrich (17) that isolated liver microsomes had changed cytochrome P450 activity after exposure to perfluoro-n-hexane seems also to suggest that there may be some combination between PFC and liver substance.

But now we know that certain PFC do not remain in the liver but are instead excreted through the lungs and the skin. These compounds contain only carbon and fluorine and are perfluorinated decalins, alkyldecalins and alkylcyclohexanes

It also appears at present, that straight chain fluorocarbons also leave the liver. Those having a C-O-C in the structure such as P11D and P1D, or a C-N-C, as FC47 are bound in the liver for the lifetime of the animal. Similar relationships are found for the spleen which contains 1/10 as much of the injected dose.

Summary. The state of the art regarding perfluorinated organic compounds (PFC) as oxygen and carbon dioxide carrying emulsions is briefly reviewed. Means for preparing and studying the usefulness of such perfusion media are described. Certain PFC form permanent bonds with liver tissue and are therefore not desirable for use as artificial blood. New classes of PFC which rapidly leave the liver open the way to artificial blood based upon perfluorocarbons.

Acknowledgements. This research is supported in part by HL-12419 NIH grant and the Southwestern Chapter of the American Heart Association. The authors wish to acknowledge the help of Christ Tamborski, Donald Denson, Eleanor Clark, Dan Steible and Dave Cohen. The freeze etch photo of the PFC emulsion was made by Dr. Wolfgang Buchhrim, Germany. We are indebted to Dr's G. Hug and M. Miller for the EM.

References

1. Clark, L.C., Jr. and F. Gollan. Science 152, 1755 (1966).
2. Gollan, F. and L.C. Clark, Jr. (Abstract) Physiologist 9, 191 (1966).
3. Sloviter, H.A. and Toshiharu, K. Nature 216, 458 (1967).
4. Geyer, R.P., R.G. Monroe and K. Taylor. In: Organ Perfusion and Preservation, Appleton Century-Croft, N.Y. 1968, p. 85.
5. Geyer, R.P., K. Taylor, E.B. Duffett and R. Eccles. Fed. Proc. 32(3), 927 (1973).
6. Gollan, F. and L.C. Clark, Jr. Ala. J. Med. Sci. 4, 336 (1967).
7. Modell, J.H., E.J. Newby and B.C. Ruiz. Fed. Proc. 29(5), 1731 (1970).
8. Sass, D.J., E.L. Ritman, P.E. Caskey, J. Greenleaf, N. Banchero, D. Mair and E.H. Wood. 27th Meeting of the Aerospace Medical Panel of AGARD NATO, Sept. 15-17, 1970 in Garmisch-Partenkirchen, Germany. In press.
9. Triner, L., M. Verosky, D.V. Habif and G.G. Nahas. Fed. Proc. 29(5), 1778 (1970).
10. Beisang, A.A., E.F. Graham, R.C. Lillehei, R.H. Dietzman and J.E. Carter. Transplant. Proc. 1(3), 862 (1969).
11. Clark, L.C., Jr., S. Kaplan, F. Becattini and G. Benzing III. Fed. Proc. 29(5), 1764 (1970).
12. Hug, George and Shirley Soukup. Personal communication.
13. Clark, L.C., Jr. Fed. Proc. 29(5), 1696 (1970).
14. Clark, L.C., Jr., F. Becattini and S. Kaplan. Triangle(Sandoz) 11, 115 (1972).
15. Rosenblum, W.I. Circulation XLVI(4), II48, Abst. #189 (1972).
16. Clark, L.C., Jr., G. Misrahy and R.P. Fox. J. App. Physiol. 13, 85 (1958).
17. Ullrich, V. and H. Diehl. Eur. J. Biochem. 20, 509 (1971).

PHARMACOLOGICAL STIMULATION OF RED BLOOD CELL METABOLISM FOR HIGH ALTITUDE PREADAPTATION

Lorna Grindlay Moore, George J. Brewer, Fred J. Oelshlegel, Jr., and Andrew M. Rose

Departments of Anthropology, Human Genetics, and Human Performance Center, University of Michigan, Ann Arbor, Michigan

A decrease in erythrocyte oxygen affinity is held to be an important component of high altitude adaptation. Ordinarily at 10-11,000 feet, a four to five day period or longer of exposure to high altitude is required to build-up the 2,3-diphosphoglycerate (DPG) levels of red cells which brings about the decrease in oxygen affinity. It is during this early period of altitude exposure that symptoms (acute mountain sickness) related chiefly to the central nervous system are frequently experienced.

Pharmacological stimulation of red cell metabolism with the administration of oral phosphate has provided us with a tool with which to explore the possibility of altitude preadaptation by raising DPG levels prior to ascent. Brain and Card (1972) have demonstrated the ability of a single dose of a combination of sodium and potassium phosphate salts to elevate DPG levels by 15%. We have been able to maintain elevated DPG levels for up to five days with continued doses of oral phosphate. Our work with phosphate has also shown that its effect on red cell glycolysis is variable. DPG responses in the same individual administered the same or nearly the same dosages at two different times have varied somewhat. This

This work has been supported in part by the Research and Development Command, Department of the Army & Navy under Contract No. DADA-17-69-C-9103, the Wenner-Gren Foundation for Anthropological Research, Grant No. 2854, and by Training Grant No. 5T01-GM-00071 from the National Institutes of Health.

variability may be due, at least in part, to the multiple mode of phosphate's effect on red cell glycolysis. Inorganic phosphate actually enters the glycolytic pathway as a cofactor at the step where glyceraldehyde phosphate (GAP) is converted to 1,3 diphosphoglycerate (1,3 DPG) by means of the enzyme glyceraldehyde-3-phosphate dehydrogenase. However, it is not likely that an increase in the level of inorganic phosphate serves to activate this step since GA-3-PD is not normally a rate-limiting enzyme. Rather, the primary points at which an increase in organic phosphate effects erythrocyte metabolism more likely are at the hexokinase (HK) and phosphofructokinase (PFK) steps. Inorganic phosphate is known to relieve the glucose-6-phosphate (G6P) inhibition of HK and to relieve the ATP inhibition of PFK (Minakami and Yoshikawa, 1966; Rapoport 1968; Rose et al., 1964). Further studies are necessary in order to confirm that the HK and PFK steps are actually those at which the addition of inorganic phosphate has its effect. If, in fact, it is the HK and PFK steps that are involved, the effect of phosphate on red cell metabolism parallels the natural effects of high altitude since we have been able to implicate HK and PFK as the red cell enzymes which are activated upon acute exposure to 10,200 feet and 14,100 feet altitudes (Moore, Brewer and Oelshlegel 1972; Moore, Brewer and Sing in preparation).

During this past winter, we undertook a field trip to Climax, Colorado, elevation 11,200 feet, with hopes of accomplishing three goals.

1)   To evaluate the ability of oral phosphate to elevate DPG levels by the 15% increases over baseline values that we have observed in previous high altitude studies.

2)   To answer the question of whether or not high DPG levels are maintained at high altitude once phosphate administration is stopped. That is, are high DPG levels retained or is there a return to baseline values and then a subsequent build-up?

3)   To use psychomotor human performance tests to evaluate the contribution of the DPG increase and concomitant decrease in erythrocyte oxygen affinity to central nervous system functioning.

The protocol for these studies involved treating five subjects with phosphate and five subjects as controls. This preliminary study was not double blind. The phosphate regimen is summarized in Table 1. The prealtitude adaptation phase of the study lasted 36 hours and was initiated with priming doses of sodium bicarbonate plus phosphate. The rationale behind the use of bicarbonate was that the effectiveness of phosphate on red cell glycolysis at the PFK step is greater the higher the pH. However, the bicarbonate

Table 1

Protocol for Phosphate Treatment

Day	Agent	Schedule
Day -1 (preceding ascent)	Phosphate solution sodium & potassium salts. pH 7.4	initial dose = 60 mmoles 30 mmoles doses continued 3 times at 4 hour intervals
	Sodium bicarbonate solution	1.25 mEqv/Kg 3 times at 4 hour intervals, first 12 hours
Day of ascent to 11,200 ft. (Climax, Colo.)	Phosphate solution	30 mmoles 3 times daily at 8 hour intervals
Day 1 at 11,200 ft.	Phosphate solution	30 mmoles 3 times daily at 8 hour intervals
Day 2 at 11,200 ft.	Phosphate solution	30 mmoles 3 times daily at 8 hour intervals
Day 3 at 11,200 ft.	Phosphate solution	30 mmoles 3 times daily at 8 hour intervals
Day 4 to 10 at 11,200 ft.	No further treatment	

was only given during the first 12 hours of preadaptation because by the time of ascent to altitude we desired a fairly normal blood pH to avoid any increase in erythrocyte oxygen affinity due to the Bohr effect. Phosphate was continued in doses of 30 millimoles three times a day after ascent to altitude for about 80 hours (or three days).

DPG increases are shown in Figure 1 and are very encouraging. An average of about a 15% increase in DPG levels in the treated group was obtained prior to ascent to altitude, which was the general area of increase we were hoping to achieve. Our second objective was also met as can be seen in Figure 1. After treatment was discontinued, DPG levels did not return to baseline values but remained generally high. The agent was tolerated relatively well. Its major side effect is the production of loose stools. The diarrhea is non-cramping in type but is a nuisance. For this,

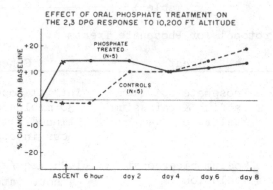

Figure 1. Phosphate administration served successfully to elevate
DPG levels in the treatment group so that DPG was already 15% above
baseline values by the time of ascent. Until Day 2, a marked contrast in DPG levels existed between the two groups.

reason, our pharmacological approach must eventually be modified,
either by agents to minimize the diarrheal effect or by addition
of new agents which permit lowering the dose of phosphate to the
point where diarrhea is not a factor.

Red blood cell glycolytic intermediates were also collected
as a part of this field study. The data is presently being subjected to reduction and statistical analysis. Qualitatively, the
metabolic profiles of the treated and untreated subjects appear
similar and in each case, stimulation of the rate-limiting HK and
PFK steps appears to be responsible for elevating the DPG levels.

Human performance tests were administered to the subjects
participating in this initial phosphate altitude trial. Members
of the University of Michigan Human Performance Laboratory collab-
orated with us on the preparation and administration of a battery
of seven tests. These were: a memory task, a reasoning task, a
balancing or critical tracking task, reaction time tasks, letter
search tasks, the Stroop test for response blocking, and a tapping
or eye-motor coordination task. This battery of tests was used to
provide an initial assessment of the usefulness, practicality, and
sensitivity of human performance testing for measuring central
nervous system function differences between preadapted and control
groups. Our preliminary results are encouraging. Figures 2 and 3
depict the results of a tapping (eye-motor coordination) and a
reasoning (cognitive function) task. In each case, the phosphate
treated group underwent greater relative improvement than did the
control group during the initial, DPG - contrast period. The presence of sizable practice effects clouds the picture and makes
more detailed analysis of these two tasks difficult. However, two

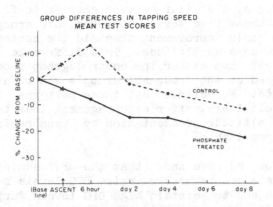

Figure 2. Over time, tapping speed steadily decreased in the phosphate group as would be expected from a practice effect. The control group scores show less overall improvement and their pronounced rise during the initial period distinguishes them from the phosphate group.

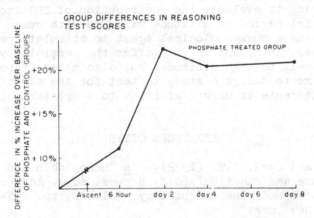

Figure 3. Scores are expressed as the percent increase over the baseline value of the phosphate group minus the percent increase of the control group. Greater overall and more pronounced improvement during the initial period characterizes the phosphate group's response.

points can be made about the seven tests taken together which offer support for ascribing advantage to the phosphate-treated group. First, on tasks where the practice function is known, the pattern of the phosphate group's practice effect conforms more to the expected practice function than does the control group's. This translates into a lesser effect of altitude on the phosphate treated compared to the control group. Second, when allowances are made for the practice effects, a similar pattern emerges in five out of the seven tests. In these five tasks, the phosphate group underwent a greater and more rapid improvement than did the control group during the first few days at altitude. This difference in rate of improvement suggest that either the control group's performance was relatively dampened or the phosphate group's was relatively stimulated at altitude. We believe that these patterns, while very tentative and preliminary, offer encouragement that the approach to evaluating acute altitude preadaptation by human performance testing may well be profitable.

In conclusion, we have shown that the oral administration of inorganic phosphate can be used to raise DPG levels before ascent to high altitudes and to maintain high DPG levels during the initial period of altitude exposure. Using phosphate as a pharmacological agent, it has been possible to preadapt the red cell metabolically to an 11,200 foot altitude. In addition, preliminary results from psychomotor human performance testing reveal group differences which generally translate into a less pronounced effect of altitude on the phosphate treated group. We are, therefore, encouraged to continue trying to evaluate the contribution of DPG increases during the initial period of altitude exposure with human performance testing. Using a pharmacological agent to stimulate red cell glycolysis serves, then, not only to offer the possibility of preadapting the red cell to high altitude, but also makes it possible for the first time to design a study to test for the adaptive significance of a decrease in oxygen affinity to short-term altitude adaptation.

REFERENCES CITED

Brain, M.C. and Card, R.T. (1972). In "Hemoglobin and Red Cell Structure and Function" (G.J. Brewer, ed.), Advances in Experimental Medicine and Biology, Vol. 28, pp. 145-154. Plenum Press, New York.
Minakami, S. and Yoshikawa, H. (1966). J. Biochem. 59, 145.
Moore, L.G., Brewer, G.J., and Oelshlegel, F.J., Jr. (1972). In "Hemoglobin and Red Cell Structure and Function" (G.J. Brewer, ed.), Advances in Experimental Medicine and Biology, Vol. 28, pp. 397-413. Plenum Press, New York.
Rapoport, S. (1968). In "Essays in Biochemistry" (P.N. Campbell and G.D. Greville, eds.), Vol. 4, pp. 69-103. Academic Press, New York.
Rose, I.A., Warms, J.V., and O'Connell, E.L. (1964). Biochem. Biophys. Research Communs. 15, 33.

CATECHOLAMINE METABOLISM AND DIGITAL CIRCULATION AFTER
HISTAMINE AND ITS ANALOGUE
Krypton Clearance and Calorimetric Blood Flow

Nosrat Eric Naftchi**, Milton Mendlowitz***,
Stefan Racoceanu***, Edward W. Lowman**

INTRODUCTION

In 1959 (1) we reported the paradoxical effects on the
digital circulation of histamine administered parenterally and of
an orally ingested histamine analogue* (2). Both histamine and
its analogue produced vasodilatation of the capillaries as
manifested by flushing contrasted with vasoconstriction of the
arteriovenous anastomoses (AVAS). It was postulated at that
time that the AVA constricted in response to heat loss produced by
histamine-mediated precapillary sphincter and hence capillary
vasodilatation. The question as to whether the AVA constriction
was a neural reflex or produced by some other mechanism such as
the release of catecholamines (CM) from nerve endings and/or
chromaffin tissues remained unanswered (3).

The biological significance of the paradoxical effect of
histamine on peripheral circulation remains obscure. In 1968
Kakiuchi and Rall (21,22) using slices of rabbit cerebellum in
which norepinephrine and histamine were both effective in raising
the level of cyclic AMP, established that both agents interact

*3-β-aminoethyl-1,2,4-triazole dihydrochloride. (Compound No.
16683, furnished by Ely Lilly and Co., Indianapolis, Indiana.)

**Laboratory of Biochemical Pharmacology, Institute of Rehabilitation
Medicine, New York University Medical Center, New York, N.Y. 10016,
Supported by the Edmond A. Guggenheim Clinical Research Endowment.
***The Department of Medicine, The Mount Sinai School of Medicine
of the City University of New York, N.Y. 10029, Supported by
U.S.P.H.S. Grants HE 06546 (CV) and HE 05802 (HTA).

with separate receptors. In an attempt to answer this question
urinary catecholamine metabolites were analyzed prior to and after
administration of histamine. In addition, sympathetic reflex
activity was inhibited by means of heat and ganglionic blockade
and also by ulnar nerve block and measurements were made of digital
blood flow (calorimetric) and capillary blood flow (isotope
disappearance rate) before and after vasodilatation and subsequent
administration of histamine or its analogue.

METHODS

In nine subjects (Table I), combined urinary metanephrine
and normetanephrine (total metanephrine, TMN) and in 11 subjects
(Table II), urinary 4-hydroxy-3-methoxymandelic acid (vanillyl-
mandelic acid, VMA) were assayed before and after parenteral
administration of histamine. Urinary TMN was assayed by the Pisano
technique (4) and urinary VMA by the two-dimensional chromato-
graphic method of Armstrong et al (5). Histamine base was infused
i.v. in a 5% dextrose in water solution or injected subcutaneously,
in doses ranging from 0.25 to 1.0 mg. In seven normal subjects an
index of capillary blood flow in the digit was obtained, from the
disappearance rate of ^{85}Kr radioactivity (9,10,11), before and
after administration of 10 mg of the histamine analogue (Table III).
Sympathetic inhibition was achieved by indirect heating alone.
Then 0.03 ml of ^{85}Kr solution containing 2.1 mCi/ml was injected
via a 27 gauge needle to a depth of approximately 3 mm into the
fingertip pulp of the 4th left digit. The terminal phalanx of the
finger was then introduced into the digital cup of a collimated
sodium iodide crystal scintillation detector placed at heart level
with a wrist and elbow support. A rate meter capable of counting
30,000 counts per minute (cpm), with a time constant of 1 second,
attached to a direct writing arithmetic linear chart, provided a
record of the disappearance rate of the radioactivity. After
subtraction of the background radiation, the cpm were plotted
against time on semilogarithmic paper. Measurements of the half-
time disappearance (T 1/2) were begun five minutes after the
injection in order to allow equilibrium to become established.
Each subject then ingested a single 10 mg dose of the histamine
analogue. Approximately one hour later, when the red flush over
the face, neck and hands had reached its peak, 0.03 ml ^{85}Kr
solution was again injected at a different site in the fourth
fingertip and its disappearance rate recorded as described above
(Table IV).

In a group of eight normal subjects complete vasodilatation
in the fifth finger was achieved with ulnar nerve blockade (12,13)
by injecting approximately 3 ml lidocaine hydrochloride (2%) into
the mesial olecranonepicondylar groove. Complete anesthesia of the
fifth finger and the external half of the fourth finger was taken

TABLE I

URINARY EXCRETION OF TOTAL METANEPHRINES AFTER HISTAMINE

Subject	Before Histamine Base (µg/mg of Creatinine)	3 Hours After Histamine Base (µg/mg of Creatinine)	% Increase
1	1.0	2.0	100
2	0.56	1.2	114
3	1.3	2.8	115
4	0.87	1.5	72
5	0.73	5.3	626
6	0.92	2.5	172
7	0.6	1.3	117
8	1.1	1.8	64
9	0.31	0.94	203
Mean $\pm$ SD*	0.82 $\pm$ 0.29	2.15 $\pm$ 1.25	176
S.E.D.**	0.43		
p***	$<10^{-2}$		

*Standard Deviation

**Standard Error of the Difference

***Probability that the Difference is due to chance

Total Metanephrines = Metanephrine plus Normetanephrine

TABLE II

URINARY EXCRETION OF 4-HYDROXY-3-METHOXYMANDELIC ACID AFTER HISTAMINE

Subject	Before Histamine Base (μg VMA/mg creatinine)	3 hrs. After Histamine Base (μg VMA/mg creatine)	% Increase
1	1.0	3.5	250
2	1.0	2.0	100
3	1.3	1.8	38
4	1.5	2.3	53
5	1.5	2.0	33
6	1.0	2.0	100
7	1.5	3.0	100
8	1.5	2.5	67
9	1.5	2.5	67
10	1.0	2.0	100
11	1.5	2.5	67
Mean $\pm$ SD	1.3 $\pm$ 0.24	2.4 $\pm$ 0.50	90

S.E.D.   0.17

P $<10^{-6}$

4-Hydroxy-3-methoxymandelic acid = vanillmandelic acid (VMA)

TABLE III

THE EFFECT OF HISTAMINE ANALOGUE ON

HALF-TIME DISAPPEARANCE [85]KRYPTON RADIOACTIVITY FROM THE

FOURTH LEFT FINGERTIP AFTER SYMPATHETIC INHIBITION

Subject	T 1/2 Disappearance of Kr Radioactivity	
	Before Histamine (sec)	After Histamine (sec)
1	345	702
2	397	660
3	270	390
4	360	690
5	350	690
6	420	660
7	370	792
Mean ± SD	359 ± 47	655 ± 125
% Increase		
S.E.D. 51		
P $< 10^{-6}$		

TABLE IV (GROUP A)

THE EFFECTS OF HISTAMINE ANALOGUE ON BLOOD PRESSURE PULSE RATE AND HALF-TIME DISAPPEARANCE OF ^{85}KRYPTON RADIOACTIVITY FROM THE FIFTH LEFT FINGER AFTER ULNAR NERVE BLOCKADE

Subject	Age	Sex	Before Histamine Analogue Blood pressure (mmHg)	Pulse rate/min.	T 1/2 ^{85}Kr (sec)	After Histamine Analogue Blood pressure (mmHg)	Pulse rate/min.	T 1/2 ^{85}Kr (sec)
1	35	M	122/81	74	440	122/81	93	510
2	30	M	124/80	76	260	124/84	83	440
3	32	M	120/80	73	345	122/81	87	510
4	38	M	120/80	77	330	120/78	92	540
5	41	F	102/72	75	195	102/60	96	329
6	27	F	103/70	76	325	102/76	80	457
Mean $\pm$ SD			115 $\pm$ 9/77 $\pm$ 4.8	75 $\pm$ 1.47	316 $\pm$ 83	115 $\pm$ 10.41/77 $\pm$ 8.62	89 $\pm$ 6.22	464 $\pm$ 76
SED			6.43/4.41	2.86	50.25			
P			NS/NS	< 0.001	< 0.01			

TABLE IV (GROUP B)

THE EFFECTS OF HISTAMINE ANALOGUE ON BLOOD PRESSURE PULSE RATE AND HALF-TIME DISAPPEARANCE OF ^{85}KRYPTON RADIOACTIVITY FROM THE FIFTH LEFT FINGER AFTER ULNAR NERVE BLOCKADE

Subject	Age	Sex	Before Histamine Analogue			After Histamine Analogue		
			Blood pressure (mmHg)	Pulse rate/min.	T 1/2 85 Kr (sec)	Blood pressure (mmHg)	Pulse rate/min.	T 1/2 85 Kr (sec)
7	42	M	120/80	76	630	96/76	83	9480
8	25	M	122/80	73	310	114/72	88	2160
Mean + SD	121 + 1.41/80 + 0		75 + 2.12	470 + 226	105+12.75/74+2.83	86+3.5	5820+5176	
SED	12.80/2.83		4.12	5180				
P	NS/<0.05		<0.01	NS				

as the criterion for total vasomotor nerve blockade. Then 0.03 ml of [85]Kr solution was injected as above into the pulp of the fifth fingertip and the disappearance rate was recorded. Ten mg of histamine analogue was then administered orally in six patients (Group A) and one hour later the [85]Kr disappearance rate was again recorded after reinjection of the fingertip at a different site (Table IV). In two cases the same procedure was carried out before and after administering 15 mg of the histamine analogue (Table IV, Group B).

In 13 of the above subjects (Table V), digital blood flow was measured calorimetrically prior to and at 30, 60 and 90 minute intervals after the administration of the histamine base without and during sympathetic nerve blockade and histamine administration. Sympathetic nerve blockade was achieved by indirect heating supplemented by i.v. infusion of trimethaphan camphorsulfonate (TMCS) (6). Heat was administered to the trunk by means of a heat cradle or an electric blanket for 45 minutes to one hour until positive heat balance was attained as manifested by generalized sweating. A single dose of 0.8 mg/Kg of TMCS was then given intravenously. Sympathetic nerve blockade was maintained by infusion of an additional 1 mg TMCS per ml of 5% dextrose in water running at the rate of 1 ml per minute.

In another group of 17 normal subjects (Figure 1), digital blood flow was measured before and at 30, 60 and 90 minute intervals after the ingestion of a histamine analogue. The histamine analogue was given in a 10 mg dose.

The digital calorimeter used in these studies consisted of an insulated cup filled with water into which a Beckmann thermometer was inserted (7). The water was stirred continually by means of a magnetic stirrer. Blood flow per cm^2 of skin surface was calculated from the formula:

$$\frac{dHV}{AK(Tm-Tc)}$$

in which dH is the rise in temperature of the water in the calorimeter per unit time adding or subtracting, respectively, the fall or rise in temperature with the finger removed; V is the volume of the water and the water equivalent of the finger, the calorimeter cup, the stirrer and the thermometer; A is the surface area of the terminal phalanx immersed in the calorimeter cup; K is the specific heat of the blood multiplied by its specific gravity determined from the hematocrit. Tm is mouth temperature minus 0.7°C and Tc is calorimeter temperature. These temperatures are assumed to represent the arterial and the venous blood temperatures, respectively, an assumption which is valid if room temperature is kept within 26° to 29°C (8).

TABLE V

CALORIMETRIC DIGITAL BLOOD FLOW AFTER HISTAMINE BASE

Digital Blood Flow (10^{-2}. cm^3/sec)

Subject	Before Histamine Base	After Histamine Base (minutes)		
		30	60	90
1	4.1		1.6	
2	5.1		0.30	
3	7.2	0.11	1.6	
4	4.7	1.9	1.6	
5	1.3	1.2		0.40
6	1.0	2.7		2.8
7	5.0	3.8	3.1	0.50
8	4.0	3.8	2.6	2.4
9	4.7	3.0	1.4	1.8
10	5.2	3.6	2.6	0.12
11	3.1	1.8	1.1	
12	5.5	4.6	2.8	1.1
13	5.8	3.0	0.23	
Mean $\pm$ SD	4.4 $\pm$ 1.7	2.7 $\pm$ 1.3	1.7 $\pm$ 0.92	1.3 $\pm$ 0.97
% Decrease		44	63	72
S.E.D.		0.60	0.54	0.59
P		$<10^{-2}$	$<10^{-2}$	$<10^{-2}$

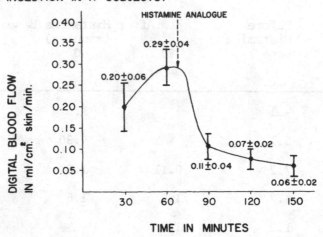

CALORIMETRIC DIGITAL BLOOD FLOW BEFORE AND AFTER SYMPATHETIC INHIBITION AND HISTAMINE ANALOGUE INGESTION IN 17 SUBJECTS.

Figure 1. Effect of sympathetic nerve blockade and histamine analogue on the digital blood flow in 17 subjects. Control digital blood flows were performed 30 minutes after resting in a temperature controlled room, after which sympathetic nerve blockade was achieved by indirect heating and infusion of histamine. Abscissa (Figure 1)

30 minutes = Control before inhibition of sympathetic nerve discharge.

60 minutes = After sympathetic blockade

90 min.; 120 min.; 150 min.= After ingestion of histamine analogue

↓ = Point of histamine Administration histamine analogue.

⊥ = Mean ± S.D.

All subjects were tested in the supine position in a quiet room. Blood pressure and radial pulse rate were measured before and during histamine infusion or after the ingestion of the histamine analogue. The room temperature was kept at 28° ± 1°C. urine samples were collected, acidified with HCl and stored.

RESULTS

Urinary catecholamine metabolite excretion. The urinary catecholamine metabolite assays for TMN and VMA, before and after the administration of the histamine base, are presented in Tables I and II. There was a significant increase in both TMN and VMA after histamine base.

Isotopic determinations of digital capillary blood flow. The half-time [85]Kr disappearance from the fourth left fingertip of seven subjects increased by 74.2% indicating a marked decrease in capillary blood flow after ingestion of histamine analogue (Table III). This decrease occurred in spite of sympathetic nerve blockade produced by indirect heating. The T 1/2 [85]Kr also increased uniformly and significantly in all six subjects in whom the ulnar nerve was blocked by lidocaine hydrochloride (Table IV, Group A), indicating again a decrease in digital capillary blood flow. The increase in disappearance rate of [85]Kr was much more marked in the two cases in which a larger dose of histamine analogue was administered (Table IV, Group B).

Calorimetric measurements of digital blood flow. The mean calorimetric digital blood flow during the administration of histamine base decreased significantly by 44%, 63%, and 72% at 30, 60 and 90 minutes respectively (Table V). In Figure 1, the increase in digital blood flow which ordinarily occurs after indirect sympathetic nerve inhibition is demonstrated. When the subjects ingested the histamine analogue the mean digital blood flow decreased as the flush progressed despite the maintenance of the sympathetic blockade (Figure 1). The flush began within 30 to 60 minutes of ingestion of the 10 mg dose and persisted for 30 to 90 minutes.

Blood pressure and radial pulse rate. When the histamine base was infused in a 5% dextrose in water solution at the rate of 0.007 mg per minute, the mean blood pressure decreased at first by 5 to 10 mm Hg as the flush appeared over the face and progressed downwards. It then increased by 10 to 20 mm Hg or more above the initial value. The mean pulse rate increased from 76 to 96 per minute. After the ingestion of the histamine analogue, the changes in pulse rate and in systolic and diastolic blood pressure were less marked (Table IV, Group A). In the two subjects receiving 15 mg of histamine analogue there was a decrease in blood pressure which may explain the great increase in the T 1/2 disappearance rate of ^{85}Kr (Table IV, Group B).

DISCUSSION

The effects of histamine on peripheral circulation have been studied with variable and often conflicting results. Usually some combination of vasodilatation and vasoconstriction was produced with variable relative intensity from region to region and species to species (14, 15, 16).

Our findings of decreased digital blood flow indicate that histamine administration produces AVA constriction and concomitant dilatation of the capillaries. The histamine-mediated capillary dilatation was accompanied by a decrease rather than an increase in capillary blood flow, due most likely, to stagnation of the blood within the dilated capillaries. Since the subjects were in positive heat balance and their reflex sympathetic activity was blocked, the more likely explanation of the AVA constriction after histamine would be the direct release of catecholamines from sympathetic neural and/or chromaffin tissues (17). This conclusion is reinforced by the ulnar nerve block studies which have been shown to block sympathetic nerve discharge (12, 13) in the 5th finger completely, thus precluding breakthrough of the neural blockade. The possibility of vasoconstriction due to stimulation by catecholamines released into the circulation, however, has not been ruled out (17, 18, 19). Our studies demonstrate that both histamine and its analogue decrease the digital blood flow and increase the urinary excretion of catecholamine metabolites. It appears, therefore, that histamine and its analogue have both a capillary dilation and an AVA constriction effect probably mediated by a direct effect on capillaries and an effect on catecholamine release interacting with separate receptors (21, 22). The increase in urinary catecholamine metabolites found after administration of histamine or its analogue could be attributable to stimulation of neural activity with indirect release (18) from the neural or chromaffin stores, or alternatively, to direct release (17) from such stores. This hypothesis would afford a more rational explanation of the effect of histamine as a provocative test for pheochromocytoma (20).

SUMMARY

The effects of histamine or its analogue on the digital circulation and on urinary catecholamine metabolite excretion were investigated both without and during sympathetic neural inhibition. Histamine or its analogue produced capillary dilatation manifested by flushing of the skin and at the same time decreased digital blood flow as indicated by calorimetric measurements and ^{85}Krypton (^{85}Kr) disappearance rate. This effect is probably attributable to the direct action of histamine in dilating precapillary sphincters and to the relative stagnation of blood within the capillaries. Histamine or its analogue also produced constriction of digital arteriovenous anastomoses (AVA) both before and after sympathetic nerve blockade and increased urinary excretion of catecholamine metabolites. The most likely explanation of these results is direct release of catecholamines from their neural and chromaffin stores by histamine or its analogue. This effect may be mediated through cyclic AMP interacting with separate histamine and catecholamine receptors and could also account for the action of histamine as a provocative test for pheochromocytoma.

REFERENCES

1. Naftchi, N.E., Mendowitz, M., Weinreb, H.L. and Gitlow, S.: Paradoxical effect of an orally ingested histamine analogue on capillaries and arteriovenous anastomoses of the human digit. J. Appl. Physiol. 14: 949, 1959.

2. Ainsworth, C., and Jones, R.G.: 1,2,4-triazole analogues of histamine. J. Amer. Chem. Soc. 75:4915, 1953

3. Naftchi, N.E. and Mendlowitz, M.: Mechanism of paradoxical effect of histamine and its analogue on capillaries and arteriovenous anastomoses of the human digit. Fed. Proc. 26:785, 1967 (abstract).

4. Pisano, J.J.: A simple analysis for normetanephrine and metanephrine in urine. Clin. Chim. Acta 5:406, 1960.

5. Armstrong, M.D., Shaw, K.N.F. and Wall, P.E.: The phenolic acids of human urine, paper chromatography of phenolic acids. J. Biol. Chem. 218:293, 1956.

6. Mendlowitz, M., Naftchi, N.E. and Wolf, R.L.: The effect of trimethaphan camphorsulfonate on the human digital circulation. Clin. Pharm. and Ther. 9:1, 1968.

7. Mendlowitz, M.: The digital circulation. Grune and Stratton, New York, 1954.

8. Eurman, G.H. and Mendlowitz, M.: The relationship between mouth, skin, and arterial blood temperature. J. Appl. Physiol. 5:579, 1953.

9. Reller, C., Sheridan, J. and Aust, J.B.: Capillary blood flow determination with Kr^{85}. Fed. Proc. 24:443, 1965.

10. Cohen, L.S., Elliot, W.C. and Gorlin, R.: Measurement of nyocardial blood flow using Krypton[85]. Am. J. Physiol. 206:997,1964.

11. Racoceanu, S., Naftchi, N.E., Suck, A., Mendlowitz, M., and Wolf, R.L.: Digital capillary blood flow in clubbing: [85]Kr studies in hereditary and acquired cases. Ann. Int. Med. 75:933, 1971.

12. Arnott, W.M. and Macfie, J.M.: Effect of ulnar nerve block on blood flow in the reflexly vasodilated digit. J. Physiol. 107:233, 1948.

13. Sarnoff, S.J. and Simeone, F.A.: Vasodilator fibers in the human skin. J. Clin. Invest. 26:453, 1947.

14. Dale, H.H. and Laidlaw, P.P.: The physiological action of -imidazolethylamine, J. Physiol. Lon. 41:318, 1910.

15. Dale, H.H.: Some chemical factors in the control of the circulation. Lancet 1:1179, 1233, 1285, 1929.

16. Sharpey-Schafer, E.P. and Ginsburg, J.: Humoral agents and venous tone: effects of catecholamines, 5-hydroxyptamine, histamine and nitrites. Lancet 2:1337, 1962.

17. Douglas, W.W., Kanno, T. and Sampson, S.R.: Effects of acetylcholine and other medullary secretagogues and antagonists on the membrane potential of adrenal chromaffin cells. J. Physiol., London 188: 107, 1967.

18. Staszewska-Barczak, J. and Vane, J.R.: The release of catecholamines from the adrenal medulla by histamine. Br. J. Pharm. Chemother. 25:728, 1965.

19. Axelrod, J., Maclean, P.D. and Wayne Albers, R., Wiessbach, H.: Regional neurochemistry. Pergamon Press, London, 307, 1961.

20. Douglas, W.W.: Histamine and antihistamines: 5-hydroxytryptamine and antagonists. In: The Pharmacological Basis of Therapeutics, edited by Goodman, L.S. and Gilman, A. London, The Macmillan Company, 1970, P.621.

21. Kakiuchi, S. and Rall, T.W.: The influence of chemical agents on the accumulation of adenosine 3',5'-phosphate in slices of rabbit cerebellum. Mol. Pharmacol. 4:367, 1968a.

22. Kakiuchi, S. and Rall, T.W.: Studies on adenosine 3',5'-phosphate in rabbit cerebral cortex. Mol. Pharmacol. 4:379, 1968b.

Augmentation of Tissue Oxygen by Dimethyl Sulfoxide and Hydrogen

Peroxide

Bert Myers and William Donovan

Dept. of Surgery LSU School of Medicine and the Vet. Admin.

Hosp. in New Orleans

Dimethyl sulfoxide (DMSO) increases the permeability of tissue and has been reported to aid in the transport of drugs through intact skin. 1, 2 These experiments were designed to see if the drug would transport oxygen in the form of hydrogen peroxide (H202) through skin. A mass spectrometer was used to measure tissue gas levels (P02, PC02, PN2).

In the rabbit, cutaneously applied DMSO with or without H202 had no effect on normal skin, but the combination elevated 02 levels in ischemic tissue. DMSO alone caused consistent decreases in the PC02 of devascularized tissue. Cutaneously applied drug had no effect on porcine skin, perhaps due to the thick epidermis, but injection of the drugs into sutured wounds generally led to a rise in the P02 in the fat up to 3.0 cm away from the incision.

Methods

14 gauge teflon catheters over 17 G. needles were inserted longitudinally into the skin on both sides of eight anesthetized white rabbits, threading the catheters between the skin and panniculus carnosus. The central needle was removed and through the cannula a teflon coated stainless steel probe was inserted. The guiding cannulas were removed and the probes were sutured in place and connected to the vacuum chamber of a medical mass spectrometer. After pump down, a 12.0 cm circular flap, based on the scapular artery and vein, was outlined with the probe in the center. The circular incision was carried through the panniculus carnosus, fulgurating all vessels except the central artery and vein, which were loosely encircled with 00 silk.

Raising the flap confirmed that the probe was superficial to the skin muscle and not exposed to air. Since it was found that exposing the undersides of the flaps to air raised O2 readings, the flaps were sutured back into their beds before devascularizing them. At this point each rabbit had two large viable island flaps, each with a probe in the center. The skin over the probes was painted with 2.0 cc of either 99% DMSO or equal parts of DMSO and 3.0% H2O2, being careful that none of the liquid spilled into the wound. After 30 min. the previously placed ligatures were tied, devascularizing the flaps. A vital dye (fluorescein or disulphine blue) was given intravenously to document the degree of ischemia. After gas levels had stabilized, the skin was again painted with the drugs, randomizing the sides.

In anesthetized pigs, 4.0 x 9.0 single pedicle dorsally based flaps were raised with a probe in the distal devascularized end, and the same drugs were painted on.

To test diffusion of O2 through the subcutaneous fat in the pig, 3.0 cm incisions were made from 1.0 to 6.0 cm away from probes. The incisions were sutured and drugs were injected into the wound cavity through a teflon catheter.

Throughout all experiments PO2, PCO2, and PN2 were recorded.

Results

Control tissue gas levels were remarkably stable, varying only with depth of anesthesia. Raising the flap and exposing the undersurface to air always led to a rise in PO2, a finding reported by Wilson, Toomey, and Owens. 3 After the flaps were sutured in place, values returned to control levels. Application of DMSO had no effect on normal skin. Ligating the blood supply to the flap led to a rapid and pronounced fall in PO2 and rise in PCO2.

Table I
Values of Gases (mm Hg) in Rabbit Flaps

	PO2		PCO2	
	mean	σ	mean	σ
Normal skin	36	6.4	46	7.5
" with DMSO	41	5.6	41	5.8
" " DMSO +H2O2	33	8.2	43	6.5
Devasc. skin	7	3.7	88	28.3
" +H2O2	8	1.5	66	17.2
" " + DMSO	24	28.7	57	23.2

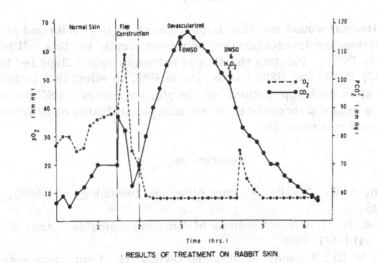

RESULTS OF TREATMENT ON RABBIT SKIN

Application of DMSO to ischemic skin led to an average fall in PCO2 of 30 mm Hg, along with an occasional slight rise in PO2. DMSO and H2O2 more consistently raised the PO2, an effect which lasted up to 60 min. after each application.

In the pig, devascularization of a skin flap led to marked rise in PCO2 and fall in PO2, but cutaneous application of DMSO and H2O2 had no effect on these values.

Also in pigs, making an incision and exposing the subcutaneous fat to air generally led to rises in tissue PO2 up to 6.0 cm away, an effect that ceased after the wound was sutured. Injecting H2O2 into sutured wounds had little effect on tissue PO2, but DMSO and H2O2 sometimes led to rises in PO2 up to 3.0 cm from the incision.

Discussion

All wounds and most skin flaps show a depressed PO2, but the exact role of hypoxia in causing tissue necrosis or wound infection is still un-certain. Investigators 4, 5 feel that raising oxygen levels in wounds and flaps would prevent infection and slough. The difficulty in delivering oxy-gen to areas without circulation is obvious, but the present experiments show that it can be done. Whether oxygen alone will be beneficial remains to be proven. The inconsistency of response to DMSO may have been due to its known deterioration on exposure to oxygen. The resulting compound, dimethyl sulfone, is non-toxic, but does not affect tissue permeability.

Summary

Experimental wound and skin flaps were made in rabbits and pigs. Making incisions or devascularizing skin consistently led to a fall in PO_2 and a rise in PCO_2. Painting the skin of ischemic rabbit flaps led to a fall in PCO_2; DMSO and H_2O_2 led to rise in PO_2, an effect that lasted up to 60 min. after each application. In the pig, cutaneous DMSO had no effect, but the drug did inconsistently enhance the diffusion of O_2 from H_2O_2 through the subcutaneous fat.

References

1. Kligman, A, M. Topical pharmacology and toxicology of DMSO, JAMA 193:140 1965.
2. Leake, C.D. Biological actions of dimethyl sulfoxide. Ann. N.Y. Acad. Sci. 141:1-671 1967.
3. Wilson, W.R., Toomey, J.M., and Owens, G. Continuous measurement of local PO_2 and PCO_2 during the creation of bipedicle skin flaps in dogs. Surg. Forum 22:495 1972.
4. Hunt, T.K., Linsey, M., Sonne, M., and Jawetz, E. Oxygen tension and wound infection. Surg. Forum 22:47 1972.
5. Bethman, W. and Schottke, C. Action of local oxygen application on healing of pedicle flaps. Zbl. Chir. 92:11235 1967.

IN VITRO RESPIRATION OF ISCHEMIC SKIN FROM AMPUTATED HUMAN LEGS

John S. Sierocki, Theodore Rosett, Raymond Penneys,
Douglas J. Pappajohn
Dept. Biochemistry, Temple University School of
Dentistry; Peripheral Vascular Sections, Philadelphia
General Hospital and Hahnemann Medical College and
Hospital

INTRODUCTION

This study was prompted by the observation that amputations of chronically ischemic, arteriosclerotic limbs must regularly be performed at a level well above any area of gangrene, if healing of the resultant surgical wound is to be expected.[1,2] The skin of such amputated legs grossly appears to be normal, except for any areas of actual gangrene which are generally located at the more distal end of the leg; the skin of these ischemic legs might play some role in maintaining the integrity of the limb, for only when it is disrupted by an injury do the visual signs of tissue breakdown and gangrene develop. Based upon these observations, the twofold purpose of this study was: 1.) to determine if the more distal, non-gangrenous sections of severely ischemic skin, shown in previous experiments by Penneys and Montgomery[3,4] to have a very low tissue oxygen tension _in vivo_, can respire actively when exposed to an adequate O supply _in vitro_; and 2.) to determine whether or not there are any quantitative differences in O_2 uptake between these more ischemic, distal, sections and the less ischemic, proximal, sections of skin removed from the same amputated limb, both endogenously and when exposed to certain intermediary metabolic substrates.

This work was done with the hope of obtaining some information which might be used in attempts to salvage a greater portion of these ischemic legs by allowing more distal amputations to be performed. Shown in Figure 1a is a leg with maximal gangrenous involvement of the entire foot and the anterolateral aspect of the leg; this leg was amputated above the knee. In sharp contrast to the specimen in Figure 1a is the leg shown in Figure 1b which has only a minimal gangrenous involvement of the second toe of the foot;

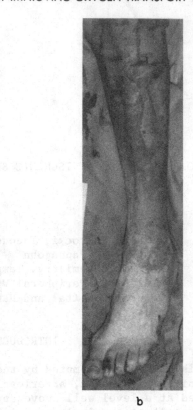

a   b

FIGURE 1.   AMPUTATED ISCHEMIC LEGS; a. maximal gangrenous involvement, and b. minimal gangrenous involvement of the toe (discolora-
tion of the leg, above the ankle, is due to merthiolate).

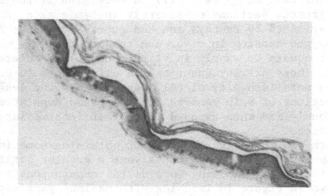

FIGURE 2.   MICROSCOPIC SECTION OF TYPICAL EPIDERMAL SHAVING

this leg was also amputated above the knee. The skin of this leg grossly appears to be normal except for the area of gangrene. If the capacity of the skin to respire was also normal, then in the presence of an adequate nutrient and oxygen supply, perhaps a more distal amputation could have been performed below the knee in order to preserve this joint which is crucial for full rehabilitation in many such amputees.

MATERIALS AND METHODS

In order to obtain specimens for study, twenty-three Phila-delphia area hospitals were contacted to determine if any legs were being amputated for arteriosclerosis obliterans with or with-out diabetes mellitus. No legs amputated for trauma, carcinoma, or sudden vascular occlusion producing acute ischemic episodes were included in this study. Immediately after amputation, the specimens selected were photographed and skin sections were taken for microscopic study from various levels on the leg, as described in Table I. Proximal (control) and ischemic, distal skin sections were then removed from the leg to include the epidermis, dermis, and as little subcutaneous fat as possible. The proximal section had an upper limit at the top of the amputated specimen, extended six centimeters distally, and encompassed the entire circumference of the leg. The distal section began at the mid-point of the ankle, extended six centimeters proximally, and likewise included the skin from the entire circumference of the leg at that level. An area of at least six centimeters always separated the proximal and distal skin sections; no skin from the foot was included in this study. These skin sections were immediately placed into an ice bath and taken directly to the biochemistry laboratory, usually within one hour after amputation, for measurement of epidermal oxygen uptake.

Preparation of the skin for study in the Warburg respirometer was done according to a modification of the method of Griesemer and Gould.[5] The skin was stretched across a dampened cellulose sponge and cleansed with EDTA buffering solution to remove any iodine or merthiolate (used as an antiseptic during surgery) which would interfere with the respiration of the epidermal cells. A rubber tube was then inserted between the skin and the sponge so that a safety razor could be used to strip away the epidermis in thin semi-cylindrical raised sections about 0.15 millimeters thick. These epidermal shavings from each section were cut into tiny pieces (about 0.5 centimeters) and divided into equivalent 500 milligram portions for study in the Warburg respirometer. Microscopic sec-tions of these shavings, as shown in Figure 2, show approximately equivalent amounts of epidermis to be present in both sections. There was a considerable amount of stratum corneum present and the epidermal to dermal ratio was about 3:1 throughout the shavings.

The epidermal shavings were placed into a Warburg respirometer along with Krebs-ringer phosphate buffer, 1500 units of penicillin

TABLE 1   TYPICAL MICROSCOPIC OBSERVATIONS OF THE SKIN FROM
                    VARIOUS LEVELS ON THE ISCHEMIC LIMB

SECTION	LOCATION ON THE LEG	MICROSCOPIC OBSERVATIONS
Proximal	Anterior tibial region 4 cm. below the upper level of amputation	A non-specific chronic inflammatory type of tissue reaction with perivascular exudation of lymphocytes and a few plasma cells; the epidermis appears normal
Inter-mediate	Midway between the proximal and distal sections on the anterior aspect of the leg	As above, a chronic non-specific type of inflammatory tissue reaction with a perivascular accumulation of lymphocytes and plasma cells; the epidermis appears normal but is perhaps slightly thicker than in the previous section
Distal	Antero-dorsal aspect of the leg at the mid-level the ankle	A definite hyperplasia of the epidermis with occasional parakeratosis and increased thickness of the stratum corneum; an increased amount of fibrosis and an increase in the vascularity of the dermis with a denser accumulation of lymphocytes and plasma cells characteristic of the chronic non-specific type of inflammatory response.

and streptomycin to suppress any bacterial contribution to O_2 utilization, and in the flasks other than the endogenous control, one of the intermediary metabolic substrates tested which included ADP, glucose-6-phosphate,alphaketoglutarate, and glutamic acid. This mixture was incubated at 37.5 degrees centigrade with readings taken every ½ hour for three hours after which the values for O_2 uptake for each specimen was calculated according to standard methods.

RESULTS

The average patient studied was 68 (53-92) years old and had had evidence of his peripheral arteriosclerotic vascular disease for 4½ (2 months to 10 years) years prior to developing gangrene. Half of these patients also had diabetes. Five amputations were above the knee, and five below. Microscopic sections of the ischemic skin, as described in Table 1, and illustrated in Figure 3, in general demonstrated a chronic non-specific inflammatory reaction with characteristic perivascular accumulations of lymphocytes and occasional plasma cells. The intensity of this reaction was comparably increased in the more ischemic, distal sections of the leg. The epidermis showed a definite and persistent hyperplasia with marked prominence of the rete ridges and rete pegs as the more ischemic distal sections of the leg was approached; this finding was absent in the less ischemic proximal portion of the leg. The stratum corneum appeared to be thicker in the distal sections when compared to the proximal sections of the leg.

A total of ten legs were studied, all of which were observed under endogenous conditions, seven in the presence of ADP, nine with alphaketoglutarate, seven with glucose-6-phosphate, and four with glutamic acid. The results of Warburg respirometry done on the proximal and distal skin sections removed from the amputated legs of these ten patients are shown in Table 2. The endogenous value for O_2 uptake by the more ischemic distal skin sections was not statistically different from that of the less ischemic proximal sections, averaging 273 microliters of O_2 absorbed per 500 mg wet weight of tissue after 3 hours for the distal sections versus 257 microliters of O_2 absorbed per 500 mg wet weight of tissue after 3 hours for the proximal sections. Both sections had a rapid uptake of O_2 and these were all direct relationships, indicating that O_2 uptake was dependent upon tissue utilization with no contribution to respiration by any bacterial contamination of the tissues studied. The QO_2 of the proximal skin was 0.17 microliters of O_2 per milligram of wet weight per hour while that for the distal skin was 0.18 microliters of O_2 per milligram of wet weight per hour, again emphasizing the similarity in respiration between the proximal and distal skin sections.

In the comparative studies between the two sections with various substrates added to the Warburg incubation flasks thus far ADP has stimulated O_2 uptake in both the proximal and distal skin sections when compared to the respective endogenous values with P

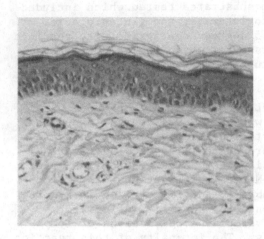
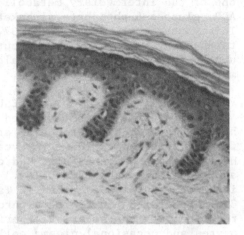

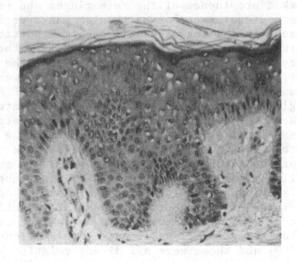

FIGURE 3. TYPICAL MICROSCOPIC SECTIONS; a. proximal,
b. intermediate, and c. distal levels of the leg.

TABLE 2　O_2 UPTAKE AFTER 180 MINUTES IN MICROLITERS PER 500 MILLIGRAMS OF SKIN

Patient	ENDOGENOUS P*	ENDOGENOUS D**	ADP P	ADP D	G-6-P P	G-6-P D	α-KG P	α-KG D	GLUTAMATE P	GLUTAMATE D
1	252	342	276	296	357	286	328	428	123	250
2	251	323	376	402	329	348	359	347	330	297
3	328	357	348	315	330	402	297	362	403	393
4	330	277	274	292	296	315	420	284	273	338
5	262	356	128	180	176	292	267	317		
6		203	280	195	225	180	180	211		
7	100	165	425	200	353	195	83	173		
8	216	75	290	295	258	200	290	195		
9	325	335		452	368	295	230	413		
10	275	295			291	452	298	355		
Average	257	273	299	291	281	286	286	309	282	319
P Value	0.40		0.75		0.75				0.40	

P * = Proximal, D** = Distal

equal to <0.01 proximally and P equal to <0.05 distally. The proximal section responded better with glucose-6-phosphate yielding a statistically significant difference between its presence and its absence endogenously, the P value being <0.05. No such statistical difference was noted distally. Compared to the endogenous value, the distal skin sections responded better with alphaketoglutarate, the P value veing <0.04 distally and >0.25 proximally.

It is pertinent to compare the effects of the various substrates upon O_2 uptake to glucose-6-phosphate as a reference instead of the endogenous value for O_2 uptake. If it is supposed that the involved distal skin underwent some sort of metabolic changes, then the endogenous respiration may not necessarily be a true refelction of the basal respiratory rate. On the assumption that all needed mechanisms and enzyme systems in the epidermal cells are present, then the addition of glucose-6-phosphate as a substrate would give a closer to normal baseline value for respiration than that observed endogenously since: 1.) glucose-6-phosphate will freely enter even diabetic epidermal cells to enter the glycolytic pathway and 2.) the more distal epidermal cells may very well have been deprived of an adequate amount of usual substrate because of the very poor or lack of circulation with inadequate tissue perfusion and consequently underwent a shift to a metabolic pathway utilizing less O_2 and/or a different substrate. Thus when the values for O_2 uptake with glucose-6-phosphate are compared with the various substrate responses, it can be seen that: 1.) alphaketoglutarate is now twenty-two percentage units higher in the distal sections than in the proximal sections compared to the eight percentage units difference when both sections were compared to their respective endogenous values, and 2.) respiration with glutamate is now fifteen percentage units higher in the distal sections than proximally compared to a difference of thirteen percentage units when both were compared to their endogenous counterparts.

DISCUSSION

The endogenous O_2 uptake for the proximal and distal skin shows that both are capable of active respiration with no significant impairment in the more ischemic distal sections when compared to the less ischemic proximal sections of the leg. In addition, no morphologic evidence of any epidermal abnormalities was noted during microscopic inspection of tissue specimens taken from various levels on the amputated legs. With these facts in mind, it could be asked why then do amputations done in the more distal sections of the leg fail to heal properly? Might the fact really be that although the distal ischemic skin is actively taking up oxygen, its respiration is different from that seen in normal non-ischemic skin? This does not seem to be the case as far as the endogenous QO_2 of chronically ischemic skin is concerned since our values were both similar to those reported by two other investigative teams. According to work done by Ohara[7], a compar-

able QO_2 for normal non-ischemic skin would be 0.12 microliters
of O_2 per milligram wet weight of tissue per hour; Evans and
Naylor reported the QO_2 value of inflamed skin to be 0.11 microliters of O_2 per milligram of wet weight of tissue per hour. Our
QO_2 values were both about 0.18 microliters of O_2 per milligram
wet weight of tissue per hour. This suggests that an abnormality
in cellular respiration is not one of the factors involved in the
failure of this ischemic distal skin to heal following amputation.
In further support of this fact is the observation that ADP stimulated O_2 uptake equally in both the proximal and distal sections
of skin, suggesting an intact oxidative phosphorylating mechanism
in the respiratory chain of the mitochondria of these epidermal
cells.

In contrast to the meaning of the endogenous values for O_2
uptake, the remainder of the comparative studies demonstrate trends
in O_2 uptake which can be used to speculate about the metabolic
activity in these chronically ischemic epidermal cells. Glucose-
6-phosphate stimulating respiration more in the proximal skin
sections might suggest a closer to "normal" metabolic state of
affairs in these cells. Glutamate and alphaketoglutarate stimulating O_2 uptake more in the distal skin sections may suggest that
an altered metabolic state, with perhaps an increase in some and
a decrease in other enzyme systems, exists in these ischemic cells.
These speculations, based upon the trends in O_2 uptake in only
a few initial exploratory experiments, could be validated or
negated by assaying for specific enzymes involved in the metabolic
pathways in these epidermal cells.

In conclusion, it should be re-emphasized that these results
confirm the fact that chronically ischemic skin is still capable
of active respiration, even in those skin sections removed from
the more ischemic distal portion of the amputated limb, and substantiate the general concept that the failure of the skin to heal
in the chronically ischemic leg is due to an inadequate supply of
nutrients and/or oxygen to these cells in vivo, while the inherent
ability of the epidermal cells to respire remains intact. In
addition, it appears that these epidermal cells are very resistant
to the chronically ischemic state. These findings encourage us
to emphasize that even in the most severely ischemic legs, which
seem to be beyond recovery, every effort should be made to surgically revascularize these ischemic tissues prior to resorting to
amputation of the limb. This should theoretically allow a more
distal amputation to be performed by providing the nutrients and
oxygen needed to heal the resultant surgical wound.

SUMMARY

This study is designed to determine if severely ischemic human skin, known to have a very low tissue pO_2 in vivo, can respire actively when offered adequate O_2 in vitro; and to observe whether or not there were any differences in respiration between the more ischemic distal, and the less ischemic proximal sections of skin removed from the same leg. This research was prompted by the observation that most of the skin in the chronically ischemic leg has a grossly normal appearance except for areas of actual gangrene. If the capacity of this skin to respire was also normal, then perhaps more distal amputations of these legs would be possible. Skin was obtained from freshly amputated, chronically ischemic, arteriosclerotic human limbs. Slices of the skin containing all of the epidermis were placed in a Warburg respirometer where O_2 uptake was measured for 3 hours. Microscopic sections showed approximately equal amounts of epidermis present in the various skin slices.

The endogenous O_2 uptake of all samples of distal skin was essentially not statistically different from that of the more proximal skin which served as the control. The O_2 uptake was quite rapid and averaged 3.03 microliters/gram of wet weight/ minute. Thus far in the comparative studies, ADP has stimulated O_2 uptake in almost all cases. Glucose-6-phosphate stimulated respiration more in the proximal skin sections while alphaketo- glutarate and glutamate stimulated respiration more in the distal skin sections. These results confirm the fact that chronically ischemic skin is still capable of active respiration, even in those skin sections removed from the more distal portion of the amputated limb.

REFERENCES

1. Romano, R.L. and Burgess, E.M. (1971) Level Selection in Lower Extremity Amputations. Clin. Orthop. 74: 177.

2. Kelly, P.J. and Janes, J.M. (1970) Criteria for Determining the Proper Level of Amputation in Occlusive Vascular Disease. J. of Bone and Joint Surg. 52: 1685.

3. Montgomery, H. and Horwitz, O. (1950) Oxygen Tension of Tissues by the Polarographic Method. I. Introduction: Oxygen Tension and Blood Flow of the Skin of Human Extremities. J. of Clin. Invest. 29: 1120.

4. Penneys, R. and Montgomery, H. (1952) Oxygen Tension of Tissues by the Polarographic Method. V. Rate of Movement of Oxygen from Peripheral Arteries to the Skin. J. of Clin. Invest. 31: 1042.

5. Rosett, T. and Fogg, J.B. (1962) Studies in the Biochemistry of Skin. I. Respiration of Adult Rat Epidermal Tissue. J. Invest. Derm. 39:395.

6. Jarrett, A. (1971) The Pentose Phosphate Pathway in Human and Animal Skin. Br. J. Derm. 84:545.

7. Ohara, K. (1951) Studies on the Oxygen Consumption of Human Skin Tissues, with Special Reference to that of the Sweat Glands. Jap. J. Physiol. 2: 1.

8. Evans, N.T.S. and Naylor, P.F.D. (1967) The Systemic Oxygen Supply to the Surface of Human Skin. Respira. Physiol. 3: 21.

ALTERING DIFFUSION RATES

John L. Gainer and Guy M. Chisolm III

University of Virginia, Charlottesville, Virginia

In a recent article concerned with the relation between glucose absorption rate and serum globulin concentration in man the authors showed a direct linear relationship between glucose absorption rate in the jejunum and the serum γ-globulin concentration, with the glucose absorption rate decreasing as the γ-globulin concentration increased. (1) We have previously shown a direct linear relationship between the diffusivity of glucose in plasma and the γ-globulin concentration (2), which is almost exactly that seen in the previously-mentioned experiments. In both cases the variation of γ-globulin was over the physiological range, although those ranges were slightly different (Zambian African subjects for the glucose absorption studies and USA Caucasians for the diffusibity work). The striking similarity between the two sets of data brings to mind an interesting question of whether diffusion through plasma is of major importance in determining the rate of uptake and release of solutes by and from tissues.

Although these studies involved glucose we have previously shown that the variations in the diffusivities of glucose and oxygen are the same when the concentrations of several of the plasma proteins are changed over the normal physiological range(3). That is, although the absolute values of the diffusion coefficients for glucose and oxygen through plasma are different, they change in the same ratios as the concentrations of the plasma proteins are varied. Thus, one might speculate that variations in the diffusivity of oxygen, due to variations in the plasma proteins, might be of importance in determining the amount of oxygen available to tissues.

We have attempted to determine what might be the reason for the variation of diffusivities with changes in concentrations of plasma proteins. For example, when the concentration of albumin in plasma is varied over its normal physiological range, the diffusivity of oxygen decreases by as much as 50%. If one does the same experiments in saline-albumin solutions, it is found that there is little decrease of the oxygen diffusivity with increasing albumin concentrations. (4) This brings up the possibility that it is the interactions of different kinds of proteins which is responsible for the different behaviors of the solutions.

First, we determined the diffusion coefficients for plasma and for a solution containing the plasma proteins dissolved in a 7.4 pH buffer solution. Varying the albumin resulted in a 50% decrease in the diffusivity in both solutions, thus it would seem unlikely that the extra constituents present in the plasma are responsible for the changes. Then, using the 7.4 pH buffer solution containing 1.75g/100 ml of α-and β-globulins we measured the diffusivities through various types of solutions. In these experiments we varied the concentration of albumin from 0 to 5 g/100 ml both with and without the presence of other plasma proteins. These results are given in Table 1.

TABLE 1

Base solution of a 7.4 pH buffer containing
1.75 g% of α-and β-globulins.

Other Proteins Present	% Decrease in Diffusivity With Increasing Albumin Concentrations
None	0
0.3 g% fibrinogen	0
1 g% γ-globulin	50
1 g% γ-globulin and 0.3 g% fibrinogen	50

These results would seem to indicate that it is some interaction between the ablumin are the globulins which is responsible for the decrease in the diffusivity.

We have also noted decreases in diffusion rates through blood plasma when various other types of substances are added. Some examples are listed in Table 2:

TABLE 2

Substance Added	Amount Added (g/100ml)	% Decrease in Diffusivity
Ethanol	0.1	10
	0.15	35
Aspirin	0.01	12
	0.1	18
Glucose	0.1	23
	0.25	47
Sodium pentathol	0.016	20
Ethyl ether	0.2	57

Thus it is possible that diffusion rates through plasma are affect by many drugs, and this may be of importance in oxygenation of tissues. In summary, if diffusion in plasma is important in the transport of oxygen to the tissues, the concentrations of the proteins and other substances in the plasma may result in very different oxygenation rates.

REFERENCES

1. Cook, G. C., Nature, 241, 284 (1973).

2. Conner, E. D. and Gainer, J. L., CEP Symposium Series, 66, No. 99, 72 (1970).

3. Navari, R. M., Gainer, J. L. and Hall, K. R., AIChE Journal, 17, 1028 (1971).

4. Navari, R. M., Gainer, J. L. and Hall, K. R., "Effect of Plasma Proteins on Diffusion", Chapter in "Blood Oxygenation", edited by D. Hershey, Plenum Press, New York, 1970.

DISCUSSION OF SESSION III

Chairmen: Dr. Lars-Erik Gelin, Dr. Konrad Messmer and
Dr. Mary P. Wiedeman

DISCUSSION OF PAPER BY J. W. IRWIN

Gelin: What material is included in your white emboli?

Irwin: We have been working on those platelets and on those
hyaline emboli for about ten years and everything we look for in
them we find. There is albumin in them that has been precipitated
out, there are the various types of globulin, there is a lot of
fibrin, and there are small platelet emboli which are separate
which I did mention. The red aggregates, at least in this reaction
which occurs quickly, do not seem to become important, although I
think this is probably tied together by some form of protein. I
always wonder why blood flows, because it really is a mess, and the
more you study it, the more you wonder why it ever flows through
anything, let alone these tiny vessels, but it does do it.

Gelin: It is probably mostly an antigen-antibody complex?

Irwin: No, it is not. It has some antibodies in it, but that is
not what most of those proteins are right out of the blood.

Gelin: How does the endothelium behave when you induce your
antigen-antibody reaction?

Irwin: The last piece of the film did show the endothelium
thickening. Yes, it does, but we have not studied that thoroughly.
You did see one vessel constricted off and also partial constriction.

Gelin: In the reaction of the tissue, I might say, that that is
one of the most early reactions.

Irwin: Another thing we have to remember is that the antigen-
antibody reaction is injury to tissue, so it is just the inflammatory process.

Gelin: From your experiments with anaphylaxis, can you differen-
tiate which is the primary effect: the precapillary constriction
in the pulmonary vessels or the forming of the blood?

Irwin: That depends on how much antibody you use. If you use a
large dose of antibody, you see fewer of the emboli, so I think

733

in that case the most important thing is constriction; whereas, when you use a more moderate amount of antibody, then the emboli become more important.

Gelin: This would mean that there is no direct action, but just a release of substances from the antibody-antigen complexes.

Irwin: I think that is true. Don't forget, they are reacting too, with the surrounding tissue.

Rubin: In the past we have been interested in what happens to the lung in severe nonthoracic sepsis, and we and others have described a lesion which has been called congested adalectisus. Hardaway, a number of years ago, described clumping of red cells using a microscopic technique in the live animal, and you have described this happening in the lung. Do you have any observation on animals that have been made previously septic as to what might happen to the pulmonary circulation during that time? We have believed that this clumping of cells and the destruction of endothelium is more or less a final common pathway of the lung's response to many insults.

Irwin: I think that is true. You are using sepsis as using bacteria. Is that what you meant by sepsis?

Rubin: Our model was appendicitis undrained in a dog that had been treated with penicillin to preclude clostridial overgrowth and we also did the same in rats.

Irwin: You are using the bacteria to produce inflammation. The only reason we prefer anaphylaxis when we make these studies is that we can control it and I do not know how to control sepsis by a degree.

Guest: Does the administration of heparin or a fibrinolytic agent cause any decrease in the degree of aggregation?

Irwin: Anticoagulants were tried in studies completed several years ago without any noticeable effect.

Gelin: Did you use cortisone to prevent any of these reactions?

Irwin: We gave the animals cortisone for a period of time. We did not prevent them, but these are not mild reactions. I gave you the doses. These are tremendous doses of antibody. You could rarely get an animal that sensitive by active anaphylaxis. You just do that because you make the antibodies some place else and just lay it in as you want it.

DISCUSSION OF PAPER BY E. H. BLOCH

Gelin: This concept brings up the possibilities of a specific or an unspecific ideology for red cell and blood cell aggregation, a concept which has been thought about. There is, as far as we know, not one single drug which can prevent all the consequences which you outlined.

Irwin: In malaria, in the various species studied, do you know of any evidence of change in oxygenation of any affected tissue?

Bloch: I have never made a measurement to determine the oxygenation of any tissue. These are phenomena which have been studied. One assumes that, because there is a linear reduction in flow, that there is an alteration deoxygenation. But there is no proof.

Gelin: Much of the biophysics involved here are, so far, poorly studied. I mean, areas like the viscous metamorphosis of the platelet. We know it is occurring, but it is very hard to determine in an exact way what really is producing it; then they suddenly become irreversibly sticky cells. Ronenmark has some nice slides on the stickiness of platelets going as single particles to the red cell surface, and thereby get them aggregated quite firmly. Lee, here in the thoracic unit, has discovered a specific protein during extracorporeal circulation which is covering, not only one cell, but several cells. They might break up if you get a water-holding colloid into the mass which is gluing them together.

Bloch: Relative to that, in a study of animals treated with phenylhydrazine, you observe the spleen and the red cell destruction. There, the red cells adhere very firmly to the wall of the reticuloendothelium in the spleen, and then you use the strongest surface activations you can think of; quinine, which is one; or heparin, and nothing budges those cells.

Wiedeman: I was interested in the pictures of the effects of alcohol on those blood vessels that you were looking at. I have personal experience that tells me that alcohol has a dehydrating effect and I just wondered if there was more to what we saw in the film.

Bloch: Yes, I would like to say immediately that this is the very top of the iceburg. In the '40's, we made a series of studies of giving known amounts of alcohol to a subject and then withdrawing the blood and doing an analytic electrofluroetic study simultaneously. Very shortly, within an hour thereafter, the albumin decreases and you have a new component, an alpha 3 globulin. You get an increase in the gamma globulin. So that is just the biochemical changes. There is some increased permeability which you cannot measure, but you can observe that. It is a very complicated study and to see the erythrocyte aggregation is only a very tiny part of it. That would be oversimplification.

Gelin: Regarding flow velocity, how much would you ascribe to low flow rate concerning aggregation power?

Bloch: This is dependent upon (which we know nothing about, again) the strength with which the particles adhere to each other. Now, if the weakest aggregates can be observed in the H-shaped bifurcation of vessels, where the courses between the two branches, balance each other out, and if the blood just flows into any other aggregate, it can just break up. That is the weakest force. In the arterial system, you have the greatest force and some of the aggregates will have physical characteristics that may be not much larger than any individual red cell, but it will be so rigid that it cannot be forced through the tip of the terminal arteriole. And so the size is not the most important part of an aggregate. It is its physical consistency. This is the reason that static pictures will give you an erroneous impression. I have seen aggregates of 50 or 60 microns squeezing, without any difficulty, through arterioles but you will have some aggregates of red cells, just red cells, for a moment, that measure only 8 micra and they will be intermittently plugging the arteriole for moments to minutes or even permanently.

Cameron: Your last comments were on the area I was interested in, namely the influence of erythrocyte deformability in this phenomenon of sludging. I wonder if you might comment on this, particularly with reference to the problems associated with sickle-cell disease.

Bloch: My colleague, before I came to this conference, asked me the same question. I looked at what I had written, and I had 12 cases and those were in 1944 and I cannot remember exactly what happened, so I cannot answer it.

Guest: I agree that no direct evidence for the adherence of fibrin on red cells is available. However, we have been able to essentially prevent sludging due to burn trauma and in decompression sickness by removing the fibrinogen from the plasma through activating the plasma fibrinolytic system. The process of sludging appears to cause a vicious cycle. Because of red cell aggregation, fewer cells traverse true capillaries. It has been demonstrated by the Todd Fibrinolytic Autograph Technique that endothelium releases an activator of profibrinolysin. Since the true capillary has a large ratio of luminal surface to cross-sectional area, the activator concentration is high in the neighborhood of red cell surfaces. We postulate that, because of the fibrinolytic activator, red cells which traverse true capillaries are cleaned or cleared of fibrin debris.

DISCUSSION OF PAPER BY L.-E. GELIN

Halmagyi: According to your thesis, the aggregation of the red cells is a more important factor in the tissue injury during shock than the flow rate. I think a good objective measure, to express quantitatively the degree of aggregation, is to measure the constant between large-vessel hematocrit and total-body hematocrit. If your thesis is correct, in all cases where there is irreversibility, where there is severe injury, you will find the greatest distortion of this constant. This has not been our experience, and as far as I understand, other people could not see the same either. I think that, for instance, the lactic acid production, which you assume is secondary to this aggregation. In shock, the lactic acid production can be completely prevented by administration of complete alpha and beta adrenergic blockade. Although this blockade does not affect the distortion of the total body and large vessel hematocrit, this is suggesting that the aggregation factor has not been effected. Yet, lactic acid release has been prevented.

Gelin: Yes, I think it is a question about the availability of oxygen for exchange. If arterial constriction is prevented by blocking agents, flow rate through the capillary and especially the off flow will be facilitated, thereby preventing stasis. I think that is one of the important aspects of flow rate, to prevent stasis. The number of cells in a specific situation does not seem to mean too much for this exchange system. I think the studies done by Shoemaker on this aspect show that a slow-flowing compartment of blood cells and a normal-flowing compartment of blood cells has to be included in the total hematocrit. But the effective dynamic hematocrit is the thing which takes part in the exchange. Prevention of stasis by lowering of arteriolar hindrance which contributes more pressure blood for vascular flow is very important in shock. In terms of whole-body hematocrit and large-vessel hematocrit, we must recall that in stasis, a large portion of red cells are not flowing, or slowly flowing, and that only the dynamic portion of the total red cell mass participates efficiently in exchange with the tissue.

DISCUSSION OF PAPER BY H. I. BICHER, D. F. BRULEY AND M. H. KNISELY

Penneys: You used the term 'anti-adhesive drugs' in the program. Many experts are constantly showing the difference between 'adhesiveness' and 'aggregation.' Do you use those terms interchangeably?

Bicher: It is just jargon. Of course, there are different function names of the platelet, but several parts of this reaction are common.

We could call them 'anti-aggregation drugs' but 'anti-adhesive' just sounds better. They are able to block both adhesiveness and aggregation, and I did not find one word to call them both.

Kessler: I wonder that you observe only small changes of local tissue pO_2 when you induce so drastic changes in arterial pO_2. Besides metabolic changes, which might partially explain your measurements, you should get a more pronounced oxygen washout.

Bicher: Yesterday, I described in some detail the tissue pO_2 response. It goes down, but we have described the plateau. If you only give nitrogen for a short period of time, it does not go down to zero. I think that what Ian Silver described on the potassium leakage of cells is very important. If we want to reproduce this time and time again in order to get new values for reoxygenation times, we can only give nitrogen for a short period of time, and then, as blocked pO_2 only approaches zero values for a very short period of time, the compensation mechanisms in our hands prevent it from reaching a real zero level. If we continue, let's say, instead of 60 seconds, for 90 seconds, then we begin to get the real drop in tissue but then we are really producing damage too.

Kessler: But you should get at least a certain washout. I agree with these metabolic things, but you should get a washout. This is what we also see with the hydrogen electrode, when we do such measurements with the hydrogen electrode.

Bicher: It is just a matter of timing. If you continue the nitrogen for a longer period, you get the washout, but we have described a delay of sometimes up to 60 seconds, between 40 and 60 seconds, between the initial fall in arterial pO_2 to the fall in tissue pO_2.

Duhm: I would like to know the effectiveness of the compounds you studied as compared to that of dipridamole. Do your compounds influence the membrane permeability of blood constituents other than the platelets? Is anion permeability of erythrocytes affected?

Bicher: Dipridamole is one of what I call a 'weak' anti-adhesive drug. In a milligram per milligram comparison, it is about 10 times weaker than, for example, Biolar 743. Also, these drugs will block all of the responses of the platelet, ADP, collagen, and so on. Dipridamole will do that, but in what is called the primary ADP response, as opposed to the secondary release response, it is much weaker than ours and there are several other drugs, aspirin, for example; I could name about 100. We have not measured permeability of ions in the red cells. All we know is that there is a change in the red cell shape, it separates, probably due to a change in permeability of water, there is some skimming and crenation, but we did not measure specific ions in the red cells.

Puffer: Have you attempted to enhance activity of the compounds by molecular modification of the compounds and, if so, do there appear to be specific physico-chemical properties related to the activity?

Bicher: These compounds were developed by mass screening. We do not know yet which is the active part of the molecule.

DISCUSSION OF PAPER BY K. MESSMER, L. GÖRNANDT, F. JESCH,
    E. SINAGOWITZ, L. SUNDER-PLASSMANN AND M. KESSLER

Gelin: I must make a comment on this very beautiful presentation stressing the significance of flow over the amount of hematocrit. The anemia, which usually occurs following injury, flow disease; any disease will have an anemia. I have tried to state that this anemia is a compensatory mechanism, a defense anemia, and should not be treated, and I think you have given all the evidences that hemodilution which nature provides, is to facilitate tissue nutrition, and I am very delighted to see your data.

Whalen: I believe you said that the low values for tissue pO_2 you found in skeletal muscle were below other published values. We have published data on rat skeletal muscle in which the histograms were quite similar to yours. Did you have much inter-individual variation? I do not feel that low values are necessarily abnormal for some animals. Perhaps these are as normal as the other measurements in other organs and dextran disturbed the mechanism(s) which was keeping the pO_2 low.

Messmer: Since most pO_2 values were found in the hypoxic range of order, whereas Kunze (1969) found a wide distribution curve, we believe that we are dealing in our dogs with impaired skeletal muscle flow. This emphasis is supported with low values for skeletal muscle flow in the surgically untouched contralateral extremity as measured with the Xenon-clearance technique. According to our opinion, we therefore are not disturbing, but improving, skeletal muscle flow distribution which is in accordance with an increase in PS (permeability-surface area product according to Renkin) values after hemodilution.

Boyan: I would like to fortify what Dr. Messmer has found in dogs with a case in a human. I had a patient who had perfuse hemorrhage and who did not have blood, and this patient received 3 liters of dextran 60. The hematocrit fell from 42 to 11 and we maintained normal volume. The patient was sustained for 2 hours with such a low hematocrit and made an uneventful recovery; so whatever Dr. Messmer has found out has been demonstrated in humans as well.

Whalen: I did not mean to say that there was anything wrong with the data or that that did not happen. What my contention is, is that you are assuming that the skeletal muscle low values were abnormal and I say that they may be normal and that the hemodilution is destroying some of the autoregulation. I do not like to say that that is what is happening, but I think that that is a possibility.

Messmer: But would you really suggest that more than 50% of the pO_2 values are found in the range from 0-12 mm Hg?

Whalen: In some animals, yes. That is why I asked whether you found a variation.

Rubin: The dose of dextran 60 which you gave your dogs was far in excess of the dose usually recommended for humans. Did you note any adverse pharmacologic effects of these large doses and was the clearance of dextran altered?

Messmer: Even though we exceed the clinical dosage of dextran by far, we never observed adverse effects (see Adv. Microcirc. 4, i-77, 1972). Kidney function and coagulation were not affected in survival studies. The dextran elimination was found normal. It is, however, recommended for clinical use not to exceed 1.5 g/kg body weight in 24 hours as dosage of clinical dextran.

DISCUSSION OF PAPER BY M. P. WIEDEMAN

Nicoll: Any time that you narrow a stream in a microvascular two-dimensional net of this variety, you begin to slow the flow up-stream from the point of narrowing. Immediately, when that occurs, you begin to get this showing of platelets to the outside and a rich platelet plasma moves down the outside. Thus, I do not think you have to have any information transmitted from the site of platelet injury back upstream to get this beginning of shoving down of the platelets to the outside for the development of this plasma-rich platelet layer.

Berman: I wish to comment on a very well designed experiment. Would you comment on the following:
    (1) Do you attribute the effect of the hemoglobin hemolysate to absorption of heat?
    (2) Was heat absorbed by Congo Red?

(3) What were the concentrations of hemoglobin and Congo Red?

(4) How pure was the hemoglobin solution?

Wiedeman: (1) I believe that the heat from the laser beam coagulates protein in the hemolysate.

(2) I have no real evidence that Congo Red absorbs heat, but in any event, it was a red solution, like blood in color, as opposed to saline, which was colorless.

(3) I do not know the concentrations of hemoglobin or Congo Red.

(4) The hemolysate was prepared by freezing washed, red blood cells and then, after the solution thawed, it was separated by centrifuging cell bodies and hemolysate.

DISCUSSION OF PAPER BY L. C. CLARK, JR., S. KAPLAN, F. BECATTINI AND V. OBROCK

Beran: We have performed push-pull exchange transfusion until a hematocrit of 3% was obtained, using emulsified fluorocarbon FC-47 prepared in a manner similar to that described by you and Geyer. Rabbits were maintained under this condition for 3 hours and then were reinfused with their own blood. During this 3-hour period, the PaO_2 was maintained at very high levels; however, a progressive decrease in $\dot{V}O_2$, $\dot{V}CO_2$ and CaO_2 (arterial oxygen content) with satisfactory ΔA-V $p\bar{O}_2$, and an increasing metabolic acidosis and unchanged PvO_2 during 100% O_2 breathing, suggests that oxygen carrying or releasing capacity of FC-47 is deficient in vivo. We postulate that the cause for this is: (1) Protein coating of fluorocarbon particles, or, (2) Retaining of fluorocarbon particles by the liver. Would you like to comment on this?

Clark: At present, it is very difficult to compare results of one laboratory with another because there is as yet no standardized way to prepare perfluorochemical emulsions. Use of purified fluoro-chemicals will partially solve this problem. We are in the process of purifying these liquids by spinning band distillation. But most important will be to characterize the final emulsion as to particle size and particle size distribution. At present a statement as to the absorbance is not too bad. All our results indicate that there is no coating on the FC particles which impedes oxygen transfer. Emulsion which has been in the body a few hours is just as good as new emulsion. There is a decrease in mixed venous pO_2, due to a decrease in cardiac output, about 2 hours after emulsion is given. The retention of FC by the liver could not explain your results because the blood FC level changes very little in three hours.

Whalen: Whenever we have used any type of artificial perfusion
media, including fluorocarbon emulsions, we have, of course, a very
high "arterial" pO_2 (500-600 mm Hg) and a high venous pO_2 (100-300
mm Hg). Yet, the tissue pO_2 is mainly zero except at intervals of
about 2 mm Hg, as if the flow were going entirely through a few
large channels.

Clark: I doubt that the high mixed venous pO_2 observed in our
animals is due to shunting the small vessels, the capillaries, or
the tissue substance. But muscle is a very mysterious tissue to
me, as far as blood flow goes. We have given large amounts of
perfluorodecalin emulsions to cannulated, awake animals. There
is really no observable effect. Cats go on playing with their
toys as if nothing were happening.

Kovach: We have used fluorocarbon in rat liver perfusion studies,
and found that the O_2 consumption is 50% higher than with hemoglobin-
free perfusion. The liver worked well for 2-3 hours. Our problem
was, that the emulsion was mildy and we could not use it because
it absorbed the light. It is very interesting to understand that
you have developed new techniques to clear it up.

DISCUSSION OF PAPER BY L. G. MOORE, G. J. BREWER, F. J. OELSHLEGEL
    AND A. M. ROSE

Duhm: Did you measure plasma and red cell inorganic phosphate
concentration? Can the increase you probably observed explain
the size of DPG levels of about 0.5 - 1 μ mole/gram RBC? Was the
blood pH elevated in the phosphate-treated group? Although you
may obtain significant results with respect to the tests of central
nervous system function, it may be a matter of caution not to draw
any conclusion from the comparison of two pairs of five persons
each with respect to an effect of a shift of the P_{50} of about 2 mm
Hg, which is probably reduced by your 15% increase of DPG levels.

Moore: First, we observed, on the average, 25% increases in the
phosphate-treated groups. Second, as I pointed out in my presentation, the in vivo effect of increases in inorganic phosphate is
probably centered on the HK and PFK enzymatic steps and depends
on altering the normally inhibitory influences on these enzymes.
Therefore, there is no reason to require the large increases in
phosphate levels found in the in vivo incubations for a significant increase in DPG levels to occur. In other words, I think
your question stems from observing the much higher increases in
phosphate observed in vitro. The in vitro and in vivo situations
in the respect to phosphate effects need not be identical. Third,
blood (whole blood and intracellular) pH was elevated in the

phosphate-treated group during altitude exposure. However, the
control group also exhibited similar pH increases. Finally, the
caution you urge was expressed in my presentation. Our study was
intended as a feasibility, pilot study whose results encourage us
to pursue further our line of reasoning.

Puffer: Are there similar correlations under hyperbaric conditions,
i.e., does a drop in DPG levels occur?

Moore: Yes, I believe work has been done on rhesus monkeys in
hyperbaric conditions in which a drop in DPG levels were seen.

DISCUSSION OF PAPER BY N. E. NAFTCHI, M. MENDLOWITZ, S. RACOCEANU
     AND E. W. LOWMAN (Paper was presented by Dr. Maurice Berard.)

Wiedeman: How does heat effect nerve blockade? There are some of
us in microcirculation who do not believe that small vessels are
innervated, if that helps in your explanation.

Berard: Generally, we managed to deplete our noradrenaline stores.
There is a question as to whether the histamine does act directly
or indirectly.

DISCUSSION OF PAPER BY B. MYERS AND W. DONOVAN

Pearce: Do you have any information on the long-term effects of
dimethyl sulfoxide and hydrogen peroxide when applied to the skin
surface?

Myers: As you know, DMSO was taken off the market; this is technical
grade. It has been used extensively in patients, apparently without
effect. These were simply acute experiments to see if we could
effect oxygen transport with them.

Rakusan: How do you explain the significant decrease in pCO_2 in
devascularized tissue after additional oxygen supply?

Myers: I am not sure, but I think we are increasing the permeability
of the skin and it is diffusing out into the air.

Goldstick: Does DMSO cause hyperemia as well as increasing skin
permeability for O_2 and CO_2?

Myers: It has been reported to, but we have not tested it for that.

DISCUSSION OF PAPER BY J. S. SIEROCKI, T. ROSETT, R. PENNEYS AND
    D. J. PAPPAJOHN

Nicoll: Have you treated your patients with the pressure increases synchronized with the normal blood pressure peak? This has been reported to be successful in many situations.

Penneys: There are many treatments for arterial disease and only a small percentage of legs with arterial disease come to amputation, I am glad to say, and there are many treatments which are helpful. These all work by increasing blood flow, but in the legs I have shown, nothing can increase their flow, including that type of instrument. Certainly there are those with borderline deprivation of flow which can be helped by this and other types of instruments.

DISCUSSION OF PAPER BY J. L. GAINER AND G. M. CHISOLM, III

Dorson: As you and I have discussed previously, there may be other explanations for your O_2 diffusion results. We have tried to repeat your work in a flowing diffusion cell, and our results agree with those of Dr. Goldstick. We used separated beef and human plasma and added both γ-globulin and albumin in independent experiments. Any activation energy process might produce the same result as yours, and it is possible that what you are observing is the absorption of protein in the micropores of the fritted disc separating your chambers. This would reduce the effective area available for diffusion and would be interpreted as a decreased diffusivity. An unrelated observation is that artificial organs are often protected by passing albumin in saline through the device before use. An appreciable albumin layer results which is both stable and reversible. You might be observing this latter phenomenon and not diffusivity reduction.

Gainer: Since the glucose diffusion data and the oxygen diffusion data were obtained using different sets of apparatus (diaphragm cell for the oxygen, microinterferometer for the glucose), and since both sets of data exhibit the same behavior with changing protein concentrations, I do not believe that our oxygen data are a result of an experimental artifact. I recognize that Dr. Dorson believes that the glucose data may be correct, and only the oxygen diffusion data is wrong, but we have shown that all solutes diffusing in a polymeric solution should behave in the same way with changing concentrations (using data other than those presented here).

<u>Goldstick</u>: As you pointed out, we have long disagreed. I have felt that your fritted-disc method is unsuitable for plasma. Is your most recent data using this method? Were you measuring oxygen diffusion?

<u>Gainer</u>: On the slides where I showed both the oxygen data and the glucose data, the oxygen data are measured with a gas liquid diffusion cell, and the glucose solution and plasma data are measured with a microinterferometric technique. So, there are two different measurement techniques. Our latest data were taken with a micro-interferometer. It is a lot easier. According to our theory, everything should decrease in the same ratio; so, we have worked with the easiest system, which is glucose.

Session IV

MATHEMATICAL STUDIES OF TISSUE OXYGENATION

PART A

Chairmen: Dr. Hermann Metzger and Dr. Carl A. Goresky

MATHEMATICAL CONSIDERATIONS FOR OXYGEN TRANSPORT TO TISSUE

(INTRODUCTORY PAPER)

Duane F. Bruley

Chemical Engineering Department

Clemson University, Clemson, South Carolina, U.S.A.

Introduction

Mathematics is the only tool available to man that can simultaneously solve the multivariable, interacting problems evolving from physiological systems analysis. The human mind is not capable of thinking in terms of n-dimensional space, therefore forcing us to rely on an organized accounting procedure as provided by applied mathematics.

Mathematical modeling can be thought of as an art as well as a science. Each individual analyzes a problem in terms of basic considerations involving heat, mass and momentum transport, chemical kinetics, thermodynamics and geometry for the formulation of meaningful functional relationships between variables. Each model, in most cases, can only approximate the real system because simplifications are generally necessary even when applying the most sophisticated solution techniques. Modeling allows the individual to be creative in his approach and solution of the problem therefore providing incentive for innovative studies.

Many examples could be sited to demonstrate the usefulness of modeling and simulation. Instead, a few general statements will be presented to illustrate why these techniques should be employed.

1. To provide practical and economical investigations of systems that would be extremely difficult to analyze experimentally and to allow faster than, equal to, or slower than real time studies as desired by the researcher.

2. To develop accurate system representations that could be
used as predictive models with possible clinical application in
closed- and open-loop control schemes.

3. To gain insight through mathematical modeling and parameter
sensitivity studies for the general understanding of physiological
processes under normal and pathological conditions.

4. To point out the need for experimental programs to determine important system parameters by means of thorough mathematical
analysis of physiological problems.

Mathematics can be beautiful. As described by Thomsen (30),
Dr. P. A. M. Dirac has formulated a "principle" centered around
his lifelong search for mathematical beauty. He has suggested that
a theoretician who works at some remove from the latest experimental
work needs a principle that will sustain his conviction in the cor-
rectness of his work during the sometimes lengthy wait for experimental confirmation. He has been criticized for his principle of
mathematical beauty on the grounds that beauty is subjective and
culturally determined. He accepts this statement with regards to
matters such as literature and painting but insists that, just as
mathematics can be understood by people of varying cultures, so
there are standards of mathematical beauty that can be appreciated
by people of differing cultural origins.

Dirac uses the Pythagorean Theorem as one example of mathematical beauty. The simplicity and exactness of the theorem can be
widely appreciated and has led to meaningful results in his own life.
Dirac worked out a theory of electrodynamics on the particle level,
quantum electrodynamics. The mathematical beauty to him was the
demand for the existance of antimatter and symmetry between matter
and antimatter. Initially the idea seemed absurd, but almost a
decade later the discovery of the positron confirmed his theory,
and today antimatter is a basic part of particle physics. It is
felt that the principle of mathematical beauty, as contrived by
Dirac, applies directly to the mathematical analysis of physiological
systems.

                       Relevant Literature

The literature sited in this writing will be incomplete by
necessity. Many papers of value to the study of tissue oxygenation
will not be referenced, however, it is not intended to mean that
this work is unimportant. The studies discussed should provide
stepping stones to contemporary work in this rapidly developing
area of research.

The famous works of August Krogh (15, 16, 17, 18) were the
first to express interest in the relation between capillary distribution and tissue oxygen supply. Krogh theorized that regardless
of the transport mechanism, the oxygen transport rate was dependent
upon the number and distribution of capillaries and the permeability
of the capillary walls and tissue. Spalteholz (18) had previously
determined the arrangement of blood vessels in striated muscle.

Krogh took the work of Spalteholz to be of high quality, but
was disappointed in both the quality and quantity of other previous
studies. Hence he undertook the experimental determination of the
distribution of capillaries in striated muscle. From his results,
Krogh concluded that open capillaries were distributed with con-
spicuous regularity in a cross-section of striated muscle. Krogh
further concluded that an annular volume of tissue surrounding a
capillary was supplied exclusively by that capillary. Krogh determined an average outer radius of this annulus from the ratio of the
total cross-sectional area to the number of capillaries in the
cross-section.

Not being a mathematician, Krogh persuaded Mr. Erlang, a Danish
mathematician, to describe oxygen transport in terms of the mathematics of the diffusion process. Despite the limitations of their
equations, the "Krogh tissue cylinder" was a major step in the study
of oxygen diffusion in living tissue (15).

Later studies concerning the distribution and flow patterns of
capillaries in tissue can be found in the work of Anrep et al. (1,
2, 3); Hillestad (11); Myers (21); Schmidt (28); Becker (4); Romanul
(26); Opitz and Schneider (23); and Knisely et al. (14).

In addition, Krogh was first to determine oxygen diffusion
rates through tissue. He developed an experimental technique for
this and was the first to report an oxygen diffusion coefficient
for tissue. Following Krogh's work, A. V. Hill (10) published a
major article which dealt with the diffusion of oxygen and lactic
acid through tissue. Hill's four-part paper was of theoretical
importance in understanding gas diffusion through living tissue.

The works of Krogh were expanded by F. J. W. Roughton (27).
Based on Krogh's assumptions, steady and unsteady-state equations
for discusion of oxygen into the "Krogh tissue cylinder" were
developed and applied. Roughton also assumed that oxygen was consumed simultaneously by zero- and first-order reactions. It was
assumed that these reactions roughly matched those for diffusion
of oxygen into myoglobin-containing red muscle tissue.

After simplification, Roughton solved his equations to give
oxygen partial pressure as a function of time and radial position

for a fixed axial position. Three separate solutions were obtained, all in the form of Bessel Functions.

Nicolson and Roughton (22) attempted to solve equations for unidirectional diffusion for a slab. Methods of numerical approximation based on the calculus of finite differences were employed for solutions. From the results, it was concluded that these equations were accurate within a 25 percent error in the first half of the process.

S. S. Kety (13) reviewed the previous work on the Krogh model and derived an equation to determine the average steady-state oxygen concentration of the capillary at any axial position. It was assumed that the oxygen content of the blood varied linearly from the arterial end to the venous end of the capillary. While this assumption is consistent with constant homogenous oxygen consumption in the tissue, it is not consistent with the relation between oxygen partial pressure and the erythrocyte oxygen content. Also, as pointed out by Kety, the omission of the effects of axial gradients can result in considerable error.

All works discussed to this point deal with oxygen supply in skeletal muscle, in the heart, in the lung and in the blood. In 1950, Opitz and Schneider (23) published the results of their experimental and theoretical research on oxygen diffusion in the brain.

Experimental findings indicated that oxygen consumption in the brain was both very high in relation to other organs and relatively constant, with little variation from periods of maximum activity to periods of complete rest. Consumption rates were measured for the whole brain and then reduced to rates for specific regions based on the general respiration ratio of one region to another.

Extensive data were also given for blood flow rates and their variation with changes in oxygen and carbon dioxide partial pressures in the arterial blood. Under normal condition, total blood flow through the brain was remarkably constant. Constant metabolic rate, all capillaries being open at all times, and subordination of vasomotor changes were given as factors affecting constant blood flow.

A wealth of data were also given on the relation between arterial and venous oxygen partial pressures. These data also indicated the effects of various arterial carbon dioxide partial pressures on venous oxygen partial pressure for a given arterial oxygen partial pressure.

Opitz and Schneider investigated conditions of cellular oxygen supply to the brain using the Krogh diffusion formula. Unable to measure the average distance between capillaries in the brain, a method was developed to estimate tissue cylinder radius based on the capillary length. Steady-state cellular oxygen pressure was studied using both Krogh's original equations and equations modified to include axial diffusion. Both studies indicated the existence of an oxygen excess in the healthy brain. The effect of the axial-gradient model was an increase of oxygen partial pressure at the point in the tissue farthest from the arterial end of the capillary, the "lethal" corner.

In 1960, Gerhard Thews (29) published a summary on his past and present work related to oxygen diffusion in the brain. Thews was apparently the first investigator to consider the existence of oxygen gradients in both the capillary and tissue xylinder. General equations to represent oxygen transport in the blood were not given. Thews stated that the nonlinearity of the oxygen dissociation curve made a mathematical analysis complicated.

Since a homogeneous arrangement of capillaries in the brain seemed unrealistic, another method was given for estimating the radius of the region of supply. Oxygen consumption in the brain was treated in a manner analogous to that of Opitz and Schneider and the same constant metabolic rate was determined for gray matter. New diffusion coefficients were measured and given in different forms.

All previous investigators assumed a flat radial oxygen profile in the capillary. Thews noted:

"It appears directly that such an O_2 pressure...cannot exist qualitatively over the whole cross-section of the capillary, since, according to diffusion laws, if the pressure gradient were zero there could be no diffusion of oxygen out of the blood."

The purpose of Thews' work was the determination of steady-state oxygen partial pressure at the "lethal" corner. Using the new expression for tissue cylinder radius and new diffusion coefficients for blood and tissue, he first calculated the oxygen partial pressure at the capillary wall. Next, he used Krogh's equation to determine the oxygen partial pressure at the "lethal" corner. He found this value to be 17 mm Hg as opposed to 30 mm Hg found by Opitz and Schneider. He thus concluded that an oxygen excess did not exist in the healthy brain.

Later, Bruley (5) derived a mathematical model which was solved by Reneau, Bruley and Knisely (25) and included a constant oxygen consumption rate and the nonlinear oxygen dissociation relationship. The effect of decreased and increased arterial oxygen partial pressure reduced blood flow rate and reduced carbon dioxide partial pressure on steady state oxygen partial pressure at the "lethal corner" of the Krogh capillary-tissue cylinder was considered. The Krogh capillary-tissue cylinder allowed for the radial diffusion distance determined by Thews. Also, Reneau, Bruley and Knisely (24) solved the unsteady state form of the model for a step change in arterial oxygen partial pressure. The model neglected a term for the accumulation of oxygen in the erythrocytes; thus the unsteady state solution did not demonstrate the actual system behavior in both the capillary and tissue. Halbert, Bruley and Knisely (9), by the discrete Monte Carlo method, solved the unsteady state model with the oxygen accumulation term included. Also, McCracken, Bruley, Reneau, Bicher and Knisely (19) solved the unsteady state model using the method of lines and considered the interacting effects of oxygen, carbon dioxide and glucose for the steady state in the capillaries and tissue of the human brain.

A single capillary and its surrounding tissue has not been the only approach for oxygen supply to tissue. Recently, researchers such as Metzger (20), Grunewald (8) and Hutten, Thews and Vaupel (12) have undertaken studies to determine if other models derived for multi-capillary systems are more descriptive of oxygen transport in the microcirculation.

Metzger developed a capillary network model which allowed inhomogeneous blood flow and analyzed the inhomogeneous flow capillary network system as well as homogeneous concurrent and counter-current flow systems by calculating tissue oxygen tension histograms for normoxia and arterial hypoxia. In normoxia, the countercurrent flow results exhibited a pronounced maximum (12 - 16%) as compared to a broad spectrum for the concurrent and inhomogeneous flow systems. On the other hand, under hypoxia, the histograms for all models were very similar.

Grunewald compared assymetric, concurrent and countercurrent capillary flow. Calculated and experimental pO_2 distributions indicated that the concurrent flow system model was 60% the most prevalent.

Hutten, Thews and Vaupel modelled an array of capillaries and tissue with a transistor and resistor analog computer from which results were calculated for various capillary configurations. It was concluded that results from an array of several capillaries were significantly different from those of a single capillary, since with an array, the supply range of oxygen in tissue was larger and and the supply conditions were better than as calculated with a single capillary.

Most of the models discussed so far consider only the steady-state. Those that have been extended to the unsteady-state are

useful for in vitro studies only, because they neglect the influence
of physiological control mechanisms. In vivo control mechanisms
play a very important role in the optimum oxygenation of tissue.

The influence of superimposed control was seen clearly in the
combined theoretical and experimental work of Bruley, Bicher, Reneau
and Knisely (6) and Bicher, Bruley, Knisely and Reneau (7). Because
of discrepancies between the predicted tissue oxygen response, using
convection-diffusion models, and experimental tests, control mechanisms were postulated. It has been suggested that a minimum of two
autoregulatory actions are functional in helping prevent neuron
damage when low blood oxygen tensions are encountered in brain.

First, blood flow rate is regulated to achieve optimal tissue
oxygenation. The exact mechanism and resulting flow rate changes
as functions of oxygen tension are unknown, however, through careful theoretical and experimental analysis, reasonable control dynamics were incorporated which simulated the temporal behavior of
tissue oxygenation for conditions of anoxic anoxia.

The second action which appears to act as a regulator is oxygen
consumption rate. Even though the brain normally operates at a
constant consumption rate, recent experimental results indicate that
the rate decreases and might even go to zero when tissue oxygen tension drops. This appears to be a protective metabolic mechanism
which prevents the self destruction of neurons.

Considerations For Further Work

There is a need to examine critically and evaluate the many
oxygen transport models that have been formulated and solved since
the initial work of Krogh in 1919. The models range from simple
ones that can be easily solved analytically to very complex models
that can only be solved using high speed electronic computing
equipment.

To determine their validity, it will be necessary to compare
calculated results with experimentally measured values. This is
a more difficult task than might be assumed. First, it is generally very difficult to design experiments and obtain meaningful
values using the complex techniques necessary in this area of
research. Second, the results from the models are necessarily
questionable because of the lack of accurate anatomical and physiological parameters. Finally, it is necessary to simplify the models
to obtain solutions thus allowing comparison for special cases only.
Providing theoretical and experimental results can be obtained, it
might still be difficult to draw conclusions about the accuracy of
calculations. This is obvious, for instance, when comparing lumped
parameter model results, which are space average values, with micro-

electrode oxygen tension recordings, which represent local or point
values. In this case, the best you can do is compare trends.

Results can be reported in many different forms, such as,
discrete point values, space averaged values, curves in the time or
frequency domain or as histograms. It is necessary to establish
what the theoretical and experimental results represent before an
analysis can be made by their comparison.

One of the greatest problems facing both the theoretician and
the experimentalist is the establishment of a meaningful geometry
and flow pattern. For instance, can a Krogh cylinder be used as a
representative sample of the whole? If not, what is the actual
histological pattern of capillaries in each organ or suborgan being
studied? This is an extremely important consideration for both
modeling the system and carrying out experimental studies. Certainly, if you want to compare point values as predicted by a distributed parameter model and the reading of the microelectrode, you
must know the precise location of the electrode tip in the capillary
network and the structure of the network being studied. A great
amount of work will be necesssry to provide the necessary geometric
relationships, capillary lengths and diameters and other anatomical
parameters needed for precise modelling.

At this time, it appears that pO_2 histograms provide the most
meaningful representation for steady-state oxygen tension distribution in tissue. Time-dependent results are best reported as time
versus point or space averaged oxygen tension.

It is apparent that more study should be directed toward the
stochastic behavior of physiological systems and the various forcing
functions. Consideration should be given to the statistical distribution of capillary lengths, capillary and tissue diameters,
flow rates, metabolic rate, arterial oxygen supply, hemodynamic
characteristics and many other variables in the capillary-tissue
system.

Finally, it is necessary that multicomponent models be investigated further for both steady-state and unsteady-state conditions.
Interaction among the necessary anabolites and metabolites and the
resulting effects on tissue survival can only be determined by a
total system analysis.

Acknowledgment

The author would like to acknowledge the financial support
from the U. S. Public Health Service for work in tissue oxygenation
(Grant NS-06957-06).

References

1. Anrep, G. V., A. Blalock, and Adli Samaan. "The Effect of Contraction Upon the Blood Flow in the Skeletal Muscle." Proc. Roy. Soc. B 114:223-245, 1933.

2. Anrep, G. V., S. Cerqua, and Adli Samaan. "The Effect of Muscular Contraction Upon the Blood Flow in the Skeletal Muscle, in the Diaphragm and in the Small Intestine." Proc. Roy. Soc. B 114:245-257, 1933.

3. Anrep, G. V. and Saalfeld. "The Blood Flow Through the Skeletal Muscle in Relation to its Contraction." Jour. Physiol., 35:375-399, 1935.

4. Becker, Ernest L., Rosemarie G. Cooper, and George D. Hataway. "Capillary Vascularization in Puppies Born at a Simulated Altitude of 20,000 Feet." Jour. Appl. Physiol.,8:166-168, 1955-56.

5. Bruley, D. F. Laboratory Research Notebook, Dept. of Chemical Engineering, Clemson University, Clemson, S. C.

6. Bruley, D. F., H. I. Bicher, D. D. Reneau and M. H. Knisely. "Autoregulatory Phenomena Related to Cerebral Tissue Oxygenation," Advances in Bioengineering, CEP Symposium Series 114, Vol. 67:195-201, 1971.

7. Bicher, H. I., D. F. Bruley, D. D. Reneau and M. H. Knisely. "Effect of Microcirculation Changes on Brain Tissue Oxygenation," J. of Physiology.,217: 689-707, 1971.

8. Grunewald, W. "Method for Comparison of Calculated and Measured Oxygen Distribution," Oxygen Supply, Urban & Schwarzenberg, Munchen, pp. 5, Feb. 1973.

9. Halberg, M., D. F. Bruley and M. H. Knisely. "Simulating Oxygen Transport in the Microcirculation by Monte Carlo Methods," Simulation, Vol. 15, N. 5, pp 206-212, November 1970.

10. Hill, A. V. "The Diffusion of Oxygen and Lactic Acid Through Tissues," Proc. Roy. Soc. B 104:39-96, 1928.

11. Hillestad, Leif. K. "The Peripheral Blood Flow in Intermittent Claudication," Acta Medica Scandinavica, 174:671-685, 1963.

12. Hutten, H., G. Thews, and P. Vaupel. "Some Special Problems Concerning the Oxygen Supply to Tissue, as Studied by an Analogue Computer," Oxygen Supply, Urban & Schwarzenberg, Munchen, pp. 25, Feb. 1973.

13. Kety, Seymour S. "Determinants of Tissue Oxygen Tension," Fed.
     Proc., 16:666-670, 1957.

14. Knisely, Melvin H., Edward H. Bloch, and Louise Warner. "Selec-
     tive Phagocytosis 1. Kobenhaun," Biologiske Skrifter.
     Bind IV, Nr. 7:43-49, 1948.

15. Krogh, August. "The Rate of Diffusion of Gases Through Animal
     Tissues with Some Remarks on the Coefficient of Invasion,"
     Jour. Physiol., 52:391-408, 1918-1919.

16. Krogh, August. "The Number and Distribution of Capillaries
     in Muscles with Calculations of the Oxygen Pressure Head
     Necessary for Supplying the Tissue," Jour. Physiol., 52:409-
     415, 1918-1919.

17. Krogh, August. "The Supply of Oxygen to the Tissues and the
     Regulations of the Capillary Circulation," Jour. Physiol.,
     52:457-474, 1918-1919.

18. Krogh, August. The Anatomy and Physiology of Capillaries.
     Yale University Press, New Haven, Con., 1 ed., 1922.

19. McCracken, T. A., D. F. Bruley, D. D. Reneau, H. I. Bicher and
     M. H. Knisely, "Systems Analysis of Transport Processes in
     Human Brain - Part II," (Systems Studies of the Simulta-
     neous Transport of Oxygen, Glucose and Carbon Dioxide in
     Human Brain) Proceedings of First Pacific Chemical Engine-
     ering Congress, Kyoto, Japan, pp 137-143, October, 1972.

20. Metzger, H. "PO_2 Histograms of Three Dimensional Systems with
     Homogeneous and Inhomogeneous Microcirculation, a Digital
     Computer Study," Oxygen Supply, Urban & Schwarzenberg,
     Munchen, pp 18, Feb. 1973.

21. Myers, Wayne W. and Carl R. Honig. "Number and Distribution
     of Capillaries as Determinants of Myocardial Oxygen
     Tension," Am. Jour. Physiol., 207:653-660, 1964.

22. Nicolson, Phyllis and F. J. W. Roughton. "A Theoretical Study
     of the Influence of Diffusion and Chemical Reaction Velo-
     city on the Rate of Exchange of Carbon Monoxide and Oxygen
     Between the Red Blood Corpuscle and the Surrounding Fluid."
     Proc. Roy. Soc. B 138:241-264, 1951.

23. Opitz, Erich and Max Schneider. "The Oxygen Supply of the
     Brain and the Mechanism of Deficiency Effects," Ergebnissee
     der Physiologie, Biologischem Chemic, und Experimentellen
     Pharmakologic., 46:126-260, 1950.

24. Reneau, D. D., D. F. Bruley, and M. H. Knisely. "A Digital Simulation of Transient Oxygen Transport in Capillaries - Tissue Systems (Cerebral Grey Matter)," Am. Inst. Chem. Engrs. Jour.,15: No. 6, 916-925, 1969.

25. Reneau, D. D., D. F. Bruley, and M. H. Knisely. "A Mathematical Simulation of Oxygen Release, Diffusion and Consumption in the Capillaries and Tissue of the Human Brain," Chemical Engineering in Medicine and Biology, ed. by D. Hershey, Plenum Press, New York, N. Y., 1967.

26. Romanul, F. C. "Distribution of Capillaries in Relation to Oxidation Metabolism of Skeletal Muscle Fibers," Nature, 201:307-308, 1964.

27. Roughton, F. J. W. "Diffusion and Chemical Reaction Velocity as Joint Factors in Determining the Rate Uptake of Oxygen and Carbon Monoxide by the Red Corpuscles," Proc. Roy. Soc. B 111:1-36, 1932.

28. Schmidt, C. F. "Central Nervous System Circulation, Fluids and Barriers - Introduction," Handbook of Physiology, Neurophysiology III, ed. by W. F. Hamilton and P. Dow, Chapter 70:1745-1750. Williams and Wilkins, Baltimore, Md., 1960.

29. Thews, Gerhard. "Oxygen Diffusion in the Brain. A Contribution to the Question of the Oxygen Supply of the Organs," Pflugers Archiv., 271:197-226, 1960.

30. Thomsen, D. E. "The Beauty of Mathematics," Science News, 103: 137-138, 1973.

GEOMETRIC CONSIDERATIONS IN MODELING OXYGEN TRANSPORT PROCESSES IN TISSUE

Hermann Metzger

Department of Physiology, Johannes Gutenberg-University Mainz, W.-Germany

SUMMARY

Numerical solution of partial differential equations describing transport processes in capillaries and tissue is used for calculation of oxygen transport to brain tissue. Calculation is based on a three-dimensional network model which covers inhomogeneities in capillary blood flow in a relation of 27:1. This study was performed in order to obtain information about the influence of the main physiological parameters on oxygen tension frequency distribution pattern. Following results were obtained: a)Increase (or decrease) of the oxygen consumption rate in tissue to about twice (or half) of the normal case for cerebral grey matter causes an extreme left (or right) shift of the oxygen tension frequency distribution = histogram with a decrease (or increase) in venous oxygen tension from 35 to 18 (or 61) mmHg. b)A decrease of capillary blood flow at the input of the network in steps of about 15% from the normal value causes a stepwise left shift of the oxygen tension frequency distribution with venous oxygen tensions from 35 to 22 (or 13) mmHg. c)An increase of the critical oxygen tension from 1 to 10 mmHg causes a considerable right shift with an increase of the venous oxygen tension from 35 to 49 mmHg.

Theoretical oxygen tension frequency distributions are compared with experimental results which have been

measured in rat brain cortex by means of oxygen microsensors. The frequently observed low oxygen tension
values are discussed on the basis of mutual parameter
changes in blood and tissue.

INTRODUCTION

Traditional closed form solutions of partial differential equations for biological transport processes
taking into account spherical or cylindrical symmetry
are not satisfactory for mathematical description of
sophisticated heterogeneous capillary-tissue systems.
Heterogeneity arises from geometrically unequal distribution of capillaries and cells in a small tissue
volume: capillaries are arranged as irregular branched network systems with unknown locations of input
and output points of the arterial blood relative to
the capillary bed (1); number and distribution of cells
diminish from the arterial to the venous end of the
capillary network, consequently, an inhomogeneous distribution of consumption rate of O_2 molecules is to be
expected in brain tissue (2).

This study was performed by means of a high speed
digital computer for simulation of oxygen transport
to a heterogeneous composed volume of brain tissue. A
small element of the total brain tissue (grey matter)
is analyzed theoretically assuming a defined morphological and physiological situation. To obtain an integral view of the oxygen supply of the whole organ, it
has to be postulated that the small tissue volume is
representative for the total brain.

Theoretical results of simulation of oxygen transport to tissue are illustrated as oxygen tension frequency distribution of all the oxygen tension values
calculated for a small tissue volume (histograms). As
described previously, oxygen tension histograms are
used for illustration of oxygen tension measurements
by use of oxygen microsensors. A comparison of experimental and theoretical results can be performed.

Experimental oxygen tension histograms. Polarographic oxygen tension measurements in brain (3, 4),
kidney and tumor tissue (5, 6) have shown the necessity for further theoretical investigations of oxygen
tension distributions in brain tissue as well as oxygen tension histograms. Some hundreds to thousands of

oxygen tension values from different test points within the organ have been measured by means of polarographic oxygen microelectrodes (tip diameter: about one
micron). The total quantity of oxygen tension values
from the same organ and animal under investigation have
been illustrated as oxygen tension histograms.

An example for oxygen tension registration is
shown in Figure 1. A small oxygen microsensor was moved
forwards from the surface into deeper layers of the
tissue in steps of 10 microns; tremendous differences
are encountered. From each experiment in the brain tissue, a high data output is obtained. To better understanding of the numerous experimental results which are
illustrated as oxygen tension frequency distribution,
shape and magnitude of the histograms are analyzed
theoretically. Typical physiological parameters are
varied within physiological ranges and their influence
on oxygen tension histograms are tested.

Model description. The capillary-tissue arrangement is represented by a small volume in which 3x3x3
capillaries (60 microns length) are arranged in regular way. The total tissue forms a cube with a side-
length of 180 microns. Input and output points to the
capillary network are located at opposite corners of
the tissue cube, i.e., 4 input and 4 output points
(arterioles and venoles) are to be found at opposite
corners of the cube. As a simplifying condition it is
assumed that the hydrodynamic resistance of all capil-

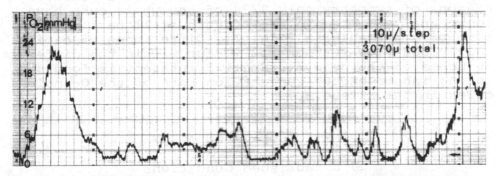

Figure 1. PO_2 registration by means of an oxygen microsensor. Stepwise movement of the manipulator in
steps of 10 microns, 3070 microns total electrode way.

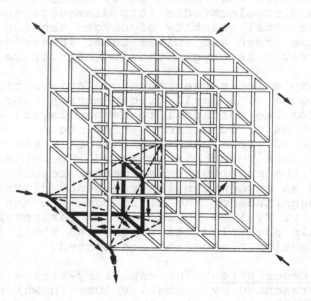

Figure 2. Capillary tissue model with tetrahedron for calculation.

laries are equal. Kirchhoffs Laws are assumed to be valid at the crosspoints of the capillaries and flow rate as well as flow directions can be calculated. By reasons of symmetry only 1/24 of the whole cube have to be analyzed, the volume of this small part has the form of a tetrahedron (fig. 2). Differences of capillary blood flow between 1:27 occur within the different branches of the network. Special interest is focussed to the inhomogeneous consumption rate of the O_2 molecules in the tissue.

METHOD OF CALCULATION

1) Established Conditions

The equations of oxygen transport to tissue are based on a number of simplifying conditions which cor-

respond to those described previously (4, 8):
a) Oxygen is transported in tissue by diffusion only
   and in capillaries by convection only.
b) Chemical reactions involving oxygen are assumed to
   take place at steady-state conditions and are described by Hills Law in the capillaries and by
   Michaelis and Mentons Law in the tissue. The reaction constant in the term is variable between 1 and
   10 mmHg.
c) Oxygen consumption in the tissue is assumed to be
   homogeneous. Inhomogeneous arrangements with high
   consumption rates at the capillary input points and
   low at the neighborhood of the venous output are
   simulated and their influence on the histograms are
   studied.
d) Flow velocity in the capillaries is calculated according to Kirchhoffs Laws and is constant over the
   capillary cross section.
e) The diffusion resistance of the capillary wall is
   small compared with that of the tissue and can be
   neglected.
f) Oxygen concentration is continuous at the capillary-tissue interface and transport across the interface can be described by Ficks First Law.
g) The analyzed tissue area continues in the same way
   to infinity; the considered tissue volume is representative for the whole organ.
h) The differences of blood flow within the different
   capillaries are 27:11:4:3:1.

   Based on the assumption (a) - (h) a mathematical
description of the capillary-tissue model is possible
and can be described by two coupled nonlinear differential equations.

2) Mathematical Equations

   The mathematical equations are transfered into
dimensionless form; numbers, characterizing capillary
and tissue qualities as well as oxygen kinetics in
blood and tissue are calculated according to the following definitions:

   a) Tissue equation:

$$\frac{\partial^2 c}{\partial x^2} + \frac{\partial^2 c}{\partial y^2} + \frac{\partial^2 c}{\partial z^2} - AG\frac{c}{CH+c} = 0$$

b) Capillary equation:

$$\frac{\partial c}{\partial s} = \frac{AK}{f(c)}\left[\left(\frac{\partial c}{\partial n}\right)_o + \left(\frac{\partial c}{\partial n}\right)_u + \left(\frac{\partial c}{\partial n}\right)_r + \left(\frac{\partial c}{\partial n}\right)_z\right]$$

with $$f(c) = \frac{c^{\gamma-1}}{(1 + K'c^\gamma)^2}$$

c) Interface conditions:

$$c\big|_{blood} = c\big|_{tissue} \qquad \frac{\partial c}{\partial n}\big|_{margin} = 0$$

$$\frac{\partial c}{\partial n}\big|_{blood} = \frac{\partial c}{\partial n}\big|_{tissue} \qquad c\big|_{input} = 1.0$$

The equations have been transformed into difference equations and solved iteratively by the method of successive overrelaxation. Relaxation parameter was calculated according known equations from literature

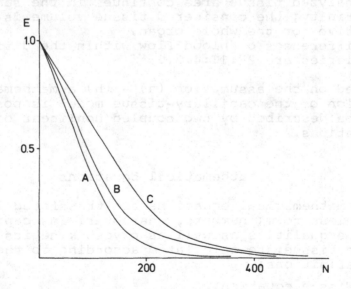

Figure 3. Error (E) of calculation as function of iteration step N. Curves: **A** = case (a) **A**; B = case (b) **A**; C = case (a) B (see table)

(7, 8). Error of calculation at the different iteration steps decrease (fig. 3). Accuracy of calculation was 1 mmHg. Computer time for one parameter combination about 10 minutes.

3) Dimensionless Parameters for Computer Simulation

a) Definitions:

Tissue parameter $\qquad AG = AP \mid P_a aD$

Capillary parameter $\qquad AK = \dfrac{DP_a a100}{dv\ 1.34\,C_{Hb}\gamma KP_a{}^{\gamma}}$

Kinetic parameters

$$CH = P_K/P_a \qquad\qquad BK = KP_a{}^{\gamma} \qquad\qquad BN = n = \gamma$$

b) Input parameters for calculation and output parameter P_v:

	AG	$AK \cdot 10^5$	CH	P_v(mmHg)
Case (a)				
A	1.824	0.083	0.01	18
B	0.910	0.079	0.01	35
C	0.456	0.079	0.01	61
Case (b)				
A	0.910	0.120	0.01	13
B	0.910	0.105	0.01	22
C	0.910	0.079	0.01	35
Case (c)				
A	0.910	0.079	0.01	35
B	0.910	0.079	0.10	49

BN = 4.0, BK = 15.4, P_a = 100 mmHg

RESULTS

A meaningful representation of the computer results can be obtained in different ways:

<u>Oxygen tension distribution</u>. This graph shows the

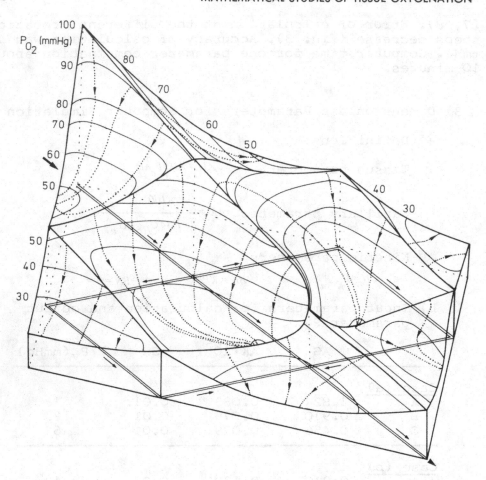

Figure 4. Isobares and O₂ flux lines for normoxia case.

iteratively calculated oxygen tension or concentrations values which are derived for each point of the capillaries and tissue.

Oxygen tension isobares. Lines of equal oxygen tension values are connected to each other. Oxygen flux lines are running at right angles to the isobares forming a rectangular net. For a special example (normoxia case) the isobares and O₂ flux lines are illustrated (fig. 4). O₂ flux lines characterize the diffusion path of the O₂ molecules into the tissue. Supply conditions of each tissue point can be easily checked, the role of shunt

diffusion of molecules from one capillary to the other
as well as the question if capillaries act as sources
or sinks can be investigated. This representation of
results has been widely used in theoretical physics and
all the physiological phenomena are completely described
by these plots.

Oxygen tension histograms. Main purpose of the investigation was the variation of the parameters AG, AK and
CH. They represent capillary blood flow and blood constants (AK) as well as tissue constants (AG, CH). As to
be seen from fig. 5 tremendous variations of the oxygen
tension frequency distribution can be expected if AG
varies from the normal value to one half or twice of

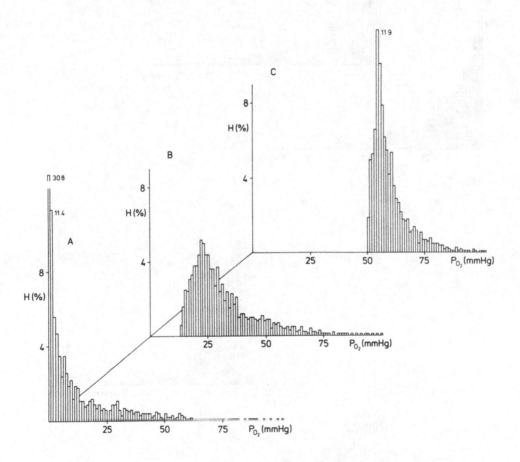

Figure 5. Oxygen tension histograms as function of parameter AG (curves A, B, C see case (a)).

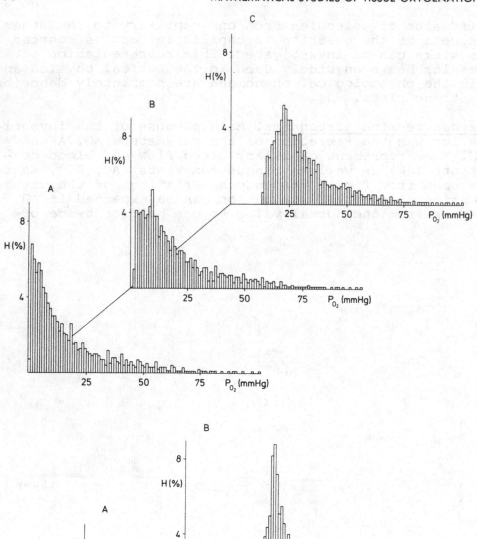

Figures 6 and 7. Top: case (b), bottom: case (c).

the normal value. A shift of the curves to the right or
left with an increase of the maximum probability occurs
under this test conditions. The parameter AG is proportional to the consumption rate of oxygen in tissue, A,
as well as to the quadrat of the capillary length, l^2.
Both, shape and distribution of the quantity of oxygen
tension values which are classified in groups of one
mmHg, are changed considerably. A change of parameter
AK between 0.120 and 0.079 results in a similiar change
in the oxygen tension histogram plot. The shift of the
histogram demonstates the tremendous influence of capillary blood flow and blood capacity; variations of
about 15% above and below the normal oxygen consumption
cause a change of venous oxygen tension between 15 and
20 mmHg compared with the normoxia case of 35 mmHg
(fig. 6, oxygen tension histograms as function of parameter AK (curves A, B, C see case (b))). A change of
the critical oxygen tension of the O_2 consumption reaction in tissue between 1 and 10 mmHg results in a considerable right shift with an increase of venous oxygen tension value to about 50 mmHg. This was to be expected because the total consumption of the tissue volume under test shows a remarkable decrease if local
oxygen tension is lowered below the critical value
(fig. 7, oxygen tension histograms as a function of
parameter CH (curves A, B see case (c))).

DISCUSSION

Polarographic oxygen microsensors are used for
studying oxygen supply conditions of different organs.
Until today the interpretation of microelectrode results is dubious: only point-shaped measurements with
the character of a random test in the heterogeneous
capillary-tissue system can be performed. This study
was to compare oxygen tension measurements in brain
tissue with oxygen tension calculations for capillaries and tissue. Experiments in the rat brain cortex
have demonstrated a considerable shift to low oxygen
tension values. By means of the inhomogeneously perfused capillary network remarkable tissue hypoxia at
the low perfused capillaries of the branched network
occurs despite the fact that venous oxygen tension is
normal. The extreme left shift of the histogram is
caused by an increase of parameter AG (consumption
rate) or AK (reciprocal of blood flow). Consequently,
both parameters can be used to fit the experimental
curves and calculate physiological constants.

NOTATION

A = O_2 consumption rate of cerebral grey matter =
0.1 ml/g min

l = capillary length of the network system = 60 (72)
microns

d = capillary diameter = 6 microns

a = solubility coefficient of O_2 molecules = 0.025
ml/ml Atm

D = diffusion coefficient in the tissue = $2 \cdot 10^{-5}$
cm^2/sec

P_a = arterial oxygen tension = 100 mmHg

K = Hill constant = $0.8575 \cdot 10^{-4}$

γ = Hill exponent = 4.04

Δs = change of saturation of oxyhemoglobin = 53%

P_K = critical oxygen tension = 1 mmHg

v = flow velocity in the capillary

c = relative oxygen tension value normalized on arterial oxygen tension

c_a = initial oxygen tension at the arterial point = 1.0

REFERENCES

(1) Lockard, I., Barham, J.R., Forlidas, N.G., and
Myeers, R.B.:J. Comp. Neurol. 112, Complete (1959).

(2) Barker, J.N.: Fed. Proc. 31 (2), 1028 (1972).

(3) Metzger, H., Erdmann, W., and Thews, G.: J. Appl.
Physiol. 31 (5), 751 (1971).

(4) Metzger, H.: Advances in Chemistry Series (in
press).

(5) Vaupel, P., Günther, H., Metzger, H., and Thews, G.:
in: Oxygen Supply, Urban & Schwarzenberg Verlag,
p. 189 (1972).

(6) Günther, H., Vaupel, P., Metzger, H., and Thews, G.:
Z. Krebsf. 77, 26 (1972).

(7) Forsythe, G.E., and Wasow, W.R.: Finite-difference
methods for partial differential equations. John
Wiley & Sons (1967).

(8) Metzger, H.: Habilitationsschrift Mainz (1971).

CAPILLARY-TISSUE EXCHANGE KINETICS: DIFFUSIONAL INTERACTIONS BETWEEN ADJACENT CAPILLARIES

Carl A. Goresky and Harry L. Goldsmith

McGill University Medical Clinic

Montreal General Hospital, Montreal, Quebec

On the basis of their blood supply, organs in the mammalian body can be classified into two types: well-perfused and poorly perfused. The well-perfused organs include liver, heart, lungs, kidneys, and brain; and the poorly perfused organs include skin and subcutaneous tissue, and resting muscle.

In order to gain an understanding of the interactions between flow and diffusion in the micro-elements of the circulation we have examined here only well-perfused organs and, in particular, have focused our attention on the liver, the lung, and the heart. In order to make the phenomena under study as simple as possible we have studied, as an analogue of oxygen, the distribution of labeled water, a substance which, unlike oxygen, is neither preferentially carried by red cells nor consumed in the tissue. The distribution of labeled water in these tissues will be regulated not only by the flow and permeability characteristics of the capillaries but also by the manner in which diffusional interaction between these elements contributes to the total.

THE LIVER AND THE LUNGS

Data from two sources were studied and compared: those coming from multiple indicator dilution studies in the intact organ (1,2), and cine-photographic studies of microcirculatory patterns in the living organ (3,4).

The dilution studies were carried out, in the liver, by rapid injection of tracer materials in blood into the portal vein, and by rapid serial sampling of hepatic venous blood; and, in the

lung, by rapid injection into the pulmonary artery, and sampling
from the carotid artery. A typical set of dilution curves from
the liver is displayed in the upper panel of Figure 1.

The labeled red cells emerge first, their outflow fraction per
ml rises to the highest and earliest peak and then decays most
rapidly. The outflow curves for each of the other compartment
labels is displaced relative to that for red cells, displaying a
diminution in peak magnitude and prolongation of transit time
which is progressively greater for each of the series: labeled
albumin, sodium, and water. In addition, for the labeled water
curve, a characteristic delay in outflow appearance is perceptible.

When the cine film of the hepatic microcirculation (3) is
studied, it becomes evident that portal venous blood leaves the
terminal portal venous radicals in a centrifugal fashion, and
drains away via a set of surrounding lobular central veins. Adjacent sinusoids are more or less parallel and the entrances to
these are adjacent, occurring along the length of the feeding
vessels. Although there is a distribution of red cell velocities
in the sinusoids, the flow is completely concurrent. There is
thus no opportunity, from the structural point of view, for a
diffusional label to bypass a vascular channel by short circuit-
ing to an adjacent vascular channel which has a staggered entrance
or exit.

The liver sinusoids are lined by a membrane pierced by large
holes and dissolved materials gain free access to the underlying
cells. The rates of flow and the dimensions are such that equilibration of labeled water would be expected to occur in the radial
direction at all points, in each tissue cylinder (6). Experimentally, the form of the labeled water curve has been found to be
independent of flow (1) and so the distribution of this label
appears to be flow limited. Perl and Chinard (5) have examined
the outflow profile expected from a single Krogh cylinder model of
capillary-tissue exchange and have found that, in this situation,
there are two asymptotic flow-limited forms. The first is that in
which no significant proportion of the label is carried in the
axial direction by flow. An impulse function introduced into the
capillary emerges as an impulse function which is delayed in time
by a factor corresponding to the ratio between the total and
vascular spaces of distribution for the label, but which is not
significantly dispersed. This is the delayed wave flow-limited
case (1). The second is that asymptotic situation in which flow
is so slow that equilibration in both axial and radial directions
occurs in the capillary-tissue cylinder, at the moment of introduction, by virtue of instantaneous diffusion along the length.
Flow then results in an exponential washout and this case may be
termed the washout flow-limited case. The lower panel of Figure 1

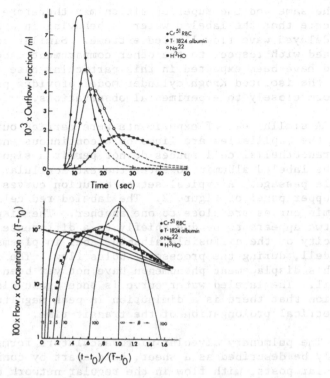

Fig. 1. Typical outflow dilution curves from an experiment
carried out in the liver. The ordinate in the upper panel is the
fraction of the injected amount per milliliter of blood. The lower
panel presents the same curves, normalized in the fashion proposed
by Perl and Chinard (5). Using a t_0 value (1) to correct for the
time spent in the large vessels, the time t of each sample is
expressed as $(t-t_0)/(\bar{t}-t_0)$, where $\bar{t}$ is the mean transit time of
that indicator. Each ordinate value is then normalized to 100 x
total flow x (outflow fraction per ml) x $(\bar{t}-t_0)$, so that the area
under each curve, if it were plotted on square grid coordinates,
would be 100. (Reprinted with permission from "Oxygen Supply",
edited by Kessler et al, Urban and Schwarzenberg, München, 1973).

demonstrates the family of dimensionless curves coming from this
Krogh cylinder model, and incorporates both the two extremes, the
delayed wave flow-limited case (with a dimensionless parameter,
the Peclet number, β, equal to ∞) and the washout flow-limited case
(β=0), and the families of intermediate cases. The dilution curves
from an organ are made up of many of these units and a vascular dispersion (large and small vessel) is superimposed. When the red
cell and labeled water curves obtained from the liver are superimposed on this kind of plot, the apparent dispersion of the two

is the same and the superimposition may therefore be taken to
indicate that the labeled water is behaving in a fashion close to
the delayed wave flow-limited extreme. Similar conclusions can be
reached with respect to the other compartment labels. This is what
would have been expected in this particular case (7), and indicates
that the isolated Krogh cylinder model provides predictions which
conform closely to experimental observations.

A similar set of experiments were carried out in the lung.
Here the capillaries are lined by a continuous endothelium and the
interendothelial cell spaces do not permit a significant fraction
of the labeled albumin to enter the extracellular space during a
single passage. A typical set of dilution curves is displayed in
the upper panel of Figure 2. The labeled red cell and labeled
albumin curves are close to one another. The displacement between
the two appears to be due chiefly to a difference in the mean
velocity of the diffusible albumin label in plasma and the labeled
red cell, during the process of bolus flow. The intimate kinetics
of this displacement phenomenon have not yet been explored in
detail. The labeled water curve is once again displaced in such a
fashion that there is a diminution in peak magnitude and a rather
symmetrical prolongation of the transit time.

The pulmonary alveolar microvasculature forms which can essentially be described as a sheet, held apart by connective tissue and
cellular posts, with flow in the regular network between (8). The
dimensions of the tissue layers adjacent to the vascular mesh and
of the connective tissue posts are such that diffusion equilibration of labeled water would be expected to occur in the adjacent
thin tissue layer, in a direction axial to the flow. Once again,
when the cine film of the microvasculature is studied (4), the
sources to each sheet are all found to be adjacent, the direction
of flow within the mesh is surprisingly uniform and one-directional,
and exits from the sheet are adjacent. No evidence of a quantitatively important proportion of countercurrent flow, or of staggered
entrances and exits, is found, on visual inspection.

Once again, the labeled water curve is found to resemble the
curves for the two vascular labels to a surprising degree when it
is superimposed upon the Perl-Chinard family of normalized curves
for a single Krogh cylinder. Again this superimposition can also
be taken to indicate that the labeled water bolus is propagating
approximately in a delayed wave flow-limited fashion. The simple
model has once more provided a major insight into the system.

In the liver, the process of oxygen consumption by the cells
may, on the basis of the foregoing, be predicted to lead to a
lobular gradient (9). The oxygen tension will be high in the area
of the sources, the portal venous and hepatic arterial blood; and
low, in the region of exits, the areas around the central vein.

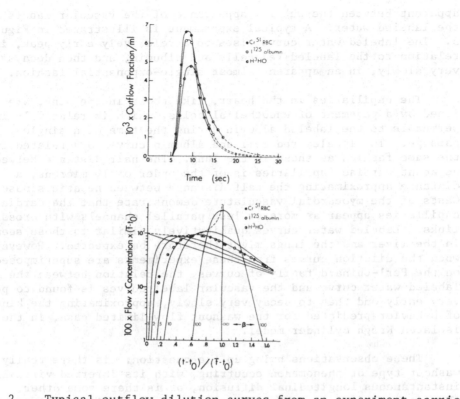

Fig. 2. Typical outflow dilution curves from an experiment carried out in the lung. The total space available for the distribution of labeled water in the alveolar sheet is very small, and so the displacement of the water curve from the vascular label is much smaller than in the case of the liver.

Histograms of oxygen tension in the liver must therefore be related to position within the hepatic lobule, if we are to come to a better understanding of the function of this organ. In the lungs the diffusion of oxygen in the gas phase will be so rapid that this phase may be regarded as well stirred. The oxygen tension in the vascular network in the alveolar sheet will therefore progressively increase, in the direction from entrance to exit. Again, if this gradient could be measured and related to flow and position, our understanding of the function of this organ would also be enhanced.

THE HEART

In dilution experiments carried out in the heart the labeled water curve has a very different shape from that seen in the experiments carried out in the liver and the lungs. No delay is

apparent between the outflow appearance of the vascular labels and
the labeled water. A typical experiment is illustrated in Figure
3. The labeled water curve rises to a relatively early peak, in
relation to the labeled red cells and albumin, and then decays
very slowly, in an apparent almost single-exponential fashion.

The capillaries in the heart, like those in the lung, are
lined by a pavement of endothelial cells, which is relatively impermeable to the labeled albumin during the time of a single
passage. The labeled red cell and albumin curves are related in
the same fashion as those in the lung. The half distance between
adjacent cardiac capillaries is of the order of 10 microns, a
distance approximating the half distance between hepatic sinusoids.
Casts of the myocardial vasculature demonstrate that the cardiac
capillaries appear as more or less parallel channels with cross
links. Labeled water curves qualitatively similar to those seen
in the liver and the lungs might therefore be expected. However,
when the dilution curves from these experiments are superimposed
on the Perl-Chinard family of curves, the relation between the
labeled water curve and the vascular label curves is found to peak
very early and then to decay very slowly, approximating the kind
of behavior predicted for the washout flow-limited case, in the
isolated Krogh cylinder model.

These observations bring up the question: is there really a
washout type of phenomenon occurring, with its inferred virtually
instantaneous longitudinal diffusion, or is there some other
phenomenon underlying these observations? Cine films (10) were
examined, in an attempt to find the answer. These show that, in
the network of cardiac capillaries, arterioles supply groups of
parallel capillaries and often act as sources, in such a fashion
that the blood supplied to opposite horizontal groups will flow in
opposing directions. Increase in flow recruits more capillaries
within each group. The sources for adjacent groups of capillaries
are not adjacent and so, at the edge of each group, there is opportunity for diffusion bypass. The labeled water can, in some
instances, leave early by means of a premature exit; and, in other
cases, may be carried in a countercurrent fashion, so that the
sojourn of label in the myocardium is prolonged. The direction of
flow in deeper and more superficial layers is not always parallel,
and the opportunity for partly cross-current connections also
arises.

Clearly the flow-structural architecture in this situation
provides the opportunity for diffusion bypass but the whole has a
rather more random than regular character. The rigid model of an
isolated Krogh cylinder with no diffusional capillary interactions
is here inadequate. The techniques needed to model this set of
circumstances are, however, at the same time not clear. Levitt (11)
has attempted to deal with this situation by continuing to use a

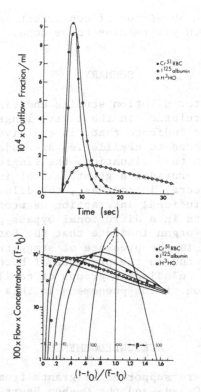

Fig. 3. Typical outflow dilution curves from an experiment carried out in the heart. (Reprinted with permission from "Oxygen Supply", edited by Kessler et al, Urban and Schwarzenberg, München, 1973).

single Krogh cylinder model, but one in which diffusion between adjacent cylinders and a random scattering of entrances and exits in adjacent units leads to a distribution in the generalized tissue cylinder which corresponds to a well-stirred extreme. This model has much merit but provides no insight into what kind of oxygen gradients would be expected in the tissue. It appears to predict only that no very high nor very low values would occur. What is needed to describe events in this tissue is some kind of probabilistic model which, when applied to the analysis of data, will provide insight into the proportion of the units which intercommunicate in a concurrent manner or in a countercurrent manner, and the way in which staggered entrances to, and exits from, capillary groups promote the diffusion bypass phenomenon. This challenge currently appears to be unanswered. Additional insight into the characteristics of this problem could be obtained if cine observations of the microvasculature of the heart were complemented by pO_2 measurements, in a situation in which the position of sources and sinks, the rates of capillary flow, and degrees and kinds of

potential interaction by virtue of concurrent and countercurrent
flows are known. Much yet remains to be done.

SUMMARY

Multiple indicator dilution studies and cine records of flow
through the microcirculation in the liver, lungs, and heart have
been examined. These indicate that, in the liver and lungs, flow
is concurrent, entrances to capillaries are adjacent and exits are
adjacent and that, in this situation, the single non-communicating
Krogh cylinder model provides a great deal of insight into exchange
processes. In the heart, in contrast, the dilution studies appear
to indicate that diffusional interaction between adjacent groups
of capillaries results in a diffusional bypass phenomenon. The
cine studies of this organ indicate that the basis for these observations lies in part in the presence of staggered capillary
entrances and exits, and in the presence of both concurrent and
countercurrent flows in adjacent groups of capillaries. The
meaningful description of the phenomena in this system provides a
major challenge.

ACKNOWLEDGEMENTS

These studies were supported by grants from the Medical
Research Council of Canada and the Quebec Heart Foundation. The
authors wish to acknowledge the superb technical assistance of
Mrs. Brita Nadeau and to thank Miss Margaret Mulherin for typing
this manuscript.

REFERENCES

1. Goresky, C.A. A linear method for determining liver sinusoidal
    and extravascular volumes. Amer. J. Physiol. 204: 626-640,
    1963.

2. Ziegler, W.H., and C.A. Goresky. Transcapillary exchange in
    the working left ventricle in the dog. Circ. Res. 29: 181-
    207, 1971.

3. Rappaport, A.M. Normal microcirculation of the mammalian
    liver (cine film). Microvasc. Res. 4: 329, 1972. The film
    was obtained from Dr. A.M. Rappaport, Department of Physiology,
    University of Toronto, Toronto, Ontario.

4. Kot, P.A. Microcirculation of the frog lung. High speed cine
    film, obtained from Dr. P.A. Kot, Georgetown Medical School,
    Washington, D.C.

5.   Perl, W., and F.P. Chinard. A convection-diffusion model of
     indicator transport through an organ. Circ. Res. 22: 273-298,
     1968.

6.   Bassingthwaighte, J.B., T.J. Knopp, and J.B. Hazelrig. A
     concurrent flow model for capillary-time exchanges. In:
     Capillary Permeability (Alfred Benzon Symposium II), ed. by
     C. Crone and N.A. Lassen. Munksgaard, Copenhagen 1970 (60-80).

7.   Goresky, C.A., W.H. Ziegler, and G.G. Bach. Capillary exchange
     modeling: barrier-limited and flow-limited distribution. Circ.
     Res. 27: 739-764, 1970.

8.   Sobin, S.S., H.T. Tremer, and Y.C. Fung. Morphometric basis
     of the sheet flow concept of the pulmonary alveolar microcir-
     culation in the cat. Circ. Res. 26: 397-414, 1970.

9.   Goresky, C.A., G.G. Bach, and B.E. Nadeau. On the uptake of
     materials by the intact liver: the transport and net removal
     of galactose. J. Clin. Invest. In Press (May 1973).

10. Hellberg, K., A. Rickart, and R. Bing. Direct observations of
     the coronary microcirculation in the arrested and beating heart
     (high speed cine). Fed. Proc. 30: 613a, 1971. The film was
     obtained from Dr. R. Bing, Huntington Memorial Hospital,
     Pasadena, California.

11. Levitt, D.G. Theoretical model of capillary exchange incorpo-
     rating interactions between capillaries. Amer. J. Physiol.
     220: 250-255, 1971.

COMPUTER CALCULATION FOR TISSUE OXYGENATION AND

THE MEANINGFUL PRESENTATION OF THE RESULTS

W. Grunewald
Max-Planck-Institut für Arbeitsphysiologie
Dortmund, West Germany

We have developed together with LÜBBERS a model for the
calculation of tissue oxygenation in organs having mostly parallel
running capillaries.
1) The model comprehends the influence of the capillary structure,
i.e., the three-dimensional arrangement of arterial inflows and
venous outflows, and
2) permits, with the aid of the oxygen partial pressure (Po_2)
frequency distribution, intercomparison between model and
experiment.

The model assumes four parallel running capillary sections of
length l seperated from each other by distance d (Fig. 1). The
arterial inflows and venous outflows are arranged on these capillary sections. The four capillary sections supply in concurrence
the tissue lying between them. Capillary sections and tissue
represent the "basic element" of the model. It is characterized by
the arrangement of the arterial and venous capillary ends, i.e.,
by the capillary structure. With known supply parameters (arterial
Po_2, venous Po_2, oxygen consumption, blood flow, HbO_2 binding
curve) and structure parameters (capillary distance, -length, and
-radius, capillary arrangement) the Po_2 course along the capillary
sections is first calculated by an iteration process reconciling
the Po_2 course with the supply area of each capillary section.
Then a meshwork of points is drawn between the capillary sections,
and the three-dimensional Po_2 field of the basic element is
approximately calculated by a relaxation process. This process
converges and approaches with diminishing distance of the points
of the meshwork the solution of the diffusion equation (Fig. 2).
In the third step Po_2 measurement in the tissue is simulated by

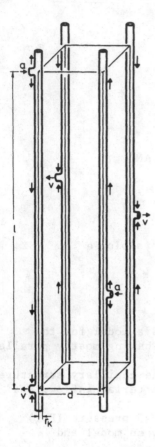

Fig. 1: Basic element of the terminal stream path.
4 straight-lined and parallel arranged capillary sections of length l and capillary radius r_K supply a tissue block with edge lengths l and d (capillary distance). To the share of a capillary section falls either an arterial inflow (a) or a venous outflow (v). In some cases both capillary ends may appear over this distance. The arrows indicate the direction of blood flow.

statistical access to the three-dimensional matrix which describes the Po_2 field. From this results the frequency distribution of each basic element.

With this model, tissue oxygenation of four capillary structures (Fig. 3) was calculated under equal supply conditions and Po_2 frequency distributions were produced:
1) of the concurrent structure according to Krogh,
2) of the countercurrent structure according to Diemer,
3) of two special asymmetric structures.

The frequency distributions of the first three structures reveal (Fig. 4) that under equal supply conditions tissue oxygenation varies for the individual structures. Asymmetric structure C_1 supplies the tissue in the most economic way, P_{min} shows the highest value. Since the shapes of the frequency distributions differ characteristically, they can be used for analysing measured Po_2 distributions. The measured Po_2 distribution $\psi_m(P)$ is then

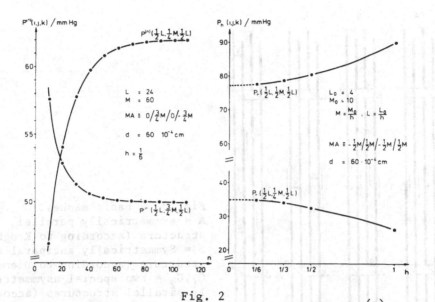

Fig. 2

Left side: Convergence of the approximated solution $P^{(n)}(i,j,k)$ at steady interval h and dependent on the number of iterations n for two points of the grid.

Right side: Convergence of the approximated solution P_h in dependence on the interval for discretization h. The number of iterations n is chosen so large that P_h is practically independent of n. The broken line indicates the extrapolation to the exact solution for the differential equation.

understood as a superposed distribution of the basic distributions φ_i (i = 1...k):

$$\psi(P) = \sum_{i=1}^{k} a_i \cdot \varphi_i(P) \quad \text{with} \quad \sum_{i=1}^{k} a_i = 1, \; a_i \geq 0 \tag{1}$$

The last square condition

$$\int_0^{P_1} \left| \psi_m(P) - \sum_{i=1}^{k} a_i \cdot \varphi_i(P) \right|^2 dP = \text{Min} \tag{2}$$

gives the weights a_i of the basic distributions in the measured distribution $\psi_m(P)$. (Here, such an analysis will not be described in detail).

The four capillary structures mentioned represent only a few selected out of a large number of other possible structures. In

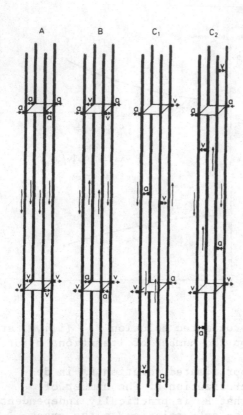

Fig. 3: 4 basic meshes.
A = symmetrically parallel
structure (according to Krogh).
B = Symmetrically antiparallel
structure (according to Diemer).
C_1, C_2 = two special asymmetrical-
ly parallel structures (according to Grunewald and Lübbers).

order to comprehend tissue oxygenation of these apparently complex
variants of structures the "twin structure model" has been developed
(Fig. 5). It consists of a basic element, as mentioned above, and
its twin element formed by prolonging the capillary sections of the
basic elements of length 1. This twin structure has the primary
advantage that with any arrangement of the capillary terminal the
end of a twin structure is identic with the beginning of the other
one. Thus, additional boundary conditions are unnecessary which, so
far, restricted the model for mathematical reasons.

With the twin structure model all systems may be examined
which resulted from shifting systematically the capillary ends
along the four sections by distance δl. Then the total of $(\frac{2l}{\delta l})^4$
variants of capillary structures are obtained. When assuming
that each of these variants appears once in the total structure
of the stream terminal, then the total inhomogeneity is covered
which is given by the different mutual positions of the capillary
ends. This inhomogeneity depends only on the interval δl. This may
be explained by an example (Fig. 6). When you choose $\delta l = \frac{1}{2}$ then
there are present $(\frac{2l}{1/2})^4 = 4^4 = 256$ variants of capillary structures.

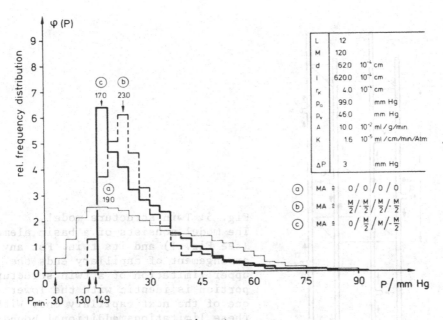

Fig. 4

Po$_2$ distribution φ(P) for (a) concurrent structure A, (b) countercurrent structure B, (c) asymmetric structure C$_1$. The shape of
these distributions shows characteristic differences. The parameters for calculation in the insert: L,M = discretization parameter; d, 1, r$_K$ = distance, length and radius of the capillary;
p$_a$, p$_v$ = arterial and venous Po$_2$; A = oxygen consumption in the
tissue, K = diffusion conductivity; ΔP = class value; MA =
localization of the arterial inflow of the capillaries of the basic
element.

By intercomparing the individual structures, the number of structures
with different tissue oxygenation can be reduced to a total of 21.
Fig. 7 shows the total frequency distribution formed by Po$_2$
distributions of the 256 structure variations. Consequently, the
twin structure model allows us, to produce Po$_2$ frequency distributions which are independent of the assumption of a defined type
of structure, as for example concurrent, countercurrent etc. Only
the "quality" of the inhomogeneity which determined the total Po$_2$
distribution is of importance. This quality may be improved for the
given example either by chosing a smaller δ1, or by chosing a
special δ1., i = 1..4. This is only a problem of the time needed
for computation. In any case, the analysis of measured Po$_2$
distributions is simplified by total frequency distributions.

We extended our model further by introducing into the twin
structure model, in addition to the structure inhomogeneity, the

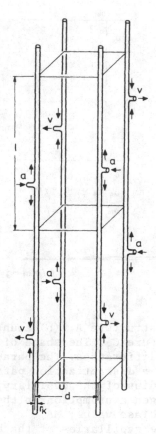

Fig. 5: Twin structure model.
The model consists of a basic element
(see Fig. 1) and its twin. For any
arrangement of capillary ends the
upper limitation of a twin structure
portion is identic with the lower
one of the next capillary end. With
these limitations additional boundary
conditions for the calculation of
the Po_2 field that strongly restricted
the general validity of the hitherto
model, can be neglected.

blood flow inhomogeneity. To each capillary inflow besides the
arterial Po_2 a blood flow value is assigned.

    For better understanding the influence of this inhomogeneity
we have compared two cases:
1) Blood flow through all the four capillary sections is equal,
that means, blood flow is homogeneous.
2) Through two capillary sections blood flow equals that of 1), for
the other two sections it is 1.5 and 0.5 times as much as in 1),
thus resulting in the same average blood flow as observed in 1).

    Comparison reveals a distinct difference between the total
Po_2 distributions. In the case of additional inhomogeneity of
blood flow there are under normoxia considerably lower Po_2 values
than are observed for the same homogeneous blood flow.

    Since with the twin model inhomogeneities having different
origins and being similar to that present in vivo can be simulated,

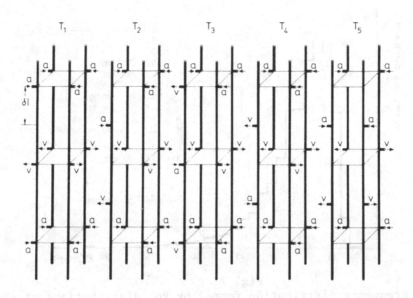

Fig. 6

Twin structures $T_1 \ldots T_5$ taken out of a total of 256 possible
structures originating by shifting systematically the capillary
ends by $1 = \frac{1}{2}$. Among the 256 twin structures some identic ones
are contained, for example, T_2 and T_4 are identic and therefore
also their Po_2 fields and Po_2 distributions, respectively. T_4
arises from T_2 by reflection on the middle plane.

we have to answer another question relating to the analysis of
measured Po_2 distributions: How large must the number of random
samplings be, i.e. the number of Po_2 values measured in tissue, to
comprehend in a sufficiently precise way the Po_2 frequency
distribution characteristic for tissue oxygenation? This question
was examined in the twin structure model with regard to a special
asymmetric structure. With the model, the case can be realized
where the tissue is punctured infinitely often and the local Po_2
is measured, i.e. where that Po_2 frequency distribution is found
which is characteristic for tissue oxygenation. When simulating
measurements with increasing Po_2 values, then the deviation of
such a Po_2 distribution from the "ideal" one can be observed.
For a Po_2 class value of 3 Torr a random sampling of 400 measured
values is necessary, and for one of 5 Torr, random sampling of 150
are needed. In both cases, the respective frequency distributions
(positive area differences among the distributions) differ by
about 15 %. (Fig. 9a+b)

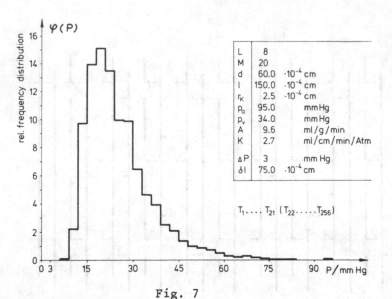

Fig. 7

Total frequency distribution formed by Po_2 distributions of the 256 twin structure variations ($T_1 \ldots T_{256}$) shown in Fig. 6 (for explanation of the symbols see Fig. 4).

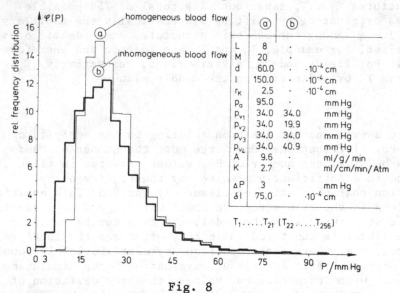

Fig. 8

Total frequency distributions, as shown in Fig. 7, for (a) homogeneous and (b) inhomogeneous blood flow. With two capillary sections of the twin structures $T_1 \ldots T_{256}$ the blood flow is 1.5 and 0.5 times as much as with the other two sections. The Po_2 values $p_{v_1} \ldots p_{v_4}$ result from the four different values of blood flow. Average blood is the same for case (a) and (b).

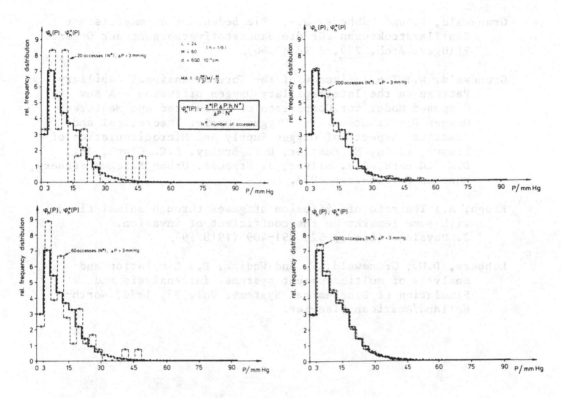

Fig. 9a, b

Po$_2$ frequency distributions.

(——φ_h(P) for a very large number (N$^+$=∞) of measured values ("ideal distribution")

--- φ_h^+(P) for an increasing number (N$^+$ = 20, 60, 200, 5000) of measured values ("statistical distribution").

φ_h^+ approaches the "ideal distribution" φ_h with increasing N$^+$.

REFERENCES

Diemer, K.: Eine verbesserte Modellvorstellung zur O$_2$-Versorgung des Gehirns. Naturwissenschaften 50, 617–618 (1963).

Grunewald, W.: Bedeutung der Kapillarstrukturen für die O$_2$-Versorgung der Organe und ihre Analyse anhand digital simulierter Modelle. Habilschrift, Bochum 1971.

Grunewald, W.: Digitale Simulation eines räumlichen Diffusions-modelles der O$_2$-Versorgung biologischer Gewebe. Pflügers Arch. 309, 266–284 (1969).

Grunewald, W. und Lübbers, D.W.: Die Bedeutung asymmetrischer
    Capillarstrukturen für die Sauerstoffversorgung der Organe.
    Pflügers Arch. 289, R 98 (1966).

Grunewald, W.: The Influence of the Threedimensional Capillary
    Pattern on the Intercapillary Oxygen Diffusion – A New
    Composed Model for Comparison of Calculated and Measured
    Oxygen Distribution. In: Oxygen Supply. Theoretical and
    Practical Aspects of Oxygen Supply and Microcirculation of
    Tissue, ed. by M. Kessler, D.F. Bruley, L.C. Clark, Jr.,
    D.W. Lübbers, I.A. Silver, J. Strauss. Urban & Schwarzenberg,
    München-Berlin-Wien 1973.

Krogh, A.: The rate of diffusion of gases through animal tissue,
    with some remarks on the coefficient of invasion.
    J. Physiol. (Lond.) 52, 391-409 (1918/19).

Lübbers, D.W., Grunewald, W. and Wodick, R.: Simulation and
    analysis of multicomponent systems. In: Analysis and
    Simulation of Biochemical Systems, Vol. 25, 1972, North-
    Holland/American Elsevier.

STOCHASTIC VERSUS DETERMINISTIC MODELS OF OXYGEN

TRANSPORT IN THE TISSUE

R. Wodick

Max-Planck-Institut für Arbeitsphysiologie

Dortmund, West Germany

PROBLEM

The difficulties in describing gas transport in the tissue are caused by the close connexion of diffusion with convection. For gas transport in the tissue the following differential equation is valid:

$$\alpha^*(p(\vec{r},t),u_i)\,\frac{\partial p(\vec{r},t)}{\partial t} = -A\,(\vec{r},t)+Q(\vec{r},t) +$$

$$\text{div}\,\{(D(\vec{r})\alpha^*(p(\vec{r},t),u_i)\cdot\text{grad } p\,(\vec{r},t)\} - \qquad (1)$$

$$\vec{v}(\vec{r},t)\cdot\text{grad}\,\{\alpha^*(p(\vec{r},t),u_i)\cdot p(\vec{r},t)\},$$

$p(\vec{r},t)$ describes the pressure of the gas depending on the site, $\vec{r}$, and the time, t. $\alpha^*(p(\vec{r},t),u_i)$ is the solubility coefficient and the apparent solubility, respectively, of the gas. The value of $\alpha^*(p(\vec{r},t),u_i)$ depends, for oxygen in tissue containing hemoglobin or myoglobin, on the pressure, $p(\vec{r},t)$, and on the parameters u_i as, for instance, Pco_2, which here are not specified in detail. $A(\vec{r},t)$ and $Q(\vec{r},t)$, respectively, describe the sinks and sources of the gas. For oxygen $A(\vec{r},t)$ means tissue respiration. $D(\vec{r})$ is the diffusion coefficient of the gas examined. $\vec{v}(\vec{r},t)$ is the flow velocity of the blood. In its general form we deal with a non-linear differential equation. One cannot help using strong simplifications for the solution, and only for certain special cases one succeeds in solving the problem though using large modern computing machines. The simplification has to be chosen in such a way that satisfying relations to reality exist, and that, on the other side, the mathematical numerical calculations can be controlled.

The processes of gas exchange and gas transport have been
studied relatively well with the Krogh cylinder model, i.e. a
capillary and the tissue cylinder surrounding it. KROGH (1),
OPITZ (2), SCHNEIDER (2), THEWS (3,4), ROUGHTON (5), BRULEY (6,7,8),
RENEAU (6,7), KNISELY (6,7,8), HALBERG (8), HUTTEN (9) and
other authors investigated on this system the connexion of diffusion with convection under various secondary conditions.

More specialized is Diemer's (9) countercurrent model where
two countercurrently perfused capillaries are assumed to be located
opposite to each other. While with the Krogh model oxygen partial
pressure values are found in the tissue (for example, at the
lethal corner) that are lower than the venous Po_2, with the
Diemer model the lowest oxygen partial pressure occurring in tissue
is identic with the venous Po_2. As examples for models including
the interaction of more than two capillaries, the capillary meshwork model according to Metzger (10), and the model supposing an
asymmetric capillary flow according to Grunewald (11)/Lübbers may
be mentioned.

A TISSUE MODEL ASSUMING STOCHASTIC CAPILLARY DISTRIBUTION

It was the purpose of our studies to detect regularities
valid for strongly varying courses of the capillaries.

We solve differential equation (1) with the approximation of
a linearized hemoglobin binding curve. The Monte-Carlo method is
very suitable for the solution, if we can assume that the gas
molecules pass through the tissue independent of the other
molecules. For the linearized case it is possible to construct an
isolated solution for special points being of particular interest.

Prior to occupying ourselves in detail with the stochastic
capillary distribution, some remarks may be made upon the Monte
Carlo Method. I think this method is familiar to all of us for
solving numerical problems. Therefore, I should like to explain
the essentials with a figure and to emphasize some primary points.
The first figure shows a particle passing through the tissue
according to random laws. On each point of the three-dimensional
grid a random number is plotted by a random generator (in the
figure shown with a die). This random number determines the further
way of the particle. In a capillary this particle gets an additional
impulse according to the flow velocity of the blood. From the
assumption that the particles pass through independently of one
another according to random laws, it follows that, if there is
no consumption by respiration and no loss of gas through the
surface, it is valid

$$\int_{-\infty}^{\infty} p(\vec{r},t) \, dt = \text{const.} \tag{2}$$

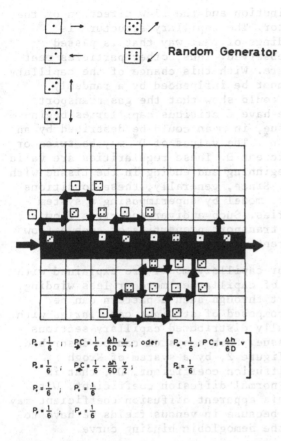

Random Generator

Fig.1: Scheme of the Monte Carlo Method. The dice represent the random generator, $P_\rightarrow$ the probability that the particle moves in the direction of the abscissa, $PC_\rightarrow$, corresponding probability in the capillary. $P_\leftarrow, P^\uparrow, P^\downarrow, P_\nearrow$, $P_\swarrow$, mean the same for the other directions of the space, PC_z, describes another possibility deviating from the first method for considering the flow through the capillary. Corresponding to the value of PC_z the particle may move several meshes Δh. D, diffusion constant, v, value of the blood flow velocity in the capillary.

This means that the value of the integral is independent of the site, $\bar{r}$, and of the diffusion constant of the gas. The validity of equation (2) may be understood by realizing that, if (2) is not valid, a saturation of the tissue perfused with constant pressure is not possible. Equation (2) is of primary importance for understanding gas transport and gas clearances.

For the Monte Carlo Method it is substantial to have a rather rapid random generator giving good random numbers. We use a generator for pseudo-random number generated by the recurrence relation in combination with a real time clock. In addition to the usual tests (as, for example, the run length test), we examined the random number generator on clearances for inert gases with Krogh cylinders, the exact result of which was known. In our opinion this method represents a particularly good procedure for applying the Monte Carlo technique to diffusion problems.

For our model of stochastic capillary distribution we

determined the capillary distribution and the flow direction of the capillaries by a random generator. The capillary structure is determined only in the surroundings of that way that is passed through by the particle to be observed. Thus, other particles meet quite another capillary structure. With this change of the capillary system the results obtained cannot be influenced by a randomly chosen vessel distribution. We could show that the gas transport through such a system, where we have fictitious capillaries begining and terminating in the tissue, in mean could be described by an apparent diffusion coefficient D^*. The values of D^* may be twice or more the usual diffusion coefficient D. These regularities are valid only in mean for capillaries beginning and ending in the tissue with a resulting total flow of zero. Since, generally, these conditions are not fulfilled, we extend the model by superimposing a system consisting of straight capillaries. Such a divergence-free flow satisfies other laws. The mean transport properties of such a flow cannot be described by an apparent diffusion coefficient.

The steps necessary for our capillary model are explained with the figure 2. We have a system of capillaries more or less winding. We could show that gas transport through such a pattern can be described in mean by a system composed of straight capillaries with a superimposition of statistically distributed capillary sections beginning and ending in the tissue. Such a system can be described, as shown in the lower part of figure 2, by a system of Krogh cylinders having an apparent diffusion coefficient, D^*, that is increased as compared with the normal diffusion coefficient. With respect to oxygen transport, this apparent diffusion coefficient may depend partly on the location, because in venous fields we have to deal with a different rise of the hemoglobin binding curve.

EXPERIMENTAL TEST OF THE MODEL

The above results obtained from model calculations according to the Monte Carlo Method have been substantiated by measurements performed by Lübbers and Leniger-Follert, and Stosseck in the cat brain[+]. For studies on transport properties of the capillary system electrochemically generated hydrogen was locally applied. The use of hydrogen guaranteed that the gas was transported only by diffusion, or by the vascular system, but not consumed by respiration. The curves shown in the third figure were obtained from measurements for which a certain quantity of gas was released at point r_1. The pressure course, $p(\vec{r},t)$, was measured on a second point, r_2, distant by 300μm. With this method reliable information is obtained upon gas transport properties of the system. According to the mean value of flow velocity we can determine the apparent

---

[+] Kessler and Thermann made simular experiments on the rat liver.

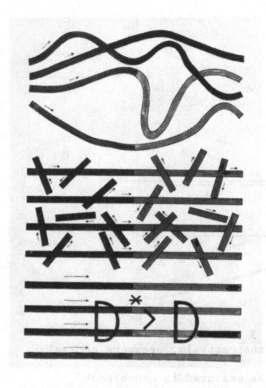

Fig. 2: Upper pannel: un-
regularly arranged capillaries;
middle pannel: capillary
arrangement corresponding to
the upper pattern with parallel
running capillaries and
stochastic distributed capillary sections; lower pannel:
capillary arrangement corresponding to the upper pattern
with parallel running
capillaries and an increased
apparent diffusion coefficient.

diffusion coefficient D^* and the value of flow velocity in
direction $(\vec{r}_1-\vec{r}_2)$. If we plot

$$z(t) = \ln \frac{p(t)}{p_o(t)}$$

against time t, then we can draw conclusions from the course of
the curves with respect to the capillary system. ($p_o(t)$ is a
respective pressure course found in dead tissue.) A course of the
curve which coincides with the x-axis suggests pure diffusion. A
largely linear course corresponds to gas transport observed in a
Krogh model. From a shift of the ordinate and a hyperbola-like
bending at the beginning, the value of **the apparent diffusion coeffi-**
cient can be concluded. Furthermore, there is another possibility
to which the stochastic capillary model can be applied, that is
for the photometric determination of the degree of oxygenation.

THE STOCHASTIC CAPILLARY MODEL FOR DETERMINING TISSUE OXYGENATION
BY REFLEXION PHOTOMETRY
(This analysis was carried out together with D.W. Lübbers)

Beside determining oxygen supply to tissue by local polarographic measurements tissue oxygenation can be measured by

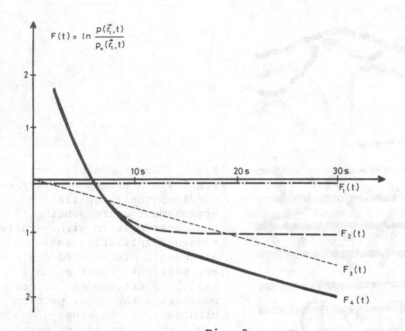

Fig. 3

$p(\vec{r}_1,t)$ measured and calculated, respectively, pressure course at point $\vec{r}_1$;
$p_0(\vec{r}_1,t)$, corresponding pressure as measured in non-blood-perfused tissue;
$F_1(t)$, pure diffusion, $F_2(t)$, apparently increased diffusion; $F_3(t)$, gas transport through parallel perfused capillaries, $F_4(t)$, combination of parallel perfused capillaries with an increased apparent diffusion coefficient. $F_4(t)$ corresponds to the course of the curve found for hydrogen in the cat brain by Lübbers and Leniger-Follert.

reflexion photometry.

Figure 4 shows the light path we have to deal with for reflexion photometry. Part of the light is reflected on the surface, another part penetrates more or less deep into the tissue. Such inhomogeneous lightpaths imply a non-linear transformation of the spectra.

For closer observation of this case, where we find beside oxygenated also desoxygenated hemoglobin in the tissue, we should like to explain the difference between additive and multiplicative dye mixing. With multiplicative dye mixing the total transmission is the product of the transmission values found for the individual components. With additive dye mixing the transmission value of the total system is obtained as the sum of the transmission of the indivi-

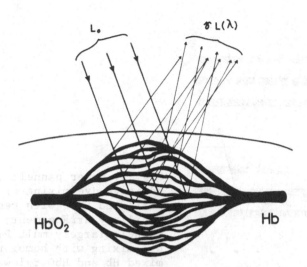

Fig. 4
Light path with reflexion photometric measurement.

dual components. In a tissue where the capillaries are regularly
arranged – as shown in the upper pannel of figure 5 – we have
largely to deal with additive dye mixing of oxygenated and desoxygenated hemoglobin. Such reflexion spectra will deviate from
the spectra where the oxygenated and desoxygenated hemoglobin is
supposed to be homogeneously mixed within the blood vessels, as
shown on the middle pannel of figure 5. On this occasion we would
find, apart from the inhomogeneous light path, a multiplicative
dye mixing of oxygenated and desoxygenated hemoglobin. Since for
most tissues our proceeding can be based neither on the upper nor
on the middle case of the distribution of hemoglobin and capillaries,
we assumed a stochastic capillary distribution. On the lower pannel
of figure 5 such a stochastic capillary distribution is shown.
This capillary distribution is calculated with the aid of a random
generator. The assumption of such a capillary distribution seems
to be reasonable as far as the accurate capillary structure of the
organ is not ascertained by morphologic investigation.

The spectra are analyzed by the "Q-analysis" method (12,
13) profiting by the fact that uniform extinction values are
uniformly transformed. When applying this method pairs of corresponding wavelengths λ^{*}_i and λ_i are selected from the measured
spectrum $G(\lambda)$. Then it is

$$G(\lambda_i) = G(\lambda^{*}_i) \text{ with } \lambda_i \neq \lambda^{*}_i$$

$$\lambda^{*}_i = f(\lambda_i)$$

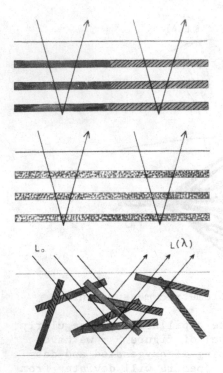

Fig. 5: Upper pannel: largely
additive dye mixing of Hb (hatched
part) and HbO_2 with regular
capillary arrangement; middle
pannel: largely multiplicative
dye mixing with homogeneously
mixed Hb and HbO_2; lower pannel:
a combination of additive and
multiplicative dye mixing with
stochastic capillary distribution.

The degree of oxygenation is found to be

$$g = \frac{\int_{\lambda_1}^{\lambda_2} (\phi_2(f(\lambda_i)) - \phi_2(\lambda_i)) \, b(\lambda) \, d\lambda}{\int_{\lambda_1}^{\lambda_2} (\phi_1(\lambda_i) - \phi_1(f(\lambda_i)) - \phi_2(\lambda_i) + \phi_2(f(\lambda_i)) b(\lambda) \, d\lambda}$$

where $\phi_1(\lambda)$ is the spectrum of the oxygenated, and $\phi_2(\lambda)$ that of
the desoxygenated hemoglobin. The proceeding is invariant against
the kind of the lightpath through the dye. $b(\lambda)$ is a weight
function.

We have found that in a tissue model with stochastically
distributed capillaries the degree of oxygenation may vary by
about $\pm 3.0\%$ at an equal quantity of oxygenated and desoxygenated
hemoglobin. This is caused by the fact that with reflexion photometry the surface layers are higher rated than the lower ones.
Leaving the capillary structure to chance may result in the
occurrence of certain differring distributions of oxygenated and
desoxygenated hemoglobin in the single layers and thus the

Fig. 6
Macroscopic Monte Carlo Simulation.

deviations of the degree of oxygenation by ± 3.0% may be explained.
In our opinion, a model of stochastic capillary distribution may
be well suited for a basic model for performing reflexion photometric investigation.

Figure 6 finally shows a macroscopic Monte Carlo simulation
of a gas transport problem in tissue. The question is asked how
long it will take for the men who leave the pub after drinking
much wine to reach the lantern. The icy street corresponds to the
rapid gas transport through the capillaries. The consumption is
produced with the macroscopic simulation by an excavation.

But you may understand the picture also as the way (please
pay attention to the icy street) an investigator has to cover to
succeed in understanding (light) the difficult problems of oxygen
transport to tissue.

REFERENCES

1. Krogh, A.: J. Physiol. 52, 391-408 and 409-415 (1918/19).

2. Opitz, E., Schneider, M.: Ergebn. Physiol. 46, 126-202 (1950).

3. Thews, G.: Acta biotheor. 10, 105-136 (1953).

4. Thews, G.: Pflügers Arch. ges. Physiol. 271, 197-226 (1960).

5. Roughton, F.R.S.: Proc. roy. Soc. B 140, 203-230 (1952).

6. Reneau, D.D., Bruley, D.F., Knisely, M.H.: A mathematical
      simulation of oxygen release, diffusion, and consumption
      in the capillaries and tissue of the human brain.
      Proceedings 33rd. Annual Chemical Engineering in Biology
      and Medicine ed. by D. Hershey, Plenum Press, New York
      135-241 (1967).

7. Reneau, D.D., Bruley, D.F., Knisely, M.H.: AICHJ 15, 916-925
      (1969).

8. Halberg, M.R., Bruley, D.F., Knisely, M.H.: Simulation 206-212
      (1970).

9. Diemer, K.: Pflügers Arch. ges. Physiol. 285, 99-108 and 109-
      118 (1965).

10. Metzger, H.: Dissertation Darmstadt (1967).

11. Grunewald, W.: Pflügers Arch. 309, 266-284 (1969).

12. Lübbers, D.W., Wodick, R.: Naturwissenschaften 59, 362-363
      (1972).

13. Lübbers, D.W., Wodick, R.: Z. Anal. Chem. 261, 271-280 (1972).

EXISTING ANATOMICAL PARAMETERS AND THE NEED FOR FURTHER

DETERMINATIONS FOR VARIOUS TISSUE STRUCTURES

Isabel Lockard

Department of Anatomy, Medical University of

South Carolina, Charleston, South Carolina, U. S. A.

Vascular patterns within the central nervous system differ considerably from one part of the brain to another but in each area are characteristic of the region (Lockard et al., 1959). Indeed the architectural pattern of the blood vessels corresponds so precisely to the pattern of the nerve cells that each small area can be identified on the basis of its blood vessels alone (figs. 1 and 2)[1]. By the use of histochemical techniques the density of blood vessels has been shown to correspond to the metabolic activity of each area (Friede, 1961).

Over the years several investigators, beginning with Craigie in the 1920's, have published figures for total capillary length per mm[3] of tissue in different parts of the brain in various species (Craigie, 1945). Although these determinations constitute one important value for each area studied, other features of vascular anatomy also provide for and limit the blood supply to neurons. Vascular beds also differ in the number and size of arterioles and venules supplying and draining each area, in the distance between capillaries, in the length of capillary segments and in the relationships which capillaries bear to the neurons they supply.

Penetrating Vessels

The brain capillary bed is a continuous network in mammals above marsupials. The penetrating arterioles, however, supply

[1]For the technique for demonstrating blood vessels and nerve cells in the central nervous system see Lockard et al., 1959.

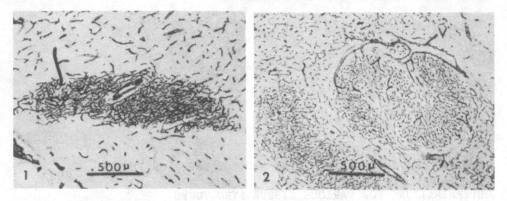

Figure 1. Supraoptic nucleus indicated by its dense capillary network. (Monkey, 20µ section. Blood vessels demonstrated by the intravascular precipitation of lead chromate.)

Figure 2. Superior olivary nucleus indicated by its capillary network. V-venule which drains the larger subdivision of the nucleus. (Cat, 30µ section. Blood vessels demonstrated with lead chromate.)

segments of the capillary bed with little or no overlap. Consequently the greater the number of arterioles supplying an area, the smaller is the volume of tissue supplied by each vessel. For example, the number of penetrating vessels supplying a segment of cerebellar cortex (fig. 3) is noticeably less than the number of vessels supplying the same amount of cerebral cortex, (fig. 4). Consequently the block of cerebral cortex supplied by each arteriole is smaller than in the cerebellar cortex.

Intercapillary Distances

Intercapillary distances are remarkably uniform within identifiable subdivisions of the capillary bed. Half the intercapillary distance corresponds roughly to the maximal distance which oxygen must travel by diffusion to reach the most distant point in the intercapillary space. From figures in the literature for total capillary length per mm^3 of tissue it is possible, using the equations on the next page, to calculate the average diameter of a hypothetical tissue cylinder (the tissue surrounding and supplied by one capillary, Krogh, 1929) in each region under consideration. The volume (V) is a constant, 1 mm^3 or 1,000,000,000$µ^3$. It is treated as a single hypothetical tissue cylinder. The total capillary length (L), for each area under consideration, is treated as the length of a single capillary running through this hypothetical tissue cylinder.

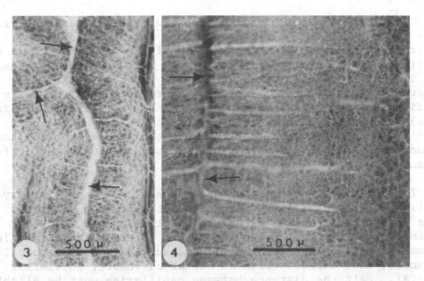

Figure 3. Blood vessels in a thick section of cat cerebellar cortex. Relatively few arterioles and venules are present connecting the surface and the underlying capillary bed. Arrows indicate the interface between two opposing surfaces. (Blood vessels demonstrated with lead chromate.)

Figure 4. Blood vessels in a thick section of cat cerebral cortex; same specimen as shown in figure 3. Notice that there are more penetrating vessels extending between the surface and the underlying capillary bed and they are larger in size than in the cerebellar cortex. Arrows indicate the interface between opposing surfaces. (Blood vessels demonstrated with lead chromate.)

$$\frac{V}{L} = A \qquad V = \text{unit volume of tissue}$$

$$A = \frac{\pi D^2}{4} \qquad L = \text{total capillary length}$$

$$D = \sqrt{\frac{4A}{3.1416}} \qquad A = \text{area of cross section of tissue cylinder}$$

$$\frac{D-d}{2} = \begin{array}{l}\text{diffusion}\\\text{distance}\end{array} \qquad D = \text{diameter of tissue cylinder}$$

$$d = \text{diameter of capillary}$$

Capillary diameter has been reported in the literature from
10μ - 6μ or even less. It frequently is not included with
measurements of capillary length. A capillary diameter of 7μ was
selected somewhat arbitrarily for use in this connection, because
that is what it usually measured on open vessels in material
prepared in this laboratory and the figure seems reasonable in
relation to the diameter of a red blood cell. If capillary diameter
($d=7\mu$) is subtracted from the diameter of the tissue cylinder (D),
the difference is the same as the average distance from one capillary
to another. Half this intercapillary distance equals the diffusion
distance.

The table on the following page consists of a preliminary
sampling of intercapillary distances (not diffusion distances) in
parts of adult cat brain, calculated in this manner from a
combination of figures in the literature for total capillary length
per mm^3 of tissue and compared with measurements of intercapillary
distances, made directly with an ocular micrometer in similar
material. Half the distance between capillaries must be slightly
less than the maximal diffusion distance, to the corners of Thews'
(1960) hexagonal tissue cylinder. Half the calculated average
intercapillary distance based on the total volume of the tissue
cylinder should be slightly less than the maximal diffusion distance
but slightly more than the measured distances. Some of the pairs
of figures are remarkably close, within a few micra of each other.
Others are significantly different (e.g. cerebral cortex lamina II,
white matter). The discrepancies may be due in part to differences
in tissue preparation or to the selection of the site for measurement
rather than, or in addition to, the manner in which the figures
were obtained. Many more determinations need to be made in these
and other areas of the nervous system.

Length of Capillary Segments

The length of capillary segments, i.e. the distance from one
junction to another, varies widely throughout the nervous system.
In some parts of the gray matter capillary segments are quite
short, as in the supraoptic nucleus (fig. 1). In other regions
they are long with few branches. The deep subcortical white
matter is a notable example (fig. 5).

In Schilder's disease degeneration of the subcortical white
matter occurs in older children and young adults. Some of the
fibers in this area do not normally myelinate until at least 21
years of age (Crosby et al., 1962). Systemic blood pressure
(Scheinker, 1954) and cerebral blood flow (Tucker et al., 1950)
are low in some patients with demyelinating diseases. Perhaps
in Schilder's disease the pattern of degeneration occurs because
the rate of flow through these long capillaries is too slow

INTERCAPILLARY DISTANCES

Cat Brain	Calculated	Measured
globus pallidus	43µ*	38µ
putamen	35	26
caudate	34	28
thalamus:		
lateral nucleus	33	35
dorsomedial nucleus	34	34
ventral nucleus	31	27
anterior nucleus	29	26
lateral geniculate nucleus	24	20
medial geniculate nucleus	31	23
parietal cortex:		
lamina I	33	26
lamina II	31	18
lamina III	30	23
lamina IV	29	21
lamina V	33	31
lamina VI	33	34
cerebellum:		
molecular layer	31	30
granular layer	27	21
red nucleus	33	26
substantia nigra:		
pars compacta	40	32
pars reticulata	44	37
white matter	51	86

*Based on figures from Campbell 1939, Dunning and Wolff 1937, Hough and Wolff 1939

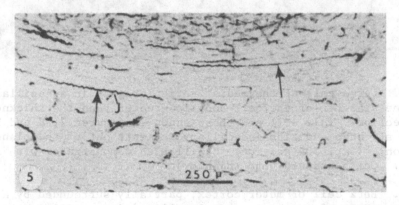

Figure 5. Long subcortical capillaries, indicated by arrows. Note capillary junction (J). (Cat, 30µ section. Capillaries demonstrated with lead chromate. Cells stained with thionin.)

to maintain the adjacent fibers at a critical time of their
development.

Capillary-neuronal Relationships

Differences occur from one part of the nervous system to
another in the relationship which capillaries bear to the neurons
they supply. In some areas individual cell bodies have closely
associated capillaries and appear to be encased in capillary
baskets. These include the large cells of the medial reticular
gray of the medulla (fig. 6) and of the lateral vestibular nucleus
(Lockard and Debacker, 1967, fig. 5), Betz cells of the cerebral
cortex (fig. 7), cells of the ventral cochlear nucleus (fig. 8),
and motor cells of the ventral horn (fig. 9). Such cells
undoubtedly get preferential treatment in the supply of oxygen and
nutrients. They may also get preferential treatment in the supply
of undesirable substances. For example, this vascular pattern may
account for the susceptibility of ventral horn cells to the virus
of poliomyelitis.

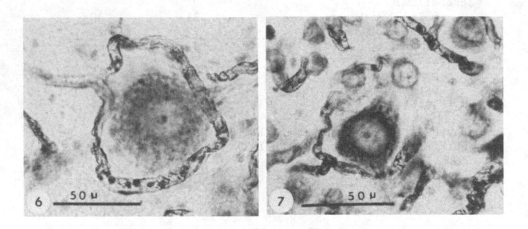

Figure 6. Large cell of the medial reticular gray in a capillary
ring. The diameter of the cell body is greater than the thickness
of the section. This cell and those shown in figures 7, 8 and 9
probably all have additional capillary segments out of the plane
of section. (Cat, 30μ section. Capillaries demonstrated with lead
chromate. Cells stained with thionin.)

Figure 7. Betz cell of motor cortex, partially surrounded by a
capillary. Part of another cortical cell and its capillary are
also visible in the upper right part of the field. (Cat, 30μ
section. Capillaries demonstrated with lead chromate. Cells
stained with thionin.)

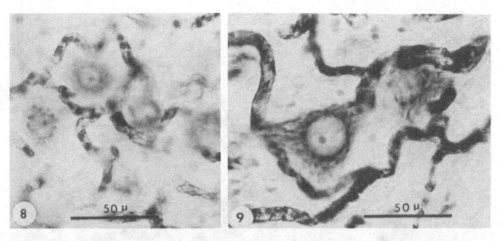

Figure 8. Cells of the ventral cochlear nucleus, and their associated capillaries. Each cell is entirely, or partly, enclosed in a capillary ring and is separated from neighboring cells by a capillary. (Cat, 30μ section. Capillaries demonstrated with lead chromate. Cells stained with thionin.)

Figure 9. Motor cell of the ventral horn, with closely associated capillaries. Another motor cell (slightly out of focus) and its associated capillaries are visible in the upper right part of the field. (Cat, 30μ section. Capillaries demonstrated with lead chromate. Cells stained with thionin.)

In other areas single capillaries supply several or many cells. Intercapillary spaces may be sparsely populated, as in the basolateral portion of the amygdala (fig. 10), or packed with cells, as in the granular layer of the cerebellum (fig. 11).

CONCLUSIONS

Although neurologic disorders resulting from insufficient blood flow through the large named arteries supplying the brain are well known (DeJong, 1967), the effects of circulatory deficiencies at the capillary level are almost unknown. The architecture of the vascular beds of the brain provides a constant upon which any number of variables can be superimposed. Some neurological disorders are of unknown etiology. In others damage occurs to neurons in selected areas or to neurons of a specific type, but the reasons for the specificity of damage are obscure. Is it possible that in some diseases this specificity, the selection of the site of damage, occurs because one or more forms of pathologic physiology, acting simultaneously or in sequence are superimposed on a set architectural pattern of small blood vessels?

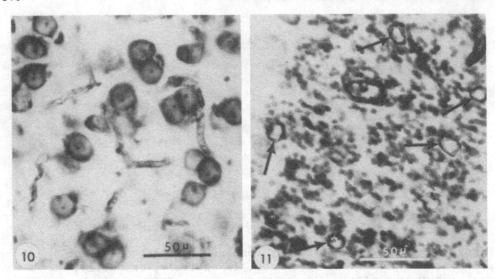

Figure 10. Amygdala (nucleus basalis, pars lateralis).
Intercapillary spaces do not contain many nerve cells. In this
region most cells are located next to capillaries. They are not
individually supplied but frequently clusters of two or three
cells are partially enclosed by a capillary. (Cat, 30µ section.
Capillaries demonstrated with lead chromate. Cells stained with
thionin.)

Figure 11. Cerebellum, granular layer, in which many cells occupy
the intercapillary spaces. (Monkey, 10µ section. Capillaries,
indicated by arrows, demonstrated with lead chromate. Cells
stained with thionin.)

Investigation of the precise architectural patterns of the small
blood vessels of the brain may provide us with an additional avenue
for study in the search for the causes of some of the unexplained
phenomena.

REFERENCES

Campbell, A. C. P. 1939 Variation in vascularity and oxidase
    content in different regions of the brain of the cat. Arch.
    Neur. Psych. 41:223-242.

Craigie, E. H. 1945 The architecture of the cerebral capillary
    bed. Biol. Rev. 20:133-146.

Crosby, E. C., T. Humphrey and E. W. Lauer 1962 Correlative
    anatomy of the nervous system. The Macmillan Co., New York.

DeJong, R. N. 1967 The neurologic examination. 3rd ed. Hoeber Med. Div., Harper and Row, Pub. New York.

Dunning, H. S. and H. G. Wolff 1937 The relative vascularity of various parts of the central and peripheral nervous system of the cat and its relation to function. J. Comp. Neur. 67:433-450.

Friede, R. L. 1961 A histochemical atlas of tissue oxidation in the brain stem of the cat. Hafner Pub. Co., New York.

Hough, H. B. and H. G. Wolff 1939 The relative vascularity of subcortical ganglia of the cat's brain; the putamen, globus pallidus, substantia nigra, red nucleus, and geniculate bodies. J. Comp. Neur. 71:427-436.

Krogh, A. 1929 The anatomy and physiology of capillaries. Yale Univ. Press, New Haven. Reprinted 1959 Hafner Pub. Co., New York.

Lockard, I., J. R. Barham, Jr., N. G. Forlidas, Jr. and R. B. Myers 1959 Simultaneous histologic demonstration of blood vessels, nerve cells and nerve fibers within the central nervous system. J. Comp. Neur. 112:169-183.

Lockard, I. and H. S. Debacker 1967 Conjunctival circulation in relation to circulatory disorders. J. South Car. Med. Assoc. 63:201-206.

Scheinker, I. M. 1954 Circulatory disturbances and management of multiple sclerosis. Ann. N. Y. Acad. Sci. 58:582-594.

Thews, G. 1960 Die Sauerstoffdiffusion im Gehirn. Ein Beitrag zur Frage der Sauerstoffversorgung der Organe. Pflüg. Arch. 271:197-226. English translation by H. Debacker, Department of Anatomy, Medical University of South Carolina, Charleston, South Carolina, USA.

Tucker, W. M., D. C. Donald, Jr. and R. A. Farmer 1950 Cerebral blood flow in multiple sclerosis. Chapter 14 in Multiple sclerosis and the demyelinating diseases. Res. Pub. Assoc. Res. Nerv. Ment. Dis. 28:203-215.

Supported by National Institutes of Health, U.S.P.H.S. Grant NB-03075.

MATHEMATICAL ANALYSIS OF OXYGEN TRANSPORT

TO TISSUE IN THE HUMAN

Eugene H. Wissler

The University of Texas at Austin

Austin, Texas 78712

A composite mathematical model which is capable of integrating thermal, respiratory, and circulatory factors in the analysis of oxygen transport is described in this paper. The model has several uses. It provides a mechanism for evaluating parameters for the human when only indirect measurements are available. It can be used also to predict responses under various pathological or stressful conditions. Much work remains to be done, but it is felt that a preliminary report should be included in this conference.

General Description of the Model. The physical system on which the equations are based consists of 15 cylindrical elements representing longitudinal segments of the head, trunk, arms, and legs. Each segment, consisting of a conglomeration of tissue, bone, fat, and skin, has a three part vascular system representing the arteries, veins, and capillaries. The circulatory path is faithfully reproduced in the sense that a tracer material introduced into an artery in a given segment will flow both into the capillaries of that segment and into the arteries of more distal segments. Blood leaving the capillaries flows into the veins where it is mixed with venous blood from other segments. The mixed venous stream at the heart flows into the pulmonary artery; exchange of both heat and mass occurs in the lungs.

Within a given segment, variables may depend on both radial position and time. Some quantities, such as density and specific heat do not depend on time, but others depend on both position and time. Local perfusion and metabolic rates are defined by physiological control equations. Temperature and the concentration of mobile species are dependent variables whose immediate values depend

813

on the previous history of the subject.

Starting with known initial conditions one can follow changes that occur in response to various imposed conditions. Fundamental are the notions of conservation of energy and mass and recognition that heat and various chemical substances move through the body under the influence of thermal and concentration gradients.

Since convection by circulating blood is a dominant transport mechanism, proper description at the capillary level is essential. Several approaches are possible, and it is not entirely clear which is best. We use the well-mixed region model for both heat and mass transfer. It is assumed that the temperature or concentration in the extra-capillary space is equal to the effluent value for blood leaving the region. For heat transfer the basic unit is a capillary and the associated tissue. For mass transfer each capillary is divided into four longitudinal sections which permits one to account for variation along the capillary.

This model was successfully employed by Pennes (11) in analyzing temperature profiles measured in the arm. It has been used subsequently by many other investigators (10,15), and seems to produce a reasonable description of thermal exchange in capillary beds. It has been employed also in the analysis of drug distribution problems (2), again producing results which are in agreement with experimental observations. Venous outflow curves from single organs are often well fitted by an exponential curve which is the expected result for the well-mixed compartment model (13).

In addition to convective transport of heat and mass by circulating blood, diffusion of heat and mass can occur in solid regions. One must account for this in analyzing thermal exchange. Whether it is important in mass transfer is questionable and probably depends on whether one is concerned with carbon dioxide or with lactic acid. Although there is evidence which indicates that one should account for diffusive transport through tissue (5), we have not included it in the present model. However, the method used to solve the heat conduction problem can be adapted with ease to solve the diffusion problem.

Equations of Change. The heat transfer model has been described previously (15). Equations are required also to account for the varying concentrations of chemical species which are involved in metabolic processes.

Since the set of metabolic reactions is very complex, we adopted a simplified scheme which does, nevertheless, include several of the more important factors. Aerobic synthesis of ATP using glucose as the substrate proceeds according to the following overall reaction.

$$C_6H_{12}O_6 + 6 O_2 + 38 \text{ ADP} + 38 P_i \rightarrow 6 CO_2 + 6 H_2O + 38 \text{ ATP}$$

(glucose)   (1)

In the absence of oxygen, synthesis of ATP proceeds as follows.

$$C_6H_{12}O_6 + 2 \text{ ADP} + 2 P_i \rightleftarrows 2 C_3H_6O_3 + 2 \text{ ATP}$$

(glucose)   (lactic acid)   (2)

Lactic acid produced in muscles is transferred to blood in the capillaries and distributed throughout the body. It can be converted to glycogen in the liver according to an aerobic mechanism which oxidizes one mole of lactic acid to carbon dioxide and water for every four moles converted into glycogen, or it can be catabolized in muscle. We assume that glucose is always available at the rate required. Hence, it is only necessary to account for three components; oxygen, carbon dioxide, and lactic acid.

Specifying the metabolic rate in a given element determines the demand for oxygen. If the rate of oxygen supply is not adequate to meet the demand, the deficit will be provided by anaerobic reactions. Since 19 moles of glucose are consumed in the anaerobic production of the same amount of ATP that is produced using one mole of glucose in the aerobic process, one can compute production rates for carbon dioxide, lactic acid, and heat for given values of demand and supply.

For accumulation of a particular component in tissue and blood, we have the following equations.

$$\varepsilon_t \frac{\partial C_t}{\partial t} = \varepsilon_t R - F_{tb} \quad \text{and} \quad \varepsilon_b \frac{\partial C_b}{\partial t} = q \frac{\partial C_b}{\partial x} = F_{tb} \quad (3)$$

in which

ε_t and ε_b = volumetric fractions of tissue and blood;
C_t and C_b = concentration; q = flux of blood through the capillaries in a unit area; x = distance along a capillary;
F_{tb} = rate of exchange between tissue and capillary per unit volume; R = rate of production per unit volume.

Although the basic equations for all components have the same form, the approximations which are appropriate in dealing with distribution of material between tissue and blood depend on the particular component involved. When dealing with oxygen, it is assumed that equilibrium exists between oxygen in myoglobin and hemoglobin. Since myoglobin is essentially saturated at oxygen tensions greater than 20 Torr, the amount of oxygen stored in myoglobin remains constant under normal conditions. These stores are released when oxygen tension falls below 20 Torr. It is assumed that equilibrium exists between carbon dioxide stores in tissue and blood, and at equilibrium the concentration in tissue is one-half of the concentration in blood. The rate of O_2 consumption is multiplied by the respiratory quotient to obtain the CO_2 production rate.

If C_{O_2} falls below a critical value in any segment, we assume that anaerobic glycolysis occurs with the production of lactic acid. The critical oxygen concentration was arbitrarily set at a constant value of 7.5 vol. percent although it should depend on the density of capillaries and the local metabolic rate, both of which change during exercise.

Equations of change are also required for lactic acid which is produced during anaerobic glycolysis. In this case, we do not assume that equilibrium exists between lactic acid in muscle and in blood. Instead, it is assumed that lactic acid is transferred from muscle to blood at a finite rate which is proportional to the difference in concentration.

The transfer of gases between pulmonary capillaries and the alveolar space must be accounted for in the lungs, but metabolic reactions and accumulation of mass in the capillaries and surrounding tissue can be neglected. Hence, the second equation in set 3 becomes

$$q(C_a - C_4) = - F_{ba} \tag{4}$$

in which C_a and C_4 = arterial and end-capillary concentrations, and F_{ba} = total rate of transfer along a capillary. Evaluation of the mass transfer rate between blood in the pulmonary capillaries and alveolar gas requires knowledge of gas tensions in blood. We use the oxygen dissociation equation and carbon dioxide buffer equations developed by Grodins, Buell, and Bart (4), and assume that the partial pressure of each component in the alveolar gas is equal to the tension of that gas in blood as it leaves the pulmonary capillary. Strictly speaking, a pressure difference is required to drive the exchange, but this difference is often negligible. It is also assumed that 2% of the pulmonary blood flow is shunted.

In developing equations of change for large arteries and veins, it is assumed that there is neither reaction nor exchange of mass with surrounding tissue.

Physiological Control Equations. The equations of change presented above permit one to compute temperatures and concentrations if physiological parameters, such as local metabolic rates, tissue perfusion rates, and the ventilation rate are known. These rates, in turn, depend on temperature, oxygen and carbon dioxide tensions, partial pressure of oxygen and carbon dioxide in the inspired gas, and the level of exercise being performed. The control equations for cutaneous blood flow, sweat rate, and rate of shivering are essentially those presented by Stolwijk and Hardy (14).

Either carbon dioxide tension or hydrogen ion concentration is usually taken as one of the dominant input variables for control of ventilation rate, but the prevailing view seems to favor use of C_{H^+}

over P_{CO_2} (7,12). It is hypothesized that three hydrogen ion concentrations are important. One is sensed by a peripheral receptor which is exposed to arterial blood. The other two are central receptors, one of which is exposed to blood while the other lies behind a barrier which is permeable to carbon dioxide but not to hydrogen ions. We assume that the central receptor responds to changes in capillary P_{CO_2} as a first order system with a time constant of 25 seconds. Hydrogen ion concentrations at the peripheral and two central receptors are weighted by factors 0.12 and 0.44, respectively, to form a single variable, CVE. The minute volume increases rapidly as CVE rises above its normal value. Arterial oxygen tensions which are outside of the normal range also affect the ventilation rate. Elevated values of P_{aO_2} attenuate the response to increasing CVE. As P_{aO_2} falls below 37 Torr during hypoxia, the minute volume increases rapidly. During exercise $\dot{V}_E$ increases rapidly, presumably due to the influence of neurogenic factors.

Cerebral blood flow varies considerably, apparently in response to varying P_{aCO_2} (8). Experimental data reported by Lambertsen (7,8)) were used to define the rate of cerebral blood flow. If the end-capillary P_{O_2} falls below 45 Torr, the base flow rate is multiplied by $1 + (45 - P_{O_2})/45$. It is assumed also that high arterial oxygen tensions ($P_{aO_2} > 500$) reduce the cerebral blood flow rate by a factor $1 - 0.00005 (P_{aO_2} - 500)$.

The perfusion rate for muscle depends on the local metabolic rate and the end-capillary oxygen tension. When $P_{O_2} < 45$, the resting value is multiplied by a factor $1 + (45 - P_{O_2})/8$. During exercise a term which is proportional to the difference between the metabolic rates in the working and resting states is added to the resting blood flow rate. It is assumed that perfusion rates change exponentially with a time constant of 30 seconds for all tissue except brain which has a time constant of 0.3 seconds.

During vigorous exercise blood flow to inactive muscle is reduced to meet the demand of active muscles without exceeding limitations on cardiac output (1). This feature was incorporated into the model by multiplying each flow rate for muscle by a factor $1 - 0.00032 \exp (0.32$ cardiac output). Since cardiac output represents the summation of local perfusion rates, solution of a transcendental equation is necessary to obtain the cardiac output.

Results. Computations performed using one set of parameters were compared with published experimental data for the following cases: inhalation of 3 and 6 percent CO_2 in air (12), inhalation of O_2 at 2 atm. (8), inhalation of hypoxic mixtures (7), sinusoidally varying P_{ICO_2} (13), and various levels of exercise (3,6). Both steady-state and transient responses were satisfactory except for

high frequency (1.2 cycles/sec) periodic drives which were very
sensitive to details of the model. Sinusoidal testing appears to
be an excellent technique for studying in vivo responses.

During exercise, excess oxygen consumption was distributed
84 percent to the legs, 8 percent to the arms and 8 percent to the
abdomen. Oxygen uptake rate was sensitive to the distribution of
blood flow. Cardiac output appears to determine the maximum oxygen
uptake rate in normal subjects; ventilation rates are more than
adequate. Maximum oxygen uptake rate was approximately 3 L/min when
the cardiac output was 17 L/min, which agrees with Fishman (3).
Lactic acid production became significant when oxygen uptake exceeded 50 percent of the maximum rate. Computed arterial lactate
concentrations agreed with reported values (6).

References

1. Bevegård, B. S. and J. T. Shepherd. J. Appl. Physiol., 21,
   123-132 (1966).
2. Bischoff, K. B., R. L. Dedrick, D. S. Zaharko, and J. A.
   Longstreth. J. Pharm. Sci., 60, 1128-1133 (1971).
3. Fishman, A. P. In Handbook of Physiology, Section 2. Circulation II, 1682, W. F. Hamilton, Sr., Editor, Am. Physiol.
   Soc., Washington, 1963.
4. Grodins, F. S., J. Buell, and A. J. Bart. J. Appl. Physiol.,
   22, 260-276 (1967).
5. Groom, A. C. and L. E. Farhi. J. Appl. Physiol., 22, 740-745
   (1967).
6. Knuttgen, H. G. and B. Saltin. J. Appl. Physiol., 32, 690-
   694 (1972).
7. Lambertsen, C. J. In Medical Physiology, V. B. Mountcastle,
   Editor, C. V. Mosby Co., St. Louis, 1968.
8. Lambertsen, C. J., R. H. Kough, D. Y. Cooper, G. L. Emmel,
   H. H. Loeschcke and C. F. Schmidt. J. Appl. Physiol., 5,
   803-813 (1953).
9. Levitt, D. G. J. Theor. Biol., 34, 103-124 (1972).
10. Mitchell, J. W., T. L. Galvez, J. Hengle, G. E. Myers, and
    K. L. Siebecker. J. Appl. Physiol., 29, 859-865 (1970).
11. Pennes, H. H. J. Appl. Physiol., 1, 93-122 (1948).
12. Reynolds, W. J., H. T. Milhorn, Jr., and G. H. Holloman, Jr.
    J. Appl. Physiol., 33, 47-54 (1972).
13. Stoll, P. J. J. Appl. Physiol., 27, 389-399 (1969).
14. Stolwijk, J. A. J. and J. D. Hardy. Pflügers Arch., 291,
    129-162 (1966).
15. Wissler, E. H. Bull. Math. Biophysics, 26, 147-166 (1964).

A MATHEMATICAL MODEL OF THE UNSTEADY TRANSPORT OF OXYGEN TO TISSUES IN THE MICROCIRCULATION*

John E. Fletcher, Ph. D.

National Institutes of Health, DCRT, Laboratory of

Applied Studies, Bldg. 12A, Rm. 2041, Bethesda, Md. 20014

INTRODUCTION

We are concerned here with the construction of a mathematical simulation model of the supply of oxygen to tissues from the microcirculation. The mathematical model describes the capillary supply to tissue as well as the distribution and consumption of oxygen within tissue.

The microscopic level at which this process takes place defies direct measurement; therefore, the basic biological characteristics can only be inferred by means of indirect (boundary) measurements. Consequently, an appropriate simulation model that can incorporate these indirect measurements and produce a complete picture of capillary-tissue exchange becomes important to the understanding of the basic biological thresholds and functional limits.

Any attempt to describe this complex physico-chemical process by means of a mathematical model clearly necessitates a number of simplifications. In attempting to formulate functional relationships among the factors involved, one considers the local character of some factors, (e.g., the value of the diffusion coefficient at a given point and in a given direction), as well as the global nature of others, (e.g., the geometrical configuration of the microcirculatory networks and the tissues they supply). The local factors essentially determine the governing differential expressions, while the global ones determine the domain of definition of the problem and the necessary boundary conditions.

*A preliminary report on the mathematical methods sketched here was presented in the Biomathematics Session of the SIAM National Meetings in Philadelphia, Pennsylvania, June 12, 1972, (see ref. 4).

The earliest published models giving a functional relation for microcirculatory flow and tissue oxygen supply are those of Krogh, (9) who used a parallel cylinder representation of the capillary bed. Krogh's single cylinder geometry is used here, but his parallel cylinder assumption may be relaxed somewhat. The essential features here are (1) that the tissue supplied by a single capillary has an approximately cylindrical configuration and (2) that the oxygen transfer across the outer boundary of this configuration is zero or essentially so, whatever the cause. To the extent these features represent an average for tissue of possibly varying local structure, the model here should be informative. A single cylinder and its notation are illustrated in figure 1.

The physical and mathematical model described in figure 1 attempts to incorporate a number of the more important features of the current state of physiological knowledge, both conceptual and experimental, into a tractable mathematical formulation. These include a consideration of the kinetics of oxygen release from hemoglobin; use of the exact form of the oxygen-hemoglobin dissociation function, $g(f)$; a nonlinear consumption rate of oxygen within the tissues, $Q(u)$; a finitely permeable capillary wall, $\alpha(z)$; and the autoregulatory effects of varying flow rates, $v(t)$. The resulting mathematical model is a nonlinear parabolic partial differential equation with derivative boundary conditions. The primary boundary condition, (i.e., at the capillary wall) is coupled to a system of nonlinear first order transport equations. These latter equations describe the intracapillary conditions. The details of the complete derivation are too extensive to be given here; however, complete details of the derivation are presented in (3) along with details and proofs of convergence of the approximating numerical scheme. The numerical scheme utilizes an unconditionally stable fractional step method which has been adapted for the approximation of the nonlinear diffusion equation in the simulation model. A similar unconditionally stable implicit method for approximating the nonlinear transport equations is developed. The combined scheme is unconditionally stable, and computes the solution directly at whole step grid points in the time and space meshes. This scheme leads directly to a system of tri-diagonal matrices of the positive type and solutions are obtained by a Gauss elimination procedure. Sample computations are presented which illustrate the model, and some applications with biological parameters are given.

A Discussion of Results

The model shown in figure 1 contains a variety of options, too numerous to detail here. Instead, we consider the most nonlinear case that might be encountered. The nonlinear hemoglobin functions $X(f)$ and $g(f)$ are determined from actual experimental data (1), (10), which are shown in figure 2. Since consistent anatomical data

THE MATHEMATICAL MODEL

TISSUE CONDITIONS :

$$\frac{\partial u}{\partial t} = \frac{1}{r}\ \frac{\partial}{\partial r}\left(r D_r\ \frac{\partial u}{\partial r}\right) + \frac{\partial}{\partial z}\left(D_z\ \frac{\partial u}{\partial z}\right) - Q(u)$$

$$\left(\frac{\partial u}{\partial r}\right)\bigg|_{r=r_t} = 0\ ;\quad D_z\left(\frac{\partial u}{\partial z}\right)\bigg|_{z=0} = 0\ ,\quad \left(D_z\ \frac{\partial u}{\partial z}\right)\bigg|_{z=z_c} = 0$$

CONSUMPTION RATE :

$$Q(u) = \begin{cases} K_1 & ,\ \text{zero order kinetics} \\ K_2 u & ,\ \text{first order kinetics, or} \\ \dfrac{K_1 u}{\mu + u} & ,\ \text{second order kinetics} \end{cases}$$

CAPILLARY WALL
PERMEABILITY :

finite : $-\left(D_r\ \dfrac{\partial u}{\partial r}\right)\bigg|_{r=r_c} = \alpha(z)\ \left[W(z,t) - u(r_c,z,t)\right]$;

infinite : $u(r_c,z,t) = W(z,t)$

$$\frac{\partial W}{\partial t} + v(t)\ \frac{\partial W}{\partial z} = \frac{2 D_r}{r_c}\left(\frac{\partial u}{\partial r}\right)\bigg|_{r=r_c} + C X(f)\left[g(f) - W(z,t)\right]$$

INTRA – CAPILLARY
CONDITIONS :

$$\frac{\partial f}{\partial t} + v(t)\ \frac{\partial f}{\partial z} = -X(f)\left[g(f) - W(z,t)\right]$$

$$f(0,t) = f_A(t)\quad W(0,t) = g(f_A(t))$$

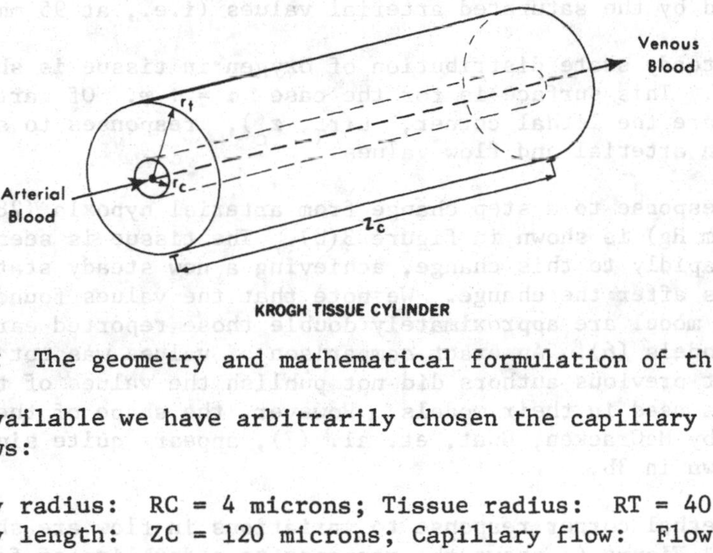

KROGH TISSUE CYLINDER

FIGURE 1: The geometry and mathematical formulation of the model.

is not available we have arbitrarily chosen the capillary parameters
as follows:

Capillary radius: RC = 4 microns; Tissue radius: RT = 40 microns
Capillary length: ZC = 120 microns; Capillary flow: Flow = 0.01
$\qquad\qquad\qquad\qquad\qquad\qquad\qquad\qquad\qquad\qquad$ vol./vol./sec.
Arterial Conditions: Normoxia = 95 mm Hg; Hypoxia = 28.5 mm Hg
Blood Capacity: C = 0.20 vol./blood vol.
Metabolic Parameters: K = 0.504 x 10^{-3} O_2 vol./Tissue vol./sec.
$\qquad\qquad\qquad\qquad\quad$ $\mu = 0.249$ x 10^{-5} O_2 vol./solute vol.
Diffusion Rates: $D_r = D_z = 0.15$ x 10^4 $micron^2$/sec.
Wall Permeability: $\alpha(z) = 1.2$ x 10^3/$micron^2$/sec. or $\alpha = +\infty$.

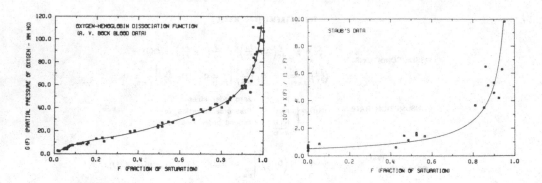

FIGURE 2: Nonlinear hemoglobin dissociation functions fitted to experimental data.

These values are chosen within ranges usually reported and used for computations in modeling the cerebral cortex, (5), (8), (9), and (11). All computational results are shown in terms of O_2 concentration normalized by the saturated arterial values (i.e., at 95 mm Hg).

The steady state distribution of oxygen in tissue is shown in figure 3a. This surface is for the case $\alpha = +\infty$. Of particular interest are the lethal corner, $u(r_T, z_c)$, responses to step changes in arterial and flow values.

The response to a step change from arterial hypoxia (28.5 mm Hg to 38.5 mm Hg) is shown in figure 3(b). The tissue is seen to respond rapidly to this change, achieving a new steady state within 10 seconds after the change. We note that the values found for this nonlinear model are approximately double those reported earlier for simpler models (6). An exact comparison of values was not possible since most previous authors did not publish the values of the parameters used in their models. However, the shape of the curves reported by McCracken, Hunt, et. al. (7), appears quite similar to those shown in 3b.

The lethal corner response to variations in flow are shown in figure 4. Figure 4a shows the response to a doubling of flow rate at steady state conditions, both for normoxia and hypoxia. Again the response is rapid, with less than 6 seconds being required to reach a new steady state. Figure 4b shows loss of oxygen supply at the lethal corner in between 5 and 6 seconds following cardiac arrest. These values are also longer than those previously reported, (6).

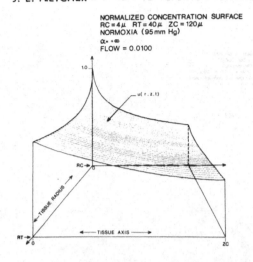

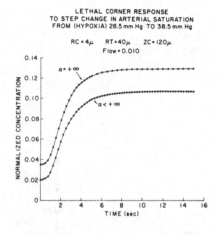

FIGURE 3: The steady state tissue distribution and lethal corner response to step increases from arterial hypoxia.

The final graph, figure 5, shows the effects of periodically varying the flow rate. This condition occurs, and the frequency used here is determined from the data of Johnson et. al., (8). In each case, the lethal corner attains a periodic variation after about 5 or 6 seconds. The upper curve is the periodic variation, the middle curve is the response to this variation about the steady flow values, while the lower curve represents a periodic variation having the steady flow value as its maximum. The contrast between these two curves illustrates that the values depend critically on the <u>nature</u> of the time-dependent flow.

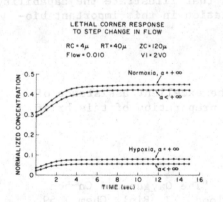

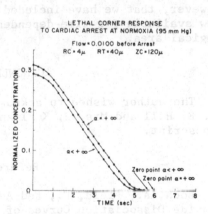

FIGURE 4: Lethal corner response to variations in flow for both the finitely permeable and infinitely permeable cases.

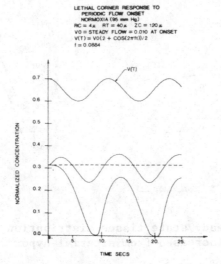

FIGURE 5: Lethal corner response to periodic flow onset at
Normoxia (95 mm Hg)

There are many other facets of this model which are now both
possible and practical to consider. We have omitted, due to space
limitations, important discussions of, (1) hemoglobin release effects,
(2) the influence of the shape of the dissociation function, and
(3) the comparative effects of neglecting axial diffusion. We hope,
however, that we have included results that illustrate the capability
now available for time-dependent simulation in this important biological area.

ACKNOWLEDGMENT

The author wishes to acknowledge the excellent assistance of
Mr. E. Hill and Mrs. J. Kuhlman in the preparation of this brief
manuscript.

References

1. Boch, A., Field, H., and Adair, G., "The Oxygen and Carbon
Dioxide Dissociation Curves of Human Blood", J. Biol. Chem., 59,
(1924), 353-378.

2. Chance, B., "Cellular Oxygen Requirements", Fed. Proc., 16
(1957), 671-691.

3. Fletcher, J. E., "On a Parabolic Boundary Value Problem of the Second Kind Describing a Biological Process", Ph. D. Thesis, Univ. of Md., 1972. (Available from University Microfilms, Ann Arbor, Mich.)

4. Fletcher, J. E., "The Effects of Time-Dependent Flow in Capillaries Supplying Substrates to Living Tissue", SIAM Review (Chronicle) Vol. 15, No. 1, 1973, p. 252.

5. Gonzalez-Fernandez, J. M., and Atta, S. E., "Transport and Consumption of Oxygen in Capillary-Tissue Structures", Mathematical Biosciences, Vol. 2, 225-262, 1968.

6. Halberg, M. R., Bruley, D. F., and Knisely, M. H., "Simulating Oxygen Transport in the Microcirculation by Monte Carlo Methods", Simulation, Vol. 15, No. 5, pp. 206-212, November, 1970.

7. Hunt, D., Bruley, D., Bicher, H., and Knisely, M., "Oxygen Transport in the Brain Microcirculation by a Hybrid Computer Nonlinear Monte Carlo Method", (To appear in proceedings of 1973 Simulation Conference, Montreal, Canada).

8. Johnson, P., and Wayland, H., "Regulation of Blood Flow in Single Capillaries", Am. J. Physiol., 212, No. 6.

9. Krogh, A., "The Number and Distribution of Capillaries in Muscles with Calculations of the Oxygen Pressure Head Necessary for Supplying the Tissue", J. Physiol., 52, 409 (1910-1919).

10. Staub, N., Bishop, J., and Forster, R., "Velocity of O_2 Uptake by Human Red Blood Cells", J. Appl. Physiol., 16, No. 2, (1961), 511-516.

11. Thews, G., "The Theory of Oxygen Transport and Its Application to Gaseous Exchange in the Lung", Oxygen Transport in Blood and Tissue, D. Lubbers, V. Luft, G. Thews, and E. Witzleb editors, George Thieme Verlag, Stuttgart, (1968), 1-20.

SIMPLIFYING THE DESCRIPTION OF TISSUE OXYGENATION

E. N. Lightfoot

Department of Chemical Engineering

University of Wisconsin

INTRODUCTION

Here as in all engineering problems first priority must go to establishing orders of magnitude, both of key system parameters and of unavoidable uncertainties. Once this is done it is possible to develop a solution providing the maximum degree of simplicity consistent with the need for accuracy. Accordingly we begin by defining a model, the modified Krogh tissue cylinder of Fig. 1, and establishing representative dimensions and other characteristics (see Fig. 1 and Table 1).

TABLE 1: CHARACTERISTIC MAGNITUDES

$$R_c = 3\mu \; ; R_T = 30\mu \; ; L = 180\mu \; ; V = 400\mu sec^{-1}$$

$$\mathcal{V} \doteq \mathcal{D} = 1.4 \times 10^{-5} \; cm^2 \; sec^{-1} \; ; k_0 = 3.7 \times 10^{-8} g\text{-}mol \; cm^{-3} \; sec^{-1}$$

$$c_{HbO_2} = 8.8 \times 10^{-6} \; g\text{-}mol \; cm^{-3} \; (\alpha^{2.66}/1 + \alpha^{2.66}) ; m_{avg} \doteq 40$$

$$\alpha = (2.68 \times 10^5 \; cm^3 \; g\text{-}mol^{-1} \chi c_{O_2}) ; \mathcal{H} = 7.4 \times 10^8 \; mmHg \; cm^3 \; g\text{-}mol^{-1}$$

$$c_{tot} \; (ART) \doteq (8.5 + 0.12) \; g\text{-}mol \; cm^{-3}$$

ANALYSIS

Tissue cylinder: We begin by assuming cylindrical symmetry and homogeneous diffusion with zero-order chemical reaction. Then in the absence of end effects the diffusion of O_2 is described by:

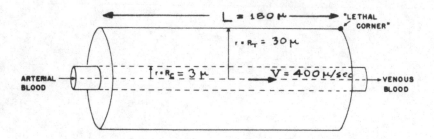

Fig. 1: The Krogh tissue cylinder used as a model here.

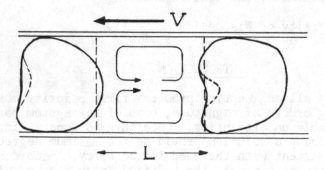

Fig. 2a: The nature of bolus flow for an observer moving with
          the red cells.

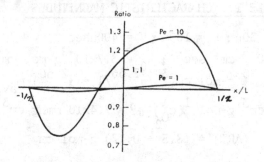

Effect of convection on local mass transfer rate

$$\text{Ratio} = \frac{\text{Local mass transfer rate (convection + diffusion)}}{\text{Local mass transfer rate (diffusion only)}}.$$

Fig. 2b: Mass transfer in the capillary bolus.

$$\frac{\partial p}{\partial t} = \mathcal{D} \, \frac{1}{r} \frac{\partial}{\partial r} \, r \, \frac{\partial p}{\partial r} + \mathcal{D} \, \frac{\partial^2 p}{\partial z^2} + k_0 \qquad (R_c < r < R_T) \qquad (1)$$

Here p denotes oxygen tension, $\mathcal{D}$ effective oxygen diffusivity, k_0 the zero-order rate constant for oxygen metabolism (moles/vol., time), and $\mathcal{H}$ is Henry's constant for oxygen solubility. Equation 1 is non-controversial in the sense that no more complex expression has been seriously proposed. For most purposes, in fact, it can be substantially simplified. It can easily be seen, especially under situations of heavy oxygen demand, that

$$(\partial^2 p/\partial z^2)/\left(\frac{1}{r} \frac{\partial}{\partial r} \, r \, \frac{\partial p}{\partial r}\right) \sim (R/L)^2 \ll 1 \qquad (2)$$

This expectation is borne out by numerical calculations of Reneau et al, and axial diffusion will therefore be neglected.

We next note that the response time t_T to changes in boundary conditions will be of the order

$$t_T \sim R_T^2/D \doteq \frac{2}{3} \text{ sec} \qquad (3)$$

This is short compared to most physiological time scales of interest here, and we will therefore assume pseudo-steady operation:

$$0 = \mathcal{D} \, \frac{1}{r} \frac{\partial}{\partial r} \, r \, \frac{\partial p}{\partial r} + k_0 \qquad (4)$$

The major exception will be fluctuations in blood flow due to incompletely dampened pulses, which may be important (see for example Crandall and Flumerfelt). We return to this point later. Integration of Eq. 4 is straightforward and gives:

$$p(R_c) - p(R_T) = \frac{R_c^2 k_0 \mathcal{H}}{4 \, \mathcal{D}} \, [\gamma^2 (\ln \gamma^2 - 1) + 1] \qquad (5)$$

$$\left. \frac{\partial p}{\partial r} \right|_{r=R_c} = - \frac{R_c^2 k_0 \mathcal{H}}{2 \, \mathcal{D}} \, (\gamma^2 - 1) \qquad (6)$$

with

$$\gamma = (R_T/R_c) \equiv \beta \quad \text{for} \quad p(R_c) \geq \beta^2 (\ln \beta^2 - 1) + 1 \qquad (7)$$

or $\quad \gamma^2(\ln\gamma^2 - 1) + 1 = p(R_c) \quad$ for $\quad p(R_c) \le \beta^2(\ln\beta^2 - 1) + 1$

$$(8)$$

Equation 8 is needed whenever there is an anoxic region at the cylinder periphery.

Capillary. The situation here is in principle much complicated, both by the two-phase nature of the flow and convective oxygen transport in the "bolus" of plasma between each pair of red cells. However the total number of red cells is large, and it has been found by Aroesty and Gross that circulation in the bolus is unimportant. This is clear from Fig. 2 if one notes that bolus Péclet numbers

$$\text{Pé}_b \equiv LV/\mathscr{D}_{O_2 p}$$

are not greater than about 10. Here L denotes bolus length, V red cell velocity, and $\mathscr{D}_{O_2 p}$ the diffusivity of oxygen in plasma. It follows that blood in the capillary may be considered a homogeneous fluid in plug flow.

It is further reasonable to assume local equilibrium between hemoglobin and dissolved oxygen (Dindorf et al; see however Longo et al) and to neglect diffusion of oxyhemoglobin (Keller et al; Spaeth). System description then takes the form

$$(1 + m)\left[\frac{\partial \bar{p}}{\partial t} + V\frac{\partial \bar{p}}{\partial z}\right] = \bar{\mathscr{D}}\left[\frac{1}{r}\frac{\partial}{\partial r}\, r\,\frac{\partial \bar{p}}{\partial r} + \frac{\partial^2 \bar{p}}{\partial z^2}\right] \qquad (9)$$

where the overline denotes conditions within the capillary and

$$m = \partial c_{HbO_2}/\partial c_{O_2} \qquad (10)$$

is the slope of the oxyhemoglobin dissociation curve. The underlined terms will henceforth be neglected as axial diffusion is almost totally negligible, and capillary response time

$$t_c \sim R_c^2(1 + m_{avg})/\bar{\mathscr{D}} \doteq \frac{1}{4}\ \text{sec} \qquad (11)$$

is very short.

We next note that oxygen flux from the capillary is to a first approximation position independent. Assuming for the moment that m is constant we find (Carslaw & Jaeger, §7.8), in the limit of large z that

$$\bar{p} - \bar{p}_0 = \left(\frac{\partial\bar{p}}{\partial r}\bigg|_{r=R}\right)\left[2\,\frac{\bar{\mathcal{D}}\,z}{(1 + m_{avg})VR_c} + \frac{r^2}{2R_c} - \frac{R_c}{4}\right] \qquad (12)$$

where $\bar{p}_0$ is the oxygen tension of entering (arterial) blood. The radial profile given by this expression is correct to within 10% for

$$z \geq 0.15\,R_c^2 V(1 + m_{avg})/\bar{\mathcal{D}} \doteq 15.4\mu \ll L \qquad (13)$$

Entrance effects are therefore very small and one may both use local values of m and relax the restriction to constant wall flux. Thus

$$\frac{\partial\langle\bar{p}\rangle}{\partial z} = \left(\frac{\partial\bar{p}}{\partial r}\right)\bigg|_{r=R_c}[2\,\bar{\mathcal{D}}/(1 + m)VR_c] \qquad (14)$$

$$\bar{p}(R_c,z) - \langle\bar{p}(z)\rangle = -(\gamma^2 - 1)(R_c^2 k_0 H/\bar{\mathcal{D}}) \qquad (15)$$

where $\langle p\rangle$ is the <u>bulk</u> or cup-mixing tension

$$\langle\bar{p}\rangle = \frac{1}{\pi R^2 V}\int_0^{2\pi}\int_0^R p(r)Vr\,dr\,d\theta \qquad (16)$$

Since m, γ, and $(\partial\bar{p}/\partial r)$ are all functions only of $\bar{p}(R)$ via the boundary condition

$$\frac{\partial p}{\partial r}\bigg|_r = (\bar{\mathcal{D}}/\mathcal{D})\,\frac{\partial\bar{p}}{\partial r} \doteq \frac{\partial\bar{p}}{\partial r} \qquad (17)$$

description is formally complete.

NUMERICAL EXAMPLE

For the conditions of Table 1

$$R_c^2 k_0 H/\mathcal{D} = (9\times10^{-8})(3.72\times7.4)/(1.4\times10^{-5}) = 0.177\text{mmHg}$$

Since there is no anoxic region here $\gamma = R_T/R_c$, and

$$p(R_c) - p(R_T) = (0.177/4)[1 + 100\,(\ln 100 - 1)] = 16.0\text{mmHg}$$

$$\langle\bar{p}\rangle - \bar{p}(R_c) = (0.177/8)(99) = 2.18\text{mmHg}$$

Total drop in oxygen tension from bulk value in blood to tissue periphery,

$$<\bar{p}> - p(R_c) = 18.2 \text{ mmHg}$$

is independent of z.

In the absence of anoxic regions the zero-order kinetics permits using a simple material balance

$$\pi R_c^2 V(<c_{tot}(0)> - <c_{tot}(L)>) = \pi(R_T^2 - R_c^2)Lk_0 \qquad (18)$$

where $<c_{tot}>$ is the bulk, or cup-mixing value of total oxygen concentration (both free and combined with hemoglobin). Equation 14 may therefore be simply replaced by Eq. 18 in the form

$$<c_{tot}(L)> = 8.60 \times 10^{-6} - (99 \times 1.8 \times 3.7 \times 10^{-8}/4) = 6.94 \times 10^{-6} \frac{\text{g-moles}}{\text{cm}^3}$$

and the relation

$$<\bar{p}(L)> = [<c_{tot}(L)> - c_{max}\alpha_L^{2.66}/(1 + \alpha_L^{2.66})] \qquad (19)$$

where $\alpha_L = <\bar{p}(L)>/ 3.73 \times 10^{-6}$ gmols/cm^3. It is then easily shown that

$$<\bar{p}(L)> \doteq 44 \text{mmHg} \quad \text{(tension in lethal corner)}$$

and

$$p(R_T,L) \doteq 44 - 18 \doteq 26 \text{mmHg (venous tension)}$$

or not much more than half the calculated oxygen tension of 44mm in the venous blood.

DISCUSSION

Even cursory comparison of the above simple development with more elaborate published expressions will show it to be equally reliable for calculating venous and lethal-corner oxygen tensions, in view of anatomical and physiological uncertainties. It may also be shown that a still simpler model, in which the capillary blood is well mixed will give identical results for these two critical tensions in the absence of local anoxia. This in turn is equivalent to the flow-limited models of Bischoff and Dedrick and suggests that

such refinements as considering velocity fluctuations is of
minor importance.

BIBLIOGRAPHY

Aroesty, J. and J. F. Gross, Microvasc. Res. 2, 247 (1970).

Crandall, E. D. and R. W. Flumerfelt, in Chemical Engineering
    in Biology and Medicine, D. Hershey, Ed., Plenum (1967).

Dindorf, J. A., E. N. Lightfoot, and K. A. Solen, Chem. Eng.
    Prog. Symp. Series # 114, pp. 75-87 (1971).

Keller, K. H. and S. K. Friedlander, Chem. Eng. Prog. Symp.
    Series, #66, 19 (1966).

Longo, L. D., G. G. Power, and R. E. Foster, II, J. Appl.
    Physiol. 26, 360 (1969).

Reneau, D. D., D. F. Bruley, and M. H. Knisely, A.I.Ch.E. J.
    15, 916-925 (1969).

Spaeth, E. E., Blood Oxygenation, CRC Critical Reviews in
    Bio-engineering (In press).

A SIMPLIFIED MODEL OF THE OXYGEN SUPPLY FUNCTION OF CAPILLARY BLOOD FLOW

William A. Hyman

Bioengineering Program
Texas A&M University
College Station, Texas   77843

ABSTRACT

An analytic model of the role of capillary blood flow in supplying oxygen to tissue has been developed. This model is less complicated than previously used models, yet may be more realistic. The fundamental point of departure here is that consideration of the radial concentration gradients in the capillary is avoided by assuming the capillary to be "well-stirred". The source-sink role of the erythrocyte with respect to oxygen is retained in a distributed sense. The inhomogeneity of blood on the scale of the capillary as well as the phenomenon of bolus flow justify this approximation. A closed form solution to the resulting equations is readily obtained and compares favorably with the results of computer based computations of axial oxygen pressure profiles. This model is therefore expected to be of significant value in further work.

INTRODUCTION

The purpose of this paper is to explain and justify an analytical representation of blood flow and oxygen transport which is less complicated yet possibly more realistic than previously used models. The geometry used is the now familiar tissue cylinder with concentric capillary which was first introduced by Krogh [1] and which has been the basis of continued study. A recent review is provided by Middleman [2].

TISSUE MODEL

Tissue is of course a non-homogeneous and probably non-iso-tropic material. Chemical species are transported through this material being either consumed or released by the cells. Although diffusion in both the radial and longitudinal direction has been considered analytically previously [3], for the current purpose of elucidating the capillary model it is convenient to consider diffu-sion in the tissue to occur only in the radial direction with a diffusion coefficient D. While this assumption does limit the capability of the model for describing the detailed oxygen concen-tration profiles in the tissue, it has been shown that the total oxygen delivery is not significantly affected.

The rate of metabolic consumption is assumed to be constant and is denoted by m. This would be true only at oxygen concentra-tions above a critical value [4] and therefore ischemic states are not properly included under this assumption. The outer radius of the tissue cylinder is R_2 and its inner radius is R_1 (which is identical to the outer radius of the capillary). The diffusion equation describing the mass balance of oxygen in the tissue under steady conditions then takes the form

$$D\frac{1}{r}\frac{d}{dr}\left(r\frac{dC^T}{dr}\right) - m = 0 \tag{1}$$

where C^T is the concentration of oxygen in the tissue and the additional assumption of axisymmetry has been included in the equa-tion. Here r is the radius at any point in the tissue, $R_1 < r < R_2$. Equation (1) can be rewritten in terms of the partial pressure of oxygen in the tissue P^T by introducing the solubility of oxygen in tissue S^T

$$D\frac{1}{r}\frac{d}{dr}\left(r\frac{dP^T}{dr}\right) - \frac{m}{S^T} = 0 \tag{2}$$

At the capillary wall resistance to the diffusion of oxygen is assumed to be negligible which requires that at the blood-tissue interface $(r = R_1)$

$$P^T = P^B \quad \text{at} \quad r = R_1 \tag{3}$$

where P^B is the partial pressure of oxygen in the blood. A balance of the mass flux at this interface requires that

$$J = -DS^T\frac{dP^T}{dr} \quad \text{at} \quad r = R_1 \tag{4}$$

where J is the mass flux of oxygen per unit area from the blood into the tissue.

Since the tissue is assumed to be served only by its central capillary a no flux boundary condition at the outer cylindrical surface is reasonable and takes the form

$$\frac{dP^T}{dr} = 0 \quad \text{at} \quad r = R_2 \tag{5}$$

CAPILLARY MODEL

Various fluid mechanical and diffusional models have been used for the purpose of considering transport processes in the flow of blood through capillaries. Many of the current models follow Reneau [5] in assuming that the capillary blood flow could be taken to be homogeneous with a distributed oxygen source-sink term and laminar flow. In such a model radial concentration gradients are included as well as axial convection. This conceptualization seems reasonable for flow in vessels with a large ratio of tube diameter to erythrocyte diameter and has been used for this purpose [6]. However for capillary flow the presence of the erythrocytes makes a homogeneous field approach somewhat illogical as well as introducing unnecessary mathematical complexity. The well documented single file flow of erythrocytes [7] with the accompanying convective motion in the plasma gaps [8], as well as the effect of these convective motions on mass transport [9] combine to make a unidirectional homogeneous flow model of questionable value. One alternative is to extend the complexity of [9]. This would however reduce the utility of these modeling procedures by requiring an extensive computational effort in obtaining results. The alternate procedure introduced below suggest instead a simplication which makes the treatment of the capillary consistent with the other approximations in the model.

In this paper a mass balance in the capillary is taken in an averaged sense over the entire capillary cross section. In this model the net convective flux of dissolved oxygen as well as the oxygen bound in the erythrocyte is balanced against the mass flux out through the capillary wall. This equation takes the form

$$V \left(\frac{\partial C^B}{\partial \chi} + N \frac{\partial \psi}{\partial \chi} \right) = \frac{2J}{R_1} \tag{6}$$

where C^B is the oxygen concentration, N is the oxygen capacity of the blood, ψ the fractional saturation of the blood with combined oxygen, V is the average velocity of the blood in the capillary and χ is the axial coordinate. It is recognized here that on the basis of Fick's first law there cannot be an outward radial flux in the absence of a radial concentration gradient. However the averaging process represented by the above equation has eliminated specific consideration of radial gradients while still allowing mass flux at the boundary. In terms of the partial pressure of oxygen in the blood equation (6) becomes

$$S^B V \left(\frac{\partial P^B}{\partial \chi} + \frac{N}{S^B} \frac{\partial \psi}{\partial \chi} \right) = \frac{2J}{R_1} \tag{7}$$

The boundary conditions applicable to the above equation are equation (3) and an inlet condition

$$P^B = P_i^B \quad \text{at} \quad \chi = 0 \tag{8}$$

MATHEMATICAL SOLUTION

Equation (2) is readily integrated and following the application of equation (5) the solution takes the familiar form

$$P^T = P_o^T - \frac{MR_2^2}{2DS^T} \ln \frac{r}{R_.} - \frac{MR_2^2}{4DS^T} \left[1 - \frac{r^2}{R_1^2} \right] \tag{9}$$

where P_o^T is the partial pressure of oxygen in the tissue at the tissue-capillary interface $r = R_1$. This value is a function of χ.

The connection between the tissue and the capillary is obtained by computing the flux into the tissue which is required to sustain the distribution of equation (9). From equations (4) and (9)

$$J = \frac{M}{2R_1} \left[R_1^2 - R_2^2 \right] \tag{10}$$

It can be seen here that this flux is independent of χ.

Substituting equation (10) into equation (7)

$$\frac{dP^B}{d\chi} + \frac{N}{S^B} \frac{d\psi}{d\chi} = \frac{M}{S^B V} \left[1 - \frac{R_2^2}{R_1^2} \right] \tag{11}$$

This equation is readily integrated with respect to χ to give

$$P^B + \frac{N}{S^B} \psi = \frac{M\chi}{S^B V} \left[1 - \frac{R_2^2}{R_2^2} \right] + B \tag{12}$$

where B is a constant. It is worth noting here that the ease with which equation (11) is integrated is the principle benefit of the current model. In fact it has not been necessary to specify the function ψ up to this point. A procedure similar to that used here had been suggested earlier [10]. However the introduction of a mass transport resistance at the capillary wall (which is not necessary for oxygen transport) made the resulting equation unintegrable in closed form.

The constant B can be determined by applying the inlet condition of equation (8)

$$B = P_i^B + \frac{N}{S^B} \psi \ (P_i^B \ ) \tag{13}$$

which is easily evaluated once ψ is specified. With B determined equation (12) represents an implicit function of P^B as a function of X. Note that this function is non-linear and depends on the dissociation function ψ. If P^B is determined at a point χ then P_o^T in equation (9) becomes known through the matching condition of equation (3). The solution is therefore complete.

NUMERICAL CALCULATIONS

It is of interest to compare the solution for the partial pressure of oxygen in the blood obtained with the present model to that obtained in [5] using the same model for the tissue and the homogeneous, laminar flow model for the capillary previously discussed. It is noteworthy in making this comparison that the work reported in [5] required a numerical finite difference approach and digital calculations. The dissociation function ψ is taken to be given by

$$\psi \ \frac{k(P^B)^n}{1+k(P^B)^n} \tag{14}$$

where k and n are constants. Their values are taken to be 0.001 and 2.2 respectively. Although it would be preferable to put equation (12) in dimensionless form this was not done in [5] and for the purpose of making a direct comparison it is convenient to use the same specific values for the individual parameters used in [5]. These values are: $R_1 = 2.5 \times 10^{-4}$ cm, $R_2 = 3.0 \times 10^{-3}$ cm, $m = 8.34 \times 10^{-4}$, $S^B = 3.42 \times 10^{-5}$, $V = 0.04$ cm/sec, $P_i^B = 95$ mm Hg, and $N = 20.4\%$.

The results of this comparison are shown in figure 1 in which the circles are for the result obtained here and the curve is taken from (5). It is apparent that the two sets of results are in close agreement.

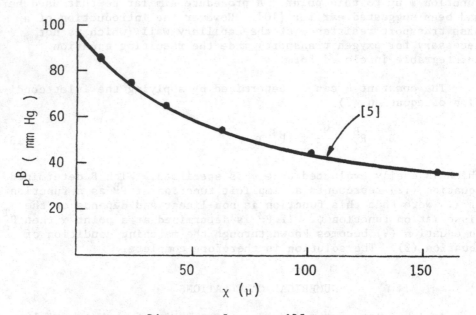

Distance along capillary

Figure 1. Comparison of present results (circles) with those
of [5] curve.

CONCLUSION

By introducing a simplification in the treatment of the oxygen
supply function of the capillary a model has been obtained which
gives essentially the same results for the axial oxygen profile in
a capillary as that obtained by far more complicated procedures.
Furthermore, the model used here is thought to represent a more
consistent level of approximation. Because of the efficiency of
the present procedure it is expected that further progress in
various problems on tissue oxygenation will be accelerated.

REFERENCES

1. Krogh, A., J. Phys. 52, 409, 1919.
2. Middleman, S., "Transport Phenomena in the Cardiovascular
        System", Wiley-Interscience: New York, 1972.
3. Blum, J., Am. J. Phys., 198, 991, 1960.

4. Jobsis, F.F., In "Handbood of Physiology" (W.O. Fenn and H. Rahn, eds), Sec. 3, 1, pp. 63-124, Amer. Phys. Soc: Washington, D.C., 1966.

5. Reneau, D.D., Ph.D. Thesis, Clemson University, Clemson, South Carolina, 1966.

6. Colton, C.K., and Drake, R.F., Chem. Eng. Symp. Series, 67, No. 114, 88, 1971.

7. Skalak, R., and Branemark, P.I., Science, 164, 717, 1969.

8. Prothero, J., and Burton, A.C., Biophys. J., 1, 565, 1961.

9. Aroesty, J., and Gross, J.F., Microvasc. Res. 2, 247, 1970.

10. Bloch, I., Bull. Math. Biophys. 5, 1, 1943.

ACKNOWLEDGEMENT

This investigation was supported, in part, by Research Grant No. 23-P-55823/6-01 from the Social and Rehabilitation Service, Department of Health, Education and Welfare to the Texas A&M Research Foundation.

OXYGEN TRANSPORT IN THE HUMAN BRAIN - ANALYTICAL SOLUTIONS

Richard R. Stewart and Catherine A. Morrazzi

Department of Chemical Engineering

Northeastern University, Boston, Ma.   02115

Reneau et al. (1966, 1967) performed a digital computer simulation of steady-state oxygen transport in the human brain. The Krough capillary-tissue cylinder geometry was used in which the blood flows axially in a cylindrical capillary of radius, R, surrounded by a layer of tissue of outer radius, R_T. The mathematical model used is given by:

Capillary: $\quad v(1 + \dfrac{NknP^{n-1}}{c(1 + kP^n)^2}) \dfrac{\partial P}{\partial z} = D \dfrac{1}{r} \dfrac{\partial}{\partial r} (r \dfrac{\partial P}{\partial r})$ $\qquad$ (1)

Tissue: $\quad D_T \dfrac{1}{r} \dfrac{\partial}{\partial r}(r \dfrac{\partial P_T}{\partial r}) - \dfrac{A}{c_T} = 0$ $\qquad\qquad$ (2)

Boundary Conditions: $\quad$ at $z = 0$, $P = $ known $P(r)$ $\qquad$ (3)

$\qquad\qquad\qquad$ at $r = 0$, $\dfrac{\partial P}{\partial r} = 0$ $\qquad$ (4)

$\qquad\qquad\qquad$ at $r = R$, $P = P_T = P_i(z)$ $\qquad$ (5)

$\qquad\qquad\qquad$ at $r = R$, $Dc \dfrac{\partial P}{\partial r} = D_T c_T \dfrac{\partial P_T}{\partial r}$ $\qquad$ (6)

$\qquad\qquad\qquad$ at $r = R_T$, $\dfrac{\partial P}{\partial r} = 0$ $\qquad$ (7)

Due to the highly nonlinear nature of Equation (1) the model comprising Equations (1)-(7) was solved numerically by Reneau. The purpose of the present study is to show that Equation (1) may be modified to yield a model which may be solved analytically for the space-averaged oxygen partial pressure, <P>, as a function of axial position, z. The approximate analytical solutions are shown to be in excellent agreement with the numerical solutions for nearly all cases.

Modification of Equation (1) begins by approximating the non-linearity which represents the derivative of the oxygen dissociation curve by a simple exponential function, that is

$$\frac{knP^{n-1}}{(1 + kP^n)^2} \cong \alpha e^{\beta P} \tag{8}$$

Figure 1 shows a comparison of these two expressions for P_{CO_2} = 40 mm Hg. Similar good agreement was obtained for P_{CO_2} = 20 mm Hg. Next, Equation (2) is space averaged by multiplying each term by $(2/R^2)r\partial r$ and integrating from r = 0 to R. This leads to

$$\frac{d<P>}{dz} + \frac{N\alpha}{c\beta} \frac{d}{dz} < e^{\beta P} > = \frac{2D}{vR} \left(\frac{\partial P}{\partial r}\right)\bigg|_R \tag{9}$$

The assumption is now made that

$$<e^{\beta P}> = e^{\beta <P>} \tag{10}$$

The derivative on the righthand side of Equation (9) is obtained by first solving Equation (2) analytically. Substitution of this derivative and integration of the result yields

$$<P> + \frac{N\alpha}{c\beta} e^{\beta <P>} = \frac{A}{vc}(1 - (\frac{R_T}{R})^2) z + <P_A> + \frac{N\alpha}{c\beta} e^{\beta <P_A>}) \tag{11}$$

A second solution of interest may be obtained by assuming that the amount of axial oxygen transport by the plasma is negligible compared with that by the erythrocytes. This assumption eliminates the first convection term in Equation (1). A mathematical develop-

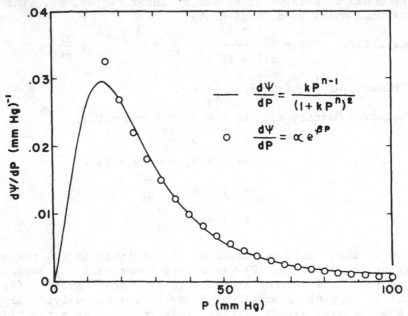

Figure 1. Derivative of O_2 dissociation curve, P_{CO_2} = 40 mm Hg.

ment similar to the above yields

$$<P> = \frac{1}{\beta} \ln \left(\frac{A\beta}{vN\alpha}(1 - (\frac{R_T}{R})^2) \ z + e^{\beta<P_A>} \right) \qquad (12)$$

Equations (11) and (12) may be solved for axial $<P>$ profiles. Equations (11) and (12) can be obtained more directly by starting with an oxygen balance on a differential length of capillary of radius, R, and introducing $<P>$ at the outset.

RESULTS AND DISCUSSION

Figure 2 shows a comparison of the numerical solution of Reneau and the approximate analytical solutions of Equations (11) and (12). At arterial pressures less than 100 mm Hg good agreement exists among the numerical and two analytical solutions. At arterial pressures above 100 mm Hg the disagreement among the three results increases as $<P_A>$ increases. The increasing disagreement between numerical and analytical solutions as $<P_A>$ increases above 100 mm Hg is due primarily to the fact that the numerical values of α and β were obtained from a nonlinear curve fit over only the P = 17-100 mm Hg range. The disagreement between the two analytical solutions for hyperbaric oxygen conditions is necessarily explained by the difference in the assumptions made in the two models. The assumptions leading to Equations (11) and (12) are identical except that Equation (12) includes the assumption of negligible convective oxygen transport by the plasma. Calculations for the model associated with Equation (11) show that, for the case of $<P_A>$ = 150 mm Hg, of the 23.417 μ^3 O_2 transported into the tissue over the first 10 μ of capillary length, approximately 50% comes from oxygen dissolved in the entering plasma. This percentage drops to 15% for $<P_A>$ = 95 mm Hg and is even less for lower values of $<P_A>$. Thus, ignoring the oxygen dissolved in the entering plasma becomes increasingly significant the higher the value of $<P_A>$ and it is not surprising that

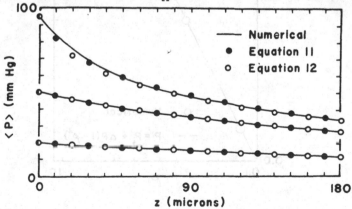

Figure 2. Comparison of numerical and analytical solutions.

axial <P> profiles given by Equations (11) and (12) differ increasingly as $<P_A>$ increases above 100 mm Hg.

   Equation (10). Equation (10) is an important assumption made in the mathematical development. Figure 3 shows a plot of radial P-profiles at the arterial and venous ends of the capillary for the 26 cases reported in Reneau (1966) which are within the scope of the present study. Only one data point per radial position is reported to avoid clutter. However, for all cases, the points are sufficiently localized so as to fit inside the circular data points of Figure 3. Thus the radial capillary profiles are parabolic and it may be shown for the parabolic case that

$$<e^{\beta P}> = e^{\beta <P>} (1 + \frac{(\frac{\beta(\Delta P)}{2})^2}{3!} + \frac{(\frac{\beta(\Delta P)}{2})^4}{5!} + \ldots) \qquad (13)$$

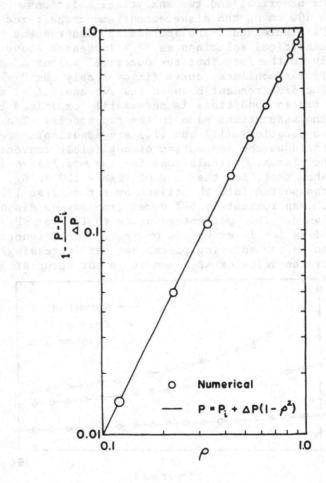

Figure 3. Radial P_{O_2} profiles in capillary from numerical solution.

That is, for a parabolic profile, Equation (10) is true to the extent that the infinite series in parentheses in Equation (13) approaches unity. For the β's and ΔP encountered in the present study this series is typically 1.003...thereby rendering Equation (10) in error by only 0.3% at worst.

Tissue Profiles. The numerical results of Reneau (1966) show that the ΔP used for the boundary condition at the arterial end of the capillary is the same ΔP found at all other axial positions. This fact coupled with the parabolicity of the radial profiles means that it should be possible to compute the oxygen partial pressure at the capillary-tissue interface from the equation

$$P_i = <P> - \frac{\Delta P}{2}. \tag{14}$$

Figure 4 compares numerically computed P_i values from Reneau (1966) with P_i values computed from $<P>$ values obtained using Equation (11). The data points in Figure 4 are representative of the 130 different radial profiles reported in Reneau (1966) which are within the scope of the present study.

Once a P_i value is obtained from Equation (14) the radial tissue profile may be computed from the analytical solution of Equation (2) which is

$$P_T = P_i(z) - \frac{AR_T^2}{2D_Tc_T} \ln \frac{r}{R} - \frac{AR^2}{4D_Tc_T}(1 - \frac{r^2}{R^2}), \quad R \le r \le R_T \tag{15}$$

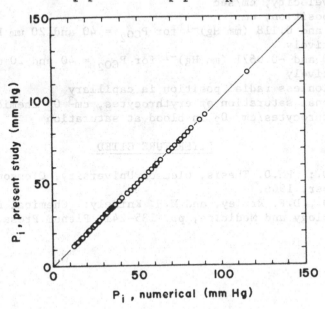

Figure 4. Comparison of P_i for numerical and present study solutions.

Sample calculations of P and P_T using Equations (11), (12), (14), (15) and the equation on Figure 3 were typically within 1 percent of the numerical results for $<P_A>$ below 100 mm Hg.

NOMENCLATURE

A = O_2 consumption rate of tissue, $cm^3 \ O_2/cm^3$ tissue-sec
c = O_2 solubility of plasma, $cm^3 \ O_2/cm^3$ blood-mm Hg
c_T = O_2 solubility of tissue, $cm^3 \ O_2/cm^3$ tissue-mm Hg
D = diffusivity of O_2 in blood, cm^2/sec
D_T = diffusivity of O_2 in tissue, cm^2/sec
k = constant in Equation (8)
n = constant in Equation (8)
N = oxygen capacity of blood, $cm^3 \ O_2$ in blood at saturation/cm^3 blood
P = oxygen partial pressure of blood, mm Hg
P_T = oxygen partial pressure of tissue, mm Hg
P_i = oxygen partial at blood-tissue interface, mm Hg
$<P>$ = space averaged oxygen partial pressure of blood, mm Hg
$<P_A>$ = space-averaged oxygen partial pressure of blood at arterial end of capillary, mm Hg
ΔP = difference between the oxygen partial pressure of blood at the centerline and at the blood-tissue interface, mm Hg
r = radial position, cm
R = radius of capillary, cm
R_T = outer radius of tissue cylinder, cm
v = blood velocity, cm/sec
z = axial position, cm
α = 0.0713 and 0.118 $(mm \ Hg)^{-1}$ for P_{CO_2} = 40 and 20 mm Hg, respectively
β = -0.0490 and -0.0671 $(mm \ Hg)^{-1}$ for P_{CO_2} = 40 and 20 mm Hg, respectively
ρ = dimensionless radial position in capillary
Ψ = fractional saturation of erythrocytes, $cm^3 \ O_2$ chemically combined in erythrocytes/$cm^3 \ O_2$ in blood at saturation

LITERATURE CITED

Reneau, D.D.: Ph.D. Thesis, Clemson University, Clemson, S.C., December, 1966.
Reneau, D.D., D.F. Bruley, and M.H. Knisely: Chemical Engineering in Biology and Medicine, pp. 135-241, Plenum Press, 1967.

CALCULATION OF CONCENTRATION PROFILES OF EXCESS ACID IN HUMAN BRAIN TISSUE DURING CONDITIONS OF PARTIAL ANOXIA

Daniel D. Reneau and Larry L. Lafitte

Department of Biomedical Engineering
Louisiana Tech University
Ruston, Louisiana 71270

The purpose of this paper is to present the initial preliminary results of a theoretical attempt to obtain a better understanding of the distribution of lactic and pyruvic acid in cerebral cortex during conditions of partial anoxia and consequent anerobic glycolysis.

The intent of the work is to begin to seek identification of parameters related to cellular metabolic products and by-products, such as pH, that might possibly be the chief factors (and not lack of oxygen) associated with the regulation of the microenvironment and changes in which may lead to nerve cell network damage, destruction and total irreversible decay. The approach to the problem is to:

(1) Initially use the Krogh capillary-tissue geometry to seek trend analyses;

(2) Use previously developed equations based on the Krogh geometry to determine the spatial distribution of oxygen and glucose in all parts of the capillary and tissue;

(3) Theoretically calculate and subsequently geometrically map anoxic tissue regions surrounding the capillary which occur under conditions of specified quantitative changes in biological functions (i.e., flow rate, metabolic rate, anoxic-anoxia);

(4) Solve interconnected equations that describe:

a) the production and diffusion of acids in the

mathematically defined anoxic region,

   b) the diffusion through the non-anoxic region,

   c) the diffusion into and removal by blood flowing through adjacent capillaries.

Mathematical simulation of the above conditions can give insight into helping answer some very important questions. For instance, reductions in rates of blood flow through capillaries can lead to the development of localized areas of cerebral tissue anoxia which become sites of anerobic glycolysis. What reduced magnitude of capillary flow rate yields minute regions of tissue anoxia but is still sufficient to remove acidic metabolic waste material? And, for each progressive reduction of determined degree below this critical flow rate, what is the magnitude and rate of increase of tissue acidosis?

MATHEMATICAL DESCRIPTION

Oxygen Diffusion. Based on the Krogh geometry, known phenomena, and certain assumptions which are outlined in Reneau, Bruley and Knisely (2), a mathematical model has been developed which decribes the change in oxygen partial pressure in capillary blood and tissue as a function of time, position, flow rate, pH, oxygen capacity, metabolic rate, and various constants such as diffusion coefficients and solubility. Certain details concerning the model are given in previous publications (2,4).

Capillary:

$$\left(1 + \frac{NknP^{n-1}}{c_1\ (1+kP^n)^2}\right)\frac{\partial P}{\partial t} = D_1\left(\frac{\partial^2 P}{\partial r^2} + \frac{1}{r}\ \frac{\partial P}{\partial r}\right) + D_1\frac{\partial^2 P}{\partial x^2} - V_x\ \frac{\partial P}{\partial x} - \quad (1)$$

$$\left(\frac{V_x NknP^{n-1}}{c_1\ (1+kP^n)^2}\right)\ \frac{\partial P}{\partial x}$$

Interface

$$P_i\Big|_{Blood} = P_i\Big|_{Tissue} \quad\quad\quad (2)$$

$$D_1 c_1\ \frac{\partial P}{\partial r}\Big|_{\substack{r=R_1 \\ Blood}} = D_2 c_2\ \frac{\partial P}{\partial r}\Big|_{\substack{r=R_2 \\ Tissue}} \quad\quad (3)$$

Tissue

$$\frac{\partial P}{\partial t} = D_2 \left[\frac{\partial^2 P}{\partial r^2} + \frac{1}{r} \frac{\partial P}{\partial r} \right] + D_2 \frac{\partial^2 P}{\partial x^2} - \frac{A}{c_2} \qquad (4)$$

Glucose Diffusion. Based on the Krogh geometry, known phenomena, and certain assumptions which are outlined in Reneau and
Knisely (3), and Reneau et al. (6), a companion mathematical
model has been developed which describes the changes in glucose
concentration in capillary blood and cerebral tissue. The
model is given below:

Capillary:

Plasma

$$\frac{\partial c}{\partial t} = D_1 \left[\frac{\partial^2 c}{\partial r^2} \quad \frac{1}{r} \frac{\partial c}{\partial r^2} \right] + D_1 \quad \frac{\partial^2 c}{\partial x^2} - V_x \frac{\partial c}{\partial x} + \qquad (5)$$

$$K \left[\frac{L_2 c'}{L_2 c' + \emptyset} - \frac{L_2 c}{L_1 c + \emptyset} \right]$$

Erythrocyte

$$\frac{\partial c'}{\partial t} = -V_x \frac{\partial c'}{\partial x} - k \left[\frac{L_2 c'}{L_2 c' + \emptyset} \quad \frac{L_1 c}{L_1 c + \emptyset} \right] \qquad (6)$$

Blood-Tissue Interface:

$$\frac{\partial c}{\partial t} = B \left[\frac{\lambda_1 c}{\lambda_1 c + \emptyset'} \right|_{} - \frac{\lambda_2 c}{\lambda_2 c + \emptyset'} \left] \right. \qquad (7)$$

$$\left. \begin{array}{l} r = R_1 \\ \text{Capillary} \\ \text{side of} \\ \text{Interface} \end{array} \right. \qquad \begin{array}{l} r = R_1 \\ \text{Tissue} \\ \text{side of} \\ \text{Interface} \end{array}$$

Tissue:

$$\frac{\partial c}{\partial t} = D_2 \left[\frac{\partial^2 c}{\partial r^2} + \frac{1}{r} \frac{\partial c}{\partial r} \right] + D_2 \frac{\partial^2 c}{\partial x^2} - A \qquad (8)$$

For this study, the above equations were reduced to steady-state and axial gradients were neglected. Nomenclature is referred to prior publications.

Acid Diffusion

Based on the Krogh geometry, known phenomena, and very idealizing assumptions including non-ionic diffusion, the following equations are presented to describe the steady-state, geometric selective, acid production, diffusion and removal by flowing capillary blood during partial anerobic glycolysis in tissue:

Tissue Equation

$$D_t \left[\frac{\partial^2 c}{\partial r^2} + \frac{1}{r} \frac{\partial c}{\partial r} \right] = -r_A \tag{9}$$

Blood Equation

$$V_x \frac{\partial c}{\partial x} = D_b \left[\frac{\partial^2 c}{\partial r^2} + \frac{1}{r} \frac{\partial c}{\partial r} \right] \tag{10}$$

Interface Equations

$$D_b \left. \frac{\partial c}{\partial r} \right| = D_t \left. \frac{\partial c}{\partial r} \right| \tag{11}$$

$$\begin{array}{ll} \text{blood side} & \text{tissue side} \\ r = \text{interface} & r = \text{interface} \end{array}$$

$$C_t \Big| = \lambda C_b \Big| \tag{12}$$

$$\begin{array}{ll} \text{tissue} & \text{blood} \\ r = \text{interface} & r = \text{interface} \end{array}$$

The interface equations describe conditions in tissue at the anerobic-aerobic interface and at the capillary wall for the blood-tissue interface.

Standard boundary conditions have been used which describe the tissue as a closed system interacting only with the capillary system.

MATHEMATICAL SIMULATION SCHEME:

The simulation procedure was as follows:

1) Under a set of quantitatively defined conditions, the oxygen equations were solved and anoxic tissue regions of the Krogh Cylinder were geometrically mapped using the solution technique outlined in Reneau et al. (5).

2) Under the same set of conditions, the glucose equations were solved to determine the distribution of glucose in anerobic and aerobic tissue regions.

3) Provided adequate glucose was available, the acid diffusion equations were solved to determine the concentration level in tissue. Production of acid was allowed to occur only in the anoxic region at a defined rate. The numerical solution technique was built around the Crank-Nicholson method. With stability, convergence was sensitive to the number of radial increments used (see Figure 1). Approximately 301 axial increments were required.

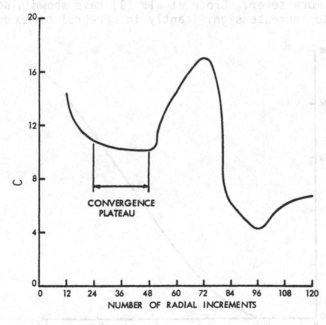

FIGURE 1.   PLOT OF CONCENTRATION AT HYPOXIC LETHAL CORNER VERSUS NUMBER OF RADIAL INCREMENTS FOR PO_2=95 mmHg and v_x=0.25 NORMAL

DISCUSSION OF RESULTS

Numerous parameter studies were conducted to study the effect of changes in different variables or combination of variables. Figures 2, 3 and 4 are representative of the significance of the effect of flow rate during partial anoxia. According to our calculations, a small region of anoxic tissue is present in the Krogh model around the lethal corner of the tissue when the PO_2 of arterial blood is 30 mmHg tension and all other factors, including flow rate, are normal. Further reductions in flow greatly increase the anoxic region in both the longitudinal and radial directions - see Reneau et al. (5). Figure 2 is a graph showing the average concentration of predicted tissue acid concentration as a function of capillary length when arterial blood PO_2 = 30 mmHg and flow rate is 25% normal. Note that the average theoretically predicted acid concentration at the lethal corner is approximately 20 mg/100 cm³ tissue. In these studies the unit rate of production of acid has been set equivalent to the rate of glucose utilization during normal conditions, and the radial diffusion coefficient for acid diffusion was set equal to a very conservative value of 10^{-5} cm²/sec. Actual conditions may be considerably more sever. Grote et al. (1) have shown glucose utilization to increase significantly in cerebral cortex during anoxia.

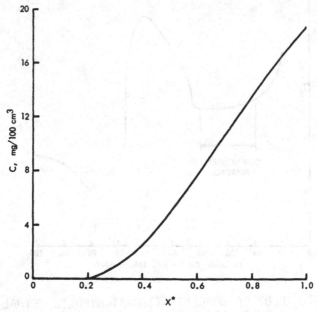

FIGURE 2. TYPICAL AXIAL CONCENTRATION PROFILE FOR
PO_2 = 30 mmHg and v_x = 0.25 NORMAL

Figure 3 represents the same conditions but the capillary flow rate has been increased to 50% of normal. This increase in flow reduces the total anoxic volume and increases the propensity for acid removal. Although the tissue region surrounding the lethal corner is still anoxic, the average concentration of acid has been reduced by a factor of 4 - from 20 mg/100 cm³ tissue to 5 mg/100 cm³ tissue. Figure 4 demonstrates the effect of an additional increase in flow rate to 75% of normal. In this case the lethal corner is still anoxic but the average concentration of acid is approximately 1.2 mg/100 cm³ tissue. Thus the theoretical acid level of the lethal corner has been reduced almost 20-fold by increasing capillary flow rate from 25% of normal to 75% of normal.

Other studies indicated a sharp sensitivity to acid production rate. Increases in the metabolic rate of acid production were accompanied by similar order of magnitude increases in tissue acid concentration.

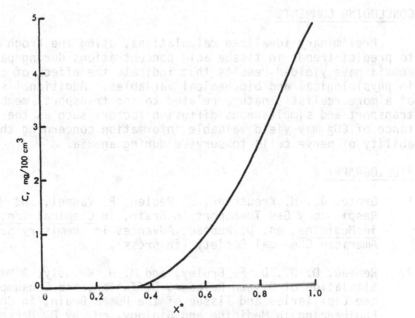

FIGURE 3.   TYPICAL AXIAL CONCENTRATION PROFILE FOR
PO_2=30 mmHg and v_x=0.50 NORMAL

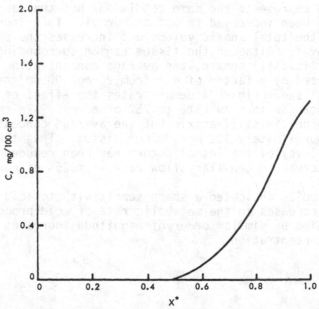

FIGURE 4.   TYPICAL AXIAL CONCENTRATION PROFILE FOR
PO_2=30 mmHg and v_x=0.75 NORMAL

CONCLUDING COMMENTS

Preliminary idealized calculations, using the Krogh system, to predict trends in tissue acid concentrations during partial anoxia have yielded results that indicate the effect of changes in physiological and biochemical variables. Additional studies of a more realistic nature related to ion transport, membrane transport and simultaneous diffusion factors such as the importance of CO_2 may yield valuable information concerning the ability of nerve cells to survive during anoxia.

BIBLIOGRAPHY

1.   Grote, J., H. Kreuscher, J. Reulen, P. Vaupel, and H. Gunther, Respiratory Gas Transport In Brain, in Chemical Engineering in Medicine, ed. D. Reneau, Advances in Chemistry Series, American Chemical Society (in press).

2.   Reneau, D. D., D. F. Bruley, and M. H. Knisely: A Mathematical Simulation of Oxygen Release, Diffusion, and Consumption in the Capillaries and Tissue of the Human Brain, in Chemical Engineering in Medicine and Biology, ed. by D. Hershey, pp. 135-241, Plenum Press, 1967.

3.   Reneau, D. D., and M. H. Knisely: A Mathematical Simulation
      of Glucose Diffusion and Consumption in Brain, Proceedings
      of the 21st Annual Conference on Engineering in Medicine and
      Biology, 10, 29.2, 1968.

4.   Reneau, D. D., D. F. Bruley, and M. H. Knisely: A Digital
      Simulation of Transient Oxygen Transport in Capillary-
      Tissue Systems (Cerebral Gray Matter), A.I.Ch.E. J., 15,
      pp 916-925, 1969.

5.   Reneau, D. D., D. F. Bruley, and M. H. Knisely: A Computer
      Simulation for Prediction of Oxygen Limitations in Cerebral
      Gray Matter, JAAMI, 4, pp. 211-223, 1970.

6.   Reneau, D. D., M. H. Knisely, H. I. Bicher, and D. F. Bruley:
      Glucose Diffusion and Consumption in the Human Brain,
      Transport Processes in Biology and Medicine, Preprint 33a,
      70th National Meeting of A.I.Ch.E., Atlantic City, New Jersey,
      August, 1971.

SIMULATING MYOCARDIUM OXYGEN DYNAMICS

Duane F. Bruley*, Daniel H. Hunt, Haim I. Bicher and

Melvin H. Knisely *Department of Chemical Engineering,

Clemson University, Clemson, S.C., U.S.A. and Department

of Anatomy, Medical University of South Carolina,

Charleston, S.C., U.S.A.

Abstract

A three-compartment model, consisting of erythrocytes, plasma, and tissue was simulated on a hybrid computer in the unsteady-state to predict myocardium oxygen tensions under normal and pathological conditions. The model includes nonequilibrium oxygen dissociation characteristics for the red cells with superimposed flow and metabolic rate changes. It was assumed that both flow velocity and metabolic rate varied as pure harmonic functions 180 degrees out of phase. Results indicate that under normal conditions, heart tissue tension remains essentially constant at a mean value rather than oscillating. It appears that this is primarily due to the frequency characteristics of the total physiological system and that myoglobin capacitance becomes important mainly at the lower heart rates.

Introduction

There are perhaps many reasons for man's investigation of natural phenomena, most principally of himself. Curiosity, self-preservation, and amazement at the simultaneous complexity and simplicity of it all have probably given researchers, (3, 4, 6, 7, 8) past and present, sufficient cause for delving into the structure and functioning of heart muscle.

A favorite approach of the research engineer and scientist is to explain a set of facts under consideration in terms of general laws deemed valid from established evidence and past experience. From the general laws, in equation form, a physical process is transformed into a mathematical model which is experimentally determined to be trustworthy for given conditions operating on the actual process.

The heart can be considered one of the most important organs of the human body. It plays a unique role in that it must function constantly to supply oxygen and other anabolites to all organs and tissue and also assist in the removal of metabolites and other contaminants from the body. Heart failure, even for short periods of time, can cause tissue anoxia and the accumulation of acid radicals which can lead to irreversible damage in most organs and tissue, most specifically brain.

Myocardial oxygen tension is an important parameter in the normal functioning of heart. Anoxic and hypoxic conditions over relatively short periods of time (5 to 10 minutes) can cause permanent damage to myocardial structure. Since the pathogenesis of many heart diseases is related to myocardial hypoxia it is obvious that an understanding of the phenomena initiating this condition would be significant.

Experimental investigation of the living heart is difficult and limited because of the delicate sensors (such as oxygen micro electrodes) necessary for oxygen measurements and the continuous movement of the beating heart. Therefore, simultaneous mathematical studies should be helpful in trying to unravel the mysteries associated with heart tissue oxygenation. This investigation is an initial attempt to mathematically examine myocardial oxygen dynamics and to understand better the significance of myoglobin in heart muscle.

Mathematical Model

A lumped parameter mathematical model based on the dimensions of a Krogh capillary-tissue unit in heart muscle was derived. (It should be noted that the anatomical dimensions can vary because of changing intercapillary distances (5)). The equations are essentially the same as those presented previously (1) except that the tissue lump considers the oxygen holdup of myoglobin.

A capillary and its surrounding tissue was separated into three compartments (figure 1). One lump represents the erythrocytes, one lump represents the plasma and another lump represents the

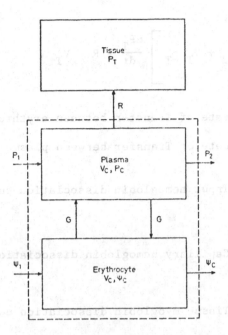

Figure 1: Three Compartment Configuration

tissue. The model considers the deoxygenation velocity of the red cell by including a velocity rate constant and a dynamic oxygen dissociation curve. Resistance to oxygen transport from the plasma to tissue is also considered by including a rate equation which is proportional to the difference in oxygen tension between the capillary plasma and tissue.

The simulation was carried out in the pseudo-dynamic state. It was assumed that both flow velocity and metabolic rate varied as pure harmonic functions, 180 degrees out of phase. The assumption of a harmonic metabolic rate is questionable (2) and can be changed easily to a constant value if desired. In either case the conclusions of this paper would not be altered.

The following set of equations were formulated.

Erythrocyte Compartment:

$$V_c N \frac{d\psi_c}{dt} = N Q \psi_1 - N Q \psi_c + G$$

Plasma Compartment:

$$b V_c c \frac{dP_c}{dt} = b Q c P_1 - b Q c P_c - G - R$$

Tissue Compartment:

$$\left[\frac{V_T \, \rho_T \, N_T \, k_T}{(1 + k_T \, P_T)^2} + V_T \, c_T \right] \frac{dP_T}{dt} = R - V_T \, A$$

where:

$G = K_1 \, (\psi_E - \psi_c)$ Rate of Transfer between erythrocyte and plasma

$R = ka \, (P_c - P_T)$ Rate of Transfer between plasma and tissue

$$\psi_1 = \frac{k_1 \, P_1^{\,n}}{1 + k_1 \, P_1^{\,n}}$$ Input hemoglobin dissociation curve

$$\psi_E = \frac{k_1 \, P_c^{\,n}}{1 + k_1 \, P_c^{\,n}}$$ Capillary hemoglobin dissociation curve

$$\psi_m = \frac{k_T \, P_T}{1 + k_T \, P_T}$$ Tissue myoglobin dissociation curve

$Q = Q_a + Q_h \sin wt$ Harmonic Flow rate

$A = (A_a + A_h \sin wt) \; f(P_T)$ Harmonic Metabolic rate

The model was simulated on a hybrid computer to handle the computational difficulties associated with the nonlinear equations. All oxygen tension output values are space averaged due to the lumped parameter nature of the model.

Results

The main variable under considerations in this simulation is myocardial tissue oxygen tension. A question of interest might be whether or not, and to what degree, the myocardium oxygen tension oscillates due to the rhythmical supply of oxygen to heart muscle? Also, what is the function of myoglobin in myocardium and what influence will it have on the temporal behavior of heart muscle oxygenation?

Considering that there is still some concern about the constancy of myocardium metabolism it has been assumed, for this simulation, that the metabolic rate oscillates, and is 180 degrees out of phase with the oscillating flow rate. These conditions represent the most extreme case of oxygen supply and consumption and therefore should produce the greatest tissue tension variations.

To demonstrate the model prediction four sets of results will be presented. The simulation was carried out using the best physiological and anatomical parameters available to the authors.

(1) Normal conditions with myoglobin at a heart rate of 60 beats per minute, Figure 1.

(2) Pathological conditions with no myoglobin present at a normal heart rate of 60 beats per minute, Figure 2.

(3) Pathological conditions (extreme) with myoglobin at a heart rate of 6 beats per minute, Figure 3.

(4) Pathological conditions (extreme) with no myoglobin present at a heart rate of 6 beats per minute, Figure 4.

Figure 1 illustrates that at a normal heart rate, with myoglobin, the tissue oxygen tension remains essentially constant, at a mean value, even with the harmonic variations in flow rate and metabolic rate. By comparing Figure 1 and Figure 2 it can be seen that the lumped parameter model predicts that myoglobin has only a slight influence in maintaining a constant tissue tension at a normal heart rate. This indicates that the physiological process is acting as a filter with the cut off frequency somewhere below 60 cycles per minute.

For the extreme pathological condition of heart rate at 6 beats per minute the effect of myoglobin can be seen more clearly. Comparing Figure 3, with myoglobin, and Figure 4, without myoglobin, illustrates that at the low heart rate the myoglobin capacitance has a definite effect on tissue oxygen tension oscillations.

Nomenclature and Parameter Values

P_1	95.	mm Hg	input O_2 tension
V_c	4.2×10^{-8}	cm^3 blood	volume of capillary
V_T	4.17×10^{-7}	cm^3 tissue	volume of tissue
c	3.42×10^{-5}	$cm^3 O_2 / cm^3$ blood-mm Hg	
N	0.204	$cm^3 O_2$ sat$/cm^3$ blood	
b	1.0	(dimensionless)	
ka	6.0×10^{-11}	$cm^3 O_2 / sec$-mm Hg	
k_1	1.08×10^{-8}	$cm^3 O_2$ sat$/sec$	
k_T	0.165	mm Hg^{-1}	
ρ_T	1.05	gm tissue$/cm^3$ tissue	
N_T	0.01	$cm^3 O_2 / gm$ tissue	

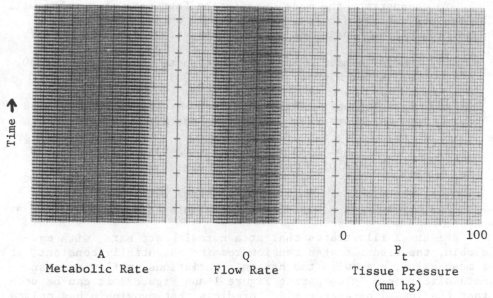

		0 100
A	Q	P_t
Metabolic Rate	Flow Rate	Tissue Pressure
		(mm hg)

Figure 1: 60 Beats per minute with myoglobin.

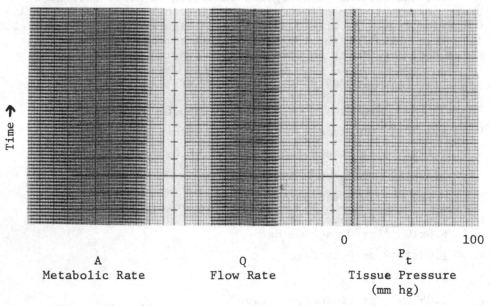

		0 100
A	Q	P_t
Metabolic Rate	Flow Rate	Tissue Pressure
		(mm hg)

Figure 2: 60 Beats per minute without myoglobin.

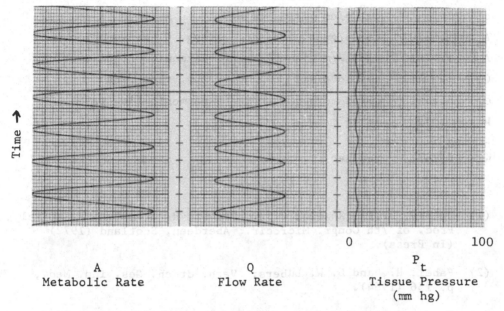

A	Q	P_t
Metabolic Rate	Flow Rate	Tissue Pressure
		(mm hg)

Figure 3: 6 Beats per minute with myoglobin.

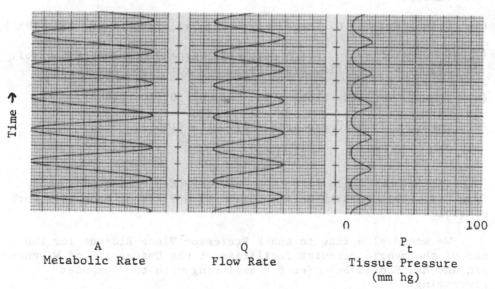

A	Q	P_t
Metabolic Rate	Flow Rate	Tissue Pressure
		(mm hg)

Figure 4: 6 Beats per minutes without myoglobin.

c_T	$2.76 \cdot 10^{-5}$ cm^3O_2/cm^3tissue-mm Hg		
A_a	1.5×10^{-3} cm^3O_2/gm tissue-sec	avg. O_2 consumption	
A_h	1.67×10^{-4} cm^3O_2/gm tissue-sec	harmonic O_2 consumption	
Q_a	1.07×10^{-8} cm^3blood/sec	avg. flow rate	
Q_h	6.45×10^{-9} cm^3blood/sec	harmonic flow rate	
k_1	0.001　　mm Hg^{-1}		
n	2.2		
w	6.283　　rad/sec		

References

(1) Bruley, D. F., Reneau, D. D., Bicher, H. I. and M. H. Knisely, Proc. of 7th Congr. Microcir., Aberdeen, Scotland (1972) (In Press).

(2) Fabel, H., and D. W. Lübers: Verh. dtsch. Ges. inn. Med., p. 156 (1964).

(3) Grote, J., and G. Thews: Pflügers Arch. 276, 142 (1962).

(4) Lübbers, D. W.: In: Heart Failure, Pathophysiology and Clinical Aspects, p. 287. Stuttgart: Thieme 1968.

(5) Martini, J. and C. R. Honig, Microvascular Research 1, 244-256 (1969).

(6) Opitz, E., and G. Thews: Arch. Kreisl.-Forsch. 18, 137 (1952).

(7) Schuchardt, S.: In: Anaesthesiology and Resuscitation, vol. 30, p. 43. Berlin, Heidelberg, New York: Springer 1969.

(8) Thews, G.: Acta biotheor. (Leiden) 10, 106 (1953).

(9) Thews, G.: Pflügers Arch. 276, 166 (1962).

Acknowledgement

The authors would like to acknowledge the financial support from U.S. Public Health Service Grant NS-06957 that made this work possible.

We would also like to thank Professor Vince Rideout for the use of the hybrid computer facilities at the University of Wisconsin and Mr. Ronald Scheafer for assisting with the computer programing.

NUMERICAL SOLUTIONS OF BLOOD FLOW IN THE ENTRANCE OF A TUBE

V. L. Shah and R. Soto

College of Engineering and Applied Science, University
of Wisconsin-Milwaukee
Medical College of Wisconsin

Hydrodynamic boundary layer predictions are obtained in the entrance region of a tube for non-Newtonian fluids like blood which obey Casson's stress-strain relation. The numerical procedure used is that of Patankar and Spalding [2]. The mathematical accuracy of the numerical procedure is demonstrated by comparing the fully developed velocity profiles with the known analytical solutions. In addition the results of the numerical solution for the case of yield stress equal to zero are compared with the available entrance flow solution for Newtonian fluids. Detailed results are presented for a wide range of yield numbers.

INTRODUCTION

During the last ten years, the major interest has been on developing capillary membrane oxygenators and hemodialysers wherein semipermeable membranes separate blood and gas phase and allow mass transfer without degradation. In the last few years, therefore, mathematical analysis of diffusion through blood flowing in a tubular membrane has been attempted by many investigators. Because of the mathematical difficulties, they have ignored the entrance region and directed their efforts in obtaining the solutions applicable to fully developed regions. In this region the velocity profile does not change.

However, when fluid enters a duct from a large plenum chamber, it enters with a uniform velocity. During its course of flow through the initial portion of the duct, the boundary layer thickens until it intercepts the boundary layer from the opposite wall. Beyond the point of intersection of these boundary layers, the shape of the velocity profile does not change, and this region is called

fully developed. The region wherein the boundary layer development occurs is called the entrance region. The length of the entrance region depends on the flow Reynolds number.

For a flow in a long tube, the effect of the entrance region
is to produce a larger pressure drop than that calculated from the
fully-developed flow solution. The detailed knowledge of the flow
development is required not only to calculate the pressure drop,
but also to solve heat and mass diffusion equations. All attention,
in the past, has been warranted towards the solution of the entrance
flow for Newtonian fluids. With the recent interest in capillary
membrane oxygenators and hemodialysers, the knowledge of blood
flow in the entrance region of a tube is of great interest.

Blood is not at all a simple fluid. A vast number of experiments have been performed on blood with varying hematocrits, anticoagulants, temperatures, and so forth, to obtain reliable values
of viscosity of blood. The majority of shear stress-shear rate
data strongly suggests non-Newtonian behavior particularly at low
rates of shear. Charm et al. [1] have fitted Casson's Equation
directly without making allowances for differences in plasma viscosity. They have come up with a relation:

$$\sqrt{\tau} = 0.166 \sqrt{\eta} + 0.33 \ (\sqrt{dynes/cm^2}) \tag{1.1}$$

which is satisfactory for shear rates between 0 and 100,000 sec^{-1}.
Here τ is a shear stress and η is a shear rate of strain.

In the present work we have used the numerical procedure of
Patankar and Spalding [2] to determine the flow development in a
tube for fluids which obey the following Casson's Stress-strain
relation:

$$\sqrt{\tau} = \sqrt{\tau_y} + K_c \ \sqrt{\eta} \tag{1.2}$$

where τ_y is the yield stress and K_c is a constant. The results have
been obtained for yield number $(\tau_y D/K_c^2 \bar{U}) = 0, 0.1, 0.3, 0.5, 1.0,$
1.5, 3.274, 5 and 10. The accuracy of the solution has been
checked by comparing the asymptotic velocity profile at a large
value of axial distance x with analytical exact solution for fully-
developed flow. In addition, the results for yield number equal to
zero are compared with the entrance flow solution for Newtonian
fluid. The results are in very close agreement.

2. GOVERNING EQUATIONS

For incompressible, laminar, axisymmetric boundary layer flow
the conservation equation of mass and momentum are:

$$\frac{\partial}{\partial x} (ru) + \frac{\partial}{\partial y} (rv) = 0 \qquad (2.1)$$

and

$$u\frac{\partial u}{\partial x} + v\frac{\partial u}{\partial r} = -\frac{1}{\rho}\frac{dp}{dx} + \frac{1}{\rho r}\frac{\partial}{\partial r} (\tau r) \qquad (2.2)$$

Here u and v are the velocity components in the axial x direction and the radial r (or normal y) direction respectively. The boundary conditions for the flow in the entrance of a tube are:

$$u = v = 0 \text{ at } r = r_0;$$

$$\frac{\partial u}{\partial r} = 0 \text{ at } r = 0, \text{ and}$$

$$u = \bar{u} \text{ at } x = 0. \qquad (2.3)$$

Here r_0 is the radius of a tube and $\bar{u}$ is the average velocity.

3.   THE SOLUTION PROCEDURE

The numerical procedure of Patankar and Spalding [2] has been used to obtain the solutions of Equation (2.2) over a wide range of yield numbers.

$$Y = \tau_y D/\bar{U} K_c^2 \qquad (3.1)$$

The coordinates employed are the dimensionless stream function ω and the distance in the flow direction x. The constant x and constant ω lines form the finite difference grid. The procedure is of the "finite difference marching integration type." Therefore, at every step in the integration, the known values of velocity for one value of x, called x_u (upstream), are used to obtain values of velocity at a slightly greater value of x called x_D (downstream). By step-wise repetition of this basic operation, the whole field of interest is investigated. For more complete descriptions of the procedure and the computer program, one may refer to Patankar and Spalding [2].

4.   SUMMARY OF RESULTS

Equation (2.2) with the boundary condition (2.3) was solved for a range of yield numbers. The values of local and apparent friction factors are presented in Figures 1 and 2. The results of the maximum velocity profile are presented in Figure 3.

The local friction coefficient f_x, defined as:

$$f_x = \frac{dp}{dx}\frac{D}{2\rho\bar{u}^2} \qquad (4.1)$$

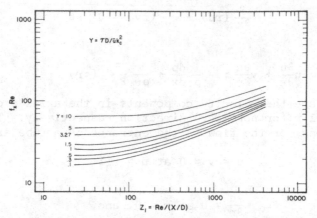

Fig. 1   Local Friction Factor in the Entrance Region of a Tube

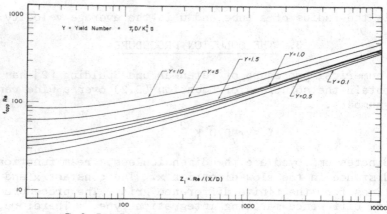

Fig. 2   Total Friction Coefficient in the Entrance Region of a Tube

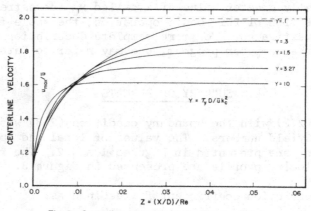

Fig. 3   Centerline Velocity in the Entrance Region

is based on the local pressure gradient at position x. Whereas
the apparent friction factor

$$f_{app} = \frac{P_0 - P_x}{\frac{1}{2} \rho \bar{u}^2 (4x/D)} \quad , \qquad (4.2)$$

is based on the total pressure drop from the entrance (x = 0) to a
position x. The pressure drop in the entrance region of a duct
arises from two sources:
  (a) The change in momentum due to change in the shape of the
       velocity profile and
  (b) The integrated effect of surface shear forces.

For more detailed results and comparison one may refer to the
reference [3].

5. CONCLUDING REMARKS

As far as we know, there are no entrance flow solutions or
experimental data available for a fluid, which obeys the Casson
stress-strain relation. It is, therefore, not possible for us to
compare our results. However, having confirmed an excellent agreement (i) in the entire region of the entrance flow for Y = 0, and
(ii) in the fully-developed region for all yield numbers, it can
be stated with confidence, that the present results are accurate
within the limits of round-off errors of computation.

It is of practical interest to know the total pressure drop
for flows in a duct. From Figure 2 and Equation (4.2), the value
of total pressure drop can be calculated for any flow condition
and position x.

6. ACKNOWLEDGEMENTS

We would like to express our deep appreciation to Mr. P. E.
Newton, Medical College of Wisconsin, Milwaukee, Wisconsin, for
introducing us to this problem, and to Dr. S. V. Patankar and
Professor D. B. Spalding, Department of Mechanical Engineering,
Imperial College, London, for many suggestions without which this
solution would not have been possible.

We are also grateful to the Wisconsin Alumni Research Foundation
for providing us with funds for the computer time through the Graduate
School, University of Wisconsin-Milwaukee.

7. NOMENCLATURE

D	Tube Diameter
f_x	Friction Factor (Equation 4.1)
f_{app}	Total Friction Factor (Equation 4.2)
K_c	Constant in the Casson Equation (1.2)
p	Pressure
r	Radius
r_0	Radius of a Tube
Re	Reynolds number (= $\rho \bar{u} D / K_c^2$)
u	Axial velocity
$\bar{u}$	Average velocity
v	Velocity in radial direction
x	Axial coordinate
y	Normal coordinate
Z	Dimensionless axial distance (= $x/D \, Re$)
Z_1	= $1/Z$

Greek Symbols

μ	Viscosity (= K_c^2)
ρ	Density
τ	Shear Stress
τ_y	Yield Stress

9. REFERENCES

1. Charm, S., and Kurland, G. S., "Viscometry of Human Blood for Shear Rates of 0-100,000 sec^{-1}," Nature (London), 206 (1965), p. 617.

2. Patankar, S. V., and Spalding, D. B., "Heat and Mass Transfer in Boundary Layer," Intertext Book, 2nd Edition, London (1970).

3. Shah, V. L. and Soto, R., "The Entry Flow for a Fluid Obeying Casson's Stress-Strain Relation," to be published.

DISCUSSION OF SESSION IV   (PART A)

Chairmen: Dr. Hermann Metzger and Dr. Carl A. Goresky

DISCUSSION OF PAPER BY H. METZGER

Iberall: There appears to be a discrepancy in your presentation
between the experimental results and your computation. The experimental results of a depth array (3000 mm) seemed to show a high
number of low pressure values (under 5 mm Hg). On the other hand,
in your computed model (above the capillary rectangle) you showed
an average of what appears to be 30 mm Hg. Can you explain this
discrepency of less than 5 mm Hg and 30 mm Hg?

Metzger: The rectangle was only one cross section of the total
tissue cube and has a high mean pO_2; going deeper into the cube,
the pO_2 decreases. All the rest of these pO_2 values are summarized
in the histogram.

Hutten: The assumptions for your model do not agree with the findings reported by Fung at the meeting of the International Society
of Physiology in Munchen (1971). The experiments performed by Fung
show that, at the branching point of two capillaries, the distribution of erythrocytes does not correspond to the different perfusion
velocities in the capillaries. Nearly all the erythrocytes will
enter the capillary with the higher perfusion velocity, nearly none
of the erythrocytes, however, will enter the capillary with the
slower perfusion velocity. Can you give an interpretation of these
findings with regard to your model?

Metzger: The model is based on the assumption of equal hydrodynamic
resistances of all capillaries; the findings of Fung are not built
into the model. It seems to me that according to the observations,
a flow dependent hydrodynamic resistance is to be built into the
model.

Lübbers: (1) The histogram you showed did not resemble the measured
pO_2-histogram of the brain. Could you comment on this?
    (2) Do you think think that the measured pO_2-histogram provides
enough data to find the correct model parameters? How many pO_2
values would be necessary?
    (3) Or what kind of data are necessary?

Metzger: (1) A fit of the experimental histograms of brain tissue
can be obtained by variation of parameters AG or AK. To find out

the right parameters, several combinations of physiological constants
are possible. Additional values have to be measured for a final fit
of experiment and theory.

    (2) That is the main problem, the histogram does not have
enough information. The number of point-shaped pO_2 measurements
should be from the same type of network from the brain. If the
type of capillary and cell arrangement changes and pO_2 measurements
by means of histograms are from different brain layers, a tremendous
increase of measurement per volume is necessary.

    (3) Especially morphological data are necessary; number and
type of capillary arrangement, number and distribution of arterial
input and venous output points, local blood flow.

Goresky: I think it is interesting to note that the first paper
has gotten into the first major difficulty that Dr. Bruley pointed
out and this is, if you have a point observation, then, if you do
not have the space parameter attached to each observation and you
try to interpret the data in terms of the model which gives you the
spatial answer, many interpretations become clouded. I think one
of the purposes of the morning session is to try to learn what kind
of interpretation there is in the data that we have and learn whether
we gather different kinds of data to come to a better understanding.

DISCUSSION OF PAPER BY C. A. GORESKY AND H. L. GOLDSMITH

Iberall: Involving your paper and Dr. Metzger's paper concerning
Krogh modeling and dimensional flow distributions, and whether there
is a winner-take-all relationship, the fact of the matter is, that
there is not a winner-take-all relationship. Hence, the flow is
not stochastic, but quite deterministic. There has to be an active
type reactance in the network. The network is not coordinated by
its resistive characteristics. It takes an actual "active reactance"
which will do the distributions.

Goresky: I can probably present a more simple point of view which
concerns a system which is in a particular state at a particular
time. We need more observations than spaces or time to understand
what is going on. If one goes from there to all of the problems
which come up with autoregulation, all of the problems which have
to do with diffusion, and the infinite varieties of metabolic conditions, the problems have even greater magnitude and are more difficult to describe. They need only to be described by visual observation rather than theoretical analysis.

Clark, J.: How important are assumptions regarding radial distribution of flow within the capillaries?

Goresky: The flow in the capillaries is of a type which has been called "bolus flow." The red cells are deformed by and fill the capillary, and move as a flexible block, by-passing a small slip layer. This has the effect of producing a mixing motion in the space between the cells, and it tends to flatten out any radial concentration distribution in the capillary. This must, however, still be considered to be a secondary effect, since the dimensions of the capillaries and the tissues involved are such that intra-capillary plasma concentrations for most small solutes can be considered to equilibrate virtually instantaneously.

Song: I am just curious about labeled water washout. I wonder whether you have tried to obtain a mixing equilibrium state in each organ too. Dr. Guyton's group has recently published a paper on the distribution of D_2O or ^3H_2O in whole body. The tissue course was over one hour but never showed any leveling off. Would you comment on this? It seems to me that the hypothesis that water passes through pores, if there are any, must be considered in comparing one organ with another, and that the length of cylindrical pores must perhaps also be considered.

Goresky: The washout curves for the whole body exhibit character-istics which emphasize the heterogeneity of tissue perfusion. I have not attempted to deal with these problems here. I have, in-stead, focused upon the distribution phenomena in a set of well-perfused organs in which the distances between the elements of the capillary which are not perfused are small. The idea of this exam-ination was to provide some kind of insight into intercapillary diffusion interactions in relatively simple systems as a background for the more complex phenomena of oxygen transport.

The second part of your comment is difficult to answer. I think that it would be wise to summarize the answer by saying that the general agreement about the area, coming out of the Benzon con-ference on capillary permeability, is that water passes, not only through pores, but also through the whole of the endothelial mem-brane lining the capillary (see Capillary Permeability, edited by C. Crone and N. Lorns, 1970).

DISCUSSION OF PAPER BY W. GRUNEWALD

Metzger: Could your result, that about 150 measuring points would be necessary for characterizing a histogram with classes of 5 mm Hg, be applied in general, or is it only valid for the model volume under consideration?

Are only 150 points necessary for a histogram in general?

Grunewald: For one basic element, if you change the structure, you change the form of the pO_2 histogram, but there is no difference in the number of pO_2 measurements.

Schuchhardt: I should like to introduce the subject of the critical distance between sinks and sources in a capillary network: the distance at which an oxygen diffusion shunt occurs. This distance depends on the given respiration. Did you calculate this distance for certain respirations with your model?

Grunewald: This is a very important question. For all of these variants of capillary structure elements, I did not prove the influence of diffusion shunt and the critical distance is not yet calculated. For the countercurrent structure, however, and for the values of oxygen consumption of the brain and the skeletal muscle, I have calculated the distance where an oxygen diffusion shunt arises. These results are published in Pflügers Archiv., (1968) and Habil. Thesis, Bochum (1971).

DISCUSSION OF PAPER BY R. WODICK

Thermann: On one of your last slides, you showed a formula to calculate the oxygenation of a given spectrum. In this formula, you only have the terms for the oxygenated and the deoxygenated hemoglobin spectra. What is it about the spectrum you really measure?

Wodick: The information of the measured spectrum is in the function $f(\lambda)$. This function describes the connection between the corresponding wave lengths of the measured spectrum.

Metzger: Would it be possible to extract physiological parameters from your analysis? For example, blood flow values from hydrogen clearance?

Wodick: Yes, we are able to calculate the blood flow with our theory.

DISCUSSION OF PAPER BY B. I. LOCKARD

Metzger: (1) Would it be possible to quantify a capillary-cellular
relationship, that means, correlate the number of cells to the
numbers of capillaries?
    (2) Do you agree with the findings of Barker (Fed. Proc.,
1972), who found a high number of cells at the arterial end of the
capillary bed and a small number at the venous end?
    (3) Is the Krogh cylinder a good basis for capillary-distance
calculations? How do you calculate capillary length?

Lockard: (1) I think it is possible, but somewhat complicated.
To determine the total number of cells around an entire capillary
segment, it would be necessary to know the intercapillary distance,
the capillary length, the numbers of cells around capillaries in
cross sections of known thicknesses, etc.
    (2) In at least some areas, where intercapillary spaces are
sparsely populated, cell bodies are located next to capillaries
(e.g., amygdala, Fig. 10) but whether or not the cells are more
plentiful near the arterial end of the capillary, I cannot say.
    (3) Yes, I believe it is. Scharrer (Quart. Rev. Biol., 1944
19:308-318) demonstrated in the opossum that, after occlusion of
a vessel in the granular layer of the cerebellum, the cells around
the affected capillaries disappeared for half the distance across
the intercapillary space.
    Capillary lengths can be measured directly on slides if they
are short enough that both ends are contained within the section
and allowance is made for the slope of the vessel. For longer
vessels, in which only one junction is seen along a capillary, as
in the subcortical capillary illustrated in Fig. 5, other techniques
will have to be devised.

Cameron: On a gross level, is there any correlation (either positive
or negative) between mean intercapillary distance and frequency of
arterioles--that is, are frequent arterioles associated with short
mean intercapillary distance or vice versa?

Lockard: I am not aware that there is a correlation between the
number of arterioles supplying a given area and the width of the
intercapillary space in the same area.

DISCUSSION OF PAPER BY J. E. FLETCHER

Guilbeau: In a similar mathematical model, Reneau and I used the Krogh tissue cylinder concept to study the response of the lethal corner to sinusoidal flow changes. We found a much longer transient decrease to the quasi-steady state response. My question is, do you think the differences could be due to differences in the frequency of the velocity changes used? What frequency in flow variation did you use? Have you made a study in which the period of oscillation was varied?

Fletcher: I have not made detailed studies of frequency effects on the lethal corner response time. I would guess, from a few preliminary computations, that the Michaelis-Menten parameters and tissue geometry parameters are more important than the frequency used. Another important factor is the form of the flow or velocity function (i.e., whether or not the flow reaches zero during a cycle). The frequency I used was determined from the data of Johnson and Wayland cited in my references. I believe the numerical value is 0.0884 cycles per second.

Cameron: I would like to return to the equilibrium vs. kinetic features of the model. In your curves for lethal corner response to flow increase, you stated that the difference derived was due to change in shape of the dissociation curve. Since this is a response to flow, would this not be a kinetic phenomenon related to the off-constant for O_2 on Hb?

It may be that the difference in response magnitude in normoxia and hypoxia is due to being below the knee of the O_2 dissociation curve in hypoxia, but still, the difference would not be due to the change in shape of the curve but to the change in the off-constant as a function of % interaction of hemoglobin.

Fletcher: Since I compute dynamically the release kinetics of hemoglobin, any change in O_2 release rate is automatically accounted for in my model. However, your second point is quite valid, and certainly related to the condition. My computations suggest that the slope of the dissociation curve is the more important factor. After all, the shape of the dissociation curve is, itself, a function of the release rate. Therefore, one cannot, in truth, completely separate the two effects.

Holland: How did you get the velocity for oxygen coming off hemoglobin? You said you computed it, but did you use any of the available measurements?

Fletcher: Yes. My modeling of the oxygen hemoglobin kinetics is based on the measurements of Staub, et al., who measured the rate constants of oxygen uptake (release) as a function of saturated hemoglobin fraction.   The equations are:

$$\frac{d[HbO_2]}{dt} = k_1[Hb][O_2] - k_2[HbO_2]$$

$$\frac{d[O_2]}{dt} = -\frac{d[HbO_2]}{dt}$$

and the release constant k_2 is the value measured by Staub, and used by me. Complete details are given in my references.

Goldstick: I was pleased to see that your calculations involved low precapillary pO_2 values, between 30 and 40 Torr. You called these 'hypoxic' but they may in fact be the normal precapillary values. Duling has repeatedly found values in this range at the proximal ends of capillaries.

Fletcher: I labeled these values as 'Hypoxic" in order to be consistent with other computations of similar type previously reported. Since my familiarity with these terms is largely from reading the literature in this area, permit me some liberties in the use of terms. My objective is to compute meaningful results and my sources of data are, in general, other papers rather than personal knowledge of physiologic states. I intend the computations as a means of suggesting possible critical values in this area.

Goresky: (1) Is the idea that there is no loss of oxygen at the borders of the Krogh cylinder not equivalent to considering a concurrent flow system with adjacent entrances and exits, one without alternate exits?
        (2) Have you included the kinetics of oxygen binding to hemoglobin and its release, as well as the form of the hemoglobin oxygen saturation curve?

Fletcher: (1) That is one way of obtaining the single Krogh cylinder; however, the single cylinder could also apply if its outer boundary were impermeable to oxygen. There may also be other local histologic reasons for adapting Krogh geometry without the specific assumption of concurrent flow.
        (2) Yes, I have included both the change in release velocity as you call it, as well as the shape of the dissociation curve itself. These are labeled respectively $X(f)$ and $g(f)$ in my model. The fitted data was shown in Figure 2.

DISCUSSION OF PAPER BY E. N. LIGHTFOOT

Fletcher: I agree with Dr. Bruley, and would add that computations with my model indicate that the neglect of axial diffusion leads to large errors during low flow conditions. These conditions, of course, are of great interest in pathologic and disease states.

Lightfoot: Any method such as mine which depends heavily on the orders of magnitudes of key parameters will be unreliable for other magnitudes. Hence, I would agree, to some degree, with your comment.

However, one must be careful to be fully consistent, and I would be very surprised if the present numerical models actually can be justified by careful comparison with experimental data. I would, for example, strongly suspect that flow pulsations, seemingly neglected, can be more important than axial dispersions.

I would like to see more critical comparisons of data with models to determine their real utility.

Guilbeau: I think that simplified mathematical analysis is often adequate for describing many complex systems, however, it is interesting to note that eloquent arguments for simplified analysis would not be nearly so convincing if the more accurate, detailed, distributed parameter solutions were not available as a basis for comparison.

DISCUSSION OF PAPER BY W. A. HYMAN

Reneau: Three comments: (1) Some of the simplified techniques used in this paper, the previous paper, and perhaps (judging from the abstracts), in future papers, were developed many years ago in Germany by Opitz and Schneider, Thews, Lübbers and many others.

(2) The comments made on the relative unimportance of radial gradients in the capillary and the relative unimportance of convective cross currents in plasma during bolus flow are in agreement with many of our studies with respect to glucose, metabolites, CO_2 and O_2. However, under conditions of pathologic-physiology, such as, where one capillary is required to do the work (oxygen supply) of its neighbors, very large radial gradients can develop, even in the capillary. Under these conditions, perhaps gradients and cross currents are of importance and your techniques would need to be modified.

(3) Under conditions of partial anoxia, where portions of the tissue are normoxic, others are hypoxic, and still others are anoxic, the total demand for oxygen by an individual capillary varies from the normal situation. Under these conditions, distributed parameter techniques are of particular use in making detailed trend studies.

DISCUSSION OF PAPER BY R. R. STEWART AND C. A. MORRAZZI

Iberall: The author must be congratulated for the absolute elegance of his presentation.

Reneau: Only a comment. We do not use space average oxygen tension values any longer except under special situations. The actual pO_2 values in blood can be obtained by space averaging saturations and then obtaining the pO_2 value from the appropriate oxygen dissociation curve. However, in this case, the capillary profiles are minor and space averaging values are quite adequate.

Bruley: I am pleased that your simplified model agrees in many cases with the solutions of our more complex models. However, considering cases of pathology, such as anemia or plasma skimming, will your model handle these problems?

Stewart: The model should be able to handle certain types of anemia through modification of parameters. The model, however, would need to be altered to handle the actual skimming process itself, since the model presented is for a single capillary and does not include branching.

Moll: How do you handle the diffusion of lactic acid to the tissue?

Stewart: It is very idealized at this point. We are making more complicated models now, but we just look at this as a homogeneous region and the regions that had no oxygen were producing the acid.

DISCUSSION OF PAPER BY D. D. RENEAU AND L. L. LAFITTE

Cameron: Were the reaction parameters in the model taken as constant in the acid producing region?

Reneau: Yes, for these initial and very preliminary studies.

DISCUSSION OF PAPER BY D. F. BRULEY, D. H. HUNT, H. I. BICHER AND
   M. H. KNISELY

Silver: Dr. N. Oshimo of the Johnson Foundation, Philadelphia, has made some observations on the relationship of myoglobin and cytochrome A which would be interesting to include in the model.
   G. Salama and I have made measurements of changes of pO_2 in the perfused beating of the heart in relation to the force of

ventricular beat and the intracellular redox state of pyridine
nucleotides. We found that, in the superficial layers of the heart,
there was virtually no change of pO_2 reasonable with floating microelectrodes when the heart was perfused with blood, but that when
oxygen capacity perfusals were used, large changes of pO_2 took
place with each beat. These changes followed the contraction cycle
with a phase shift of about 0.1 sec and disappeared as soon as the
O_2 capacity of the perfusal was increased over a certain level.

Bruley: I agree that it would be interesting to include your
findings in a model of heart muscle. Maybe we should discuss this
further to see if it is possible.

Lübbers: I would like to comment on the assumption of phasic
myocardium oxygen consumption made by Dr. Bruley. By reflection
spectro-photometric measurements, we were able to show that neither
the myoglobin nor the cytochromes (a + a_3, b and c) changed the
state of oxygenation respectively of oxidation during the heart
cycle. The question arose about the function of myoglobin: It
seems to be a short time oxygen store for a few contractions. I
think, furthermore, it could help to transport the oxygen to the
tissue at a larger rate during rythmical blood flow. Assuming that
the pO_2-distribution curve, which S. Schuchhardt has measured, is
correct, a large part of the myoglobin is not O_2 saturated. During
the circulation stop or reduction in the inner part of the muscular
wall, the myoglobin will deliver its oxygen to the respiratory
chain. Later on, if during the heart cycle, the circulation increases, the capillary will receive new blood with a sufficiently
high pO_2. In spite of delivering oxygen to the tissues, in the
capillary, the pO_2 will remain relatively constant because of the
hemoglobin being a high tension oxygen store.
    Since myoglobin can bind oxygen at low oxygen tensions (low
tension oxygen store), by its binding of oxygen in the resaturation
phase, the local tissue pO_2 can remain low. Thus, the gradient
between capillary and tissue remains steep over a longer time than
without the myoglobin. Steep gradients mean a high transport rate
of oxygen.

Bruley: The assumption of cyclic metabolism came from early work
done in the field. The model can be altered easily to consider
constant metabolism. The main point is, that this change would
not affect the conclusions drawn from the simulation; in fact, it
would substantiate the results.
    The main function of myoglobin has not yet been determined,
but the model does show that it is only a short time capacitance
(about one to two heart cycles). Your analysis of the oxygen
gradient could be meaningful with regard to transport of oxygen.

<u>Dorson</u>: It comes to mind, Duane, that this particular problem may
be more sensitive to the tissue diffusion time. Your lumped system
would damp out any flow-diffusion lags, and the flow period and
diffusion lag are of the same order.

<u>Bruley</u>: It is true that a distributed parameter model would provide
a better analysis by the inclusion of flow-diffusion lags. I am
very much in favor of a distributed parameter approach, however,
with the necessary superimposed autoregulation, the models become
difficult to handle. This work represents a first try to understand
better oxygen transport in, and the needs of, heart muscle and will
certainly be extended. I think that the model is good enough to
establish meaningful conclusions regarding the influence of myoglobin, etc., in heart muscle.

Session IV

MATHEMATICAL STUDIES OF TISSUE OXYGENATION

PART B

Chairmen: Dr. Reinhard Wodick and Dr. Wolfgang Grunewald

Section IV

MATHEMATICAL STUDIES OF TISSUE OSMOLALITY

Chairmen: Drs. Reinhard Bock and Dr. Wolfgang Günther

DISTRIBUTED MODEL SOLUTION TECHNIQUES FOR CAPILLARY-TISSUE SYSTEMS

Daniel H. Hunt*, Duane F. Bruley, Haim I. Bicher** and
Melvin H. Knisely**

Department of Chemical Engineering, Clemson University,
Clemson, South Carolina, U.S.A.
**Department of Anatomy, Medical University of South
Carolina, Charleston, South Carolina, U.S.A.

I'd like to begin with the remark that distributed parameter
models are so-called "tigers-by-the-tail" as several people in the
audience would testify, but if one can tame it, or at least hang on,
the results are usually worth the effort. In effect, one is trying
to come as close to reality as possible with such a model.

Back in 1967, Reneau, Bruley and Knisely (4) presented a distributed parameter model which describes oxygen release, diffusion
and consumption in the capillaries and tissue of the human brain.
I don't have time to go into detail on the model, but point out that
the model is a system of equations based upon the Krogh capillary-
tissue cylinder where an equation is written for the capillary,
another for the tissue, and both equations are connected by an
interface equation. Each equation, the capillary, the tissue, and
the interface, is interacting, i.e., values of oxygen tension in
the capillary are dependent upon values of oxygen tension in the
tissue and vice versa. Also, the system is nonlinear, due to the
term describing oxygen release from the erythrocytes. I should
point out too, that the model has two independent space variables,
x and r, and is time-dependent (an unsteady state model). All of
these properties can and do add difficulty to the method of solution.

The equations were solved, however, by not one, but four investigators, for reasons I will explain. The first investigator was
Reneau, who modified what is known as the alternating direction
implicit technique (ADI), i.e., each equation derivative was finite-
differenced in both an explicit and implicit manner with respect to
the x-direction and the r-direction independent variables. Then
the x-direction explicit and r-direction implicit formula equations

887

were applied and solved for a given time step by the Thomas method.
While on the next time step, the direction of the explicit and
implicit formulas was alternated by applying and solving x-direction
implicit and r-direction explicit equations by the Thomas method.
It was necessary to adapt the interface equation to the scheme,
since it involved only one space dimension, and also it was necessary to generate an approximate value for the nonlinear term. When
he programmed the technique on the digital computer, Reneau found
that, as a result of the nonlinearity, stability of the solution
method limited the magnitude of the time step. With the largest
feasible time step, ten hours of digital computer time on an IBM
7094 was required to calculate a transient response of 1.5 seconds
duration.

       Ten hours is much time and money, especially in regards to a
digital computer. Hence, in order to use the unsteady state (time-
dependent) model, a faster solution technique needed to be found.
A very desirable method would be a real-time simulation, or faster
than real-time simulation, for predicting the result of an event in
progress.

       Another approach was developed by McCracken (3), who called the
method 'digital integration explicit' (DIE), but it is also known as
'the method of lines.' Basically, the DIE method parallels the
analog computer technique for solving partial differential equations
where the discrete-differencing is done in the space dimension yielding a set of equations which are integrated simultaneously in the
time domain. However, McCracken carried the calculation one step
further, in the sense that the discrete-differencing was done in two
space dimensions, producing a set of simultaneous ordinary differential equations (in two-dimensional space), which were numerically
integrated in time on a digital computer. An advantage of the method
is that the nonlinear coefficient term was handled as easily as a
linear coefficient. Also, the stability of the solution method was
limited only by the stability of the numerical integration formula,
which, in turn, limited the magnitude of the time step. Quite an
improvement in calculation time was achieved with the DIE method.
Only 50 minutes on an IBM 360/model 50 was required for the 1.5-
second transient response.

       In the meantime, Halberg (1) investigated the Monte Carlo method
which offers some unique advantages over other techniques for the
solution of linear PDE. These are that memory requirements are minimal, solution for a single point is exclusively calculated, and
addition of independent variables to the partial differential equation adds no complexity to the method of solution. On the other
hand, the Monte Carlo method is inherently applicable solely to
linear differential equations, and difficulty arises when nonlinear-
ities are encountered. Thus, to solve the nonlinear, distributed

parameter model equations, modification of the Monte Carlo method
was necessary.

With that in mind, let me say that the Monte Carlo method has
been divided into two basic approaches, one discrete and the other
continuous. Halberg investigated the discrete Monte Carlo method,
but he found that the digital computer time for generation of a
sufficient number of random walks precluded any attempt to modify
the method for use with nonlinear differential equations.

However, the present author (2) employed a hybrid computer,
which is more suited to handle random walk generation. It was thus
possible to modify the Monte Carlo method for nonlinear equations
by using the continuous approach.

Let me retrace and explain the basic Monte Carlo method. It
is applied to differential equations defined on an open-bounded
region with a closed boundary, as prescribed by the boundary conditions. The solution for a single point is obtained by generating
random walks originating from this point, and a tally is kept of
the value associated with the final point reached by the random walk,
for instance, a boundary value for the boundary reached, etc. For
time-dependent equations, a random walk is allowed to proceed for
the time increment for which the solution is desired whence the
initial condition is tallied or, if in the meantime a boundary is
reached, the boundary value is tallied.

As pointed out earlier, it is necessary to modify this basic
approach to account for the nonlinearity. In order to accomplish
this, the solution for the bounded region of the combined tissue,
interface, capillary regions must be found. Thus the bounded region
is divided into a discrete mesh where the solution at any mesh point
can be obtained by the Monte Carlo method. For the solution to nonlinear, time-dependent equations, the initial condition at time zero,
$t_s - \Delta t = 0$, is satisfied by the steady state solution of the mesh
points. Furthermore, the nonlinear coefficient is approximated at
time zero by substituting the steady state solution values into its
nonlinear function.

To find a solution for each desired mesh point at a time which
is a small time increment Δt different from the previous known solution mesh, random walks are generated backward in time for a Δt
period, in this case, from t_s to $t_s - \Delta t$. In this way, the entire
mesh is solved at t_s. The nonlinear coefficient is updated by
substituting the solution mesh point values just calculated at time
t_s, and the next solution mesh can now be solved for a time $t_s + \Delta t$
by random walking backward to time t_s. Consequently, it follows
that the entire solution mesh may be continually advanced in time,
using only the previous solution mesh, until the total desired time

solution is achieved.

As described, the nonlinear coefficient is continually updated
by approximation with the previous solution mesh values. The question of whether interative correction is necessary or not will
depend upon the size of the Δt time increment chosen for the random
walks. In the present instance, a small enough time increment was
used not to require interative correction.

The uniqueness of the Monte Carlo method of calculating single-
point solutions allows freedom in choosing mesh point locations.
Also, instead of acquiring Monte Carlo solutions at every mesh point,
mesh points may be selected such that the remaining mesh-point
solutions can be calculated directly from discrete-difference equations or some other method.

The nonlinear Monte Carlo computation time was 56 minutes, on a
PDP15/EAI680 hybrid computer, which is slightly more than the DIE
method. On the positive side, it is considered that improvement in
hardware, such as noise-generation methods, and in software, such
as use of machine language instead of Fortran IV, would lower computation time considerably. Thus the nonlinear Monte Carlo method can
prove to be a powerful technique for the solution of sophisticated
mathematical models.

REFERENCES

1. Halberg, M., D. F. Bruley and M. H. Knisely. Simulation 15,
    no. 5, 206 (1970).

2. Hunt, D. H., Ph.D. Dissertation, Clemson University, Clemson,
    South Carolina (1971).

3. McCracken, T. A., D. F. Bruley, D. D. Reneau, H. I. Bicher and
    M. H. Knisely. Proceedings of First Pacific Chemical Engineering
    Congress, Kyoto, Japan, p. 137 (1972).

4. Reneau, D. D., D. F. Bruley and M. H. Knisely, Chemical
    Engineering in Medicine and Biology, p. 135, Plenum Press (1967).

* Present address: Department of Anatomy, Medical University of
  South Carolina, Charleston, South Carolina, U.S.A.

HETEROGENEOUS MODELS OF OXYGEN TRANSPORT IN TISSUE SLICES

Ronald C. Tai and Hsin-kang Chang

Department of Civil Engineering

State University of New York at Buffalo

Recent development of the oxygen microelectrode (Whalen,1967) has made possible more accurate measurements of tissue oxygen transport. Such measurements could greatly benefit and indeed demand the development of more sophisticated mathematical models. Since tissue is heterogeneous both in structure and in transport properties, heterogeneous tissue models appear to be in order.

In this paper a two-phase tissue composed of a cellular phase and an extracellular phase is studied. Mathematical expressions for the permeability ratio between the extracellular phase and the cellular phase are given for three heterogeneous models of tissue slice, this geometry (Fig. 1A) being chosen because most of the available data have been obtained from tissue slices. The experimental data of Gore and Whalen (1968) and of Ganfield et al (1970) are then analyzed in light of these three heterogeneous models.

NOTATION

b	Effective radius of a cell
J	Total oxygen out-flux from the tissue surface
K_c	Permeability (diffusivity x solubility) of oxygen in cellular material
K_e	Permeability in extracellular fluid
m	Metabolic rate of tissue
P_A	Ambient oxygen partial pressure at one surface of tissue slice
P_B	Ambient oxygen partial pressure at other surface of tissue slice
ΔP	Oxygen partial pressure drop across tissue slice

Q   Oxygen uptake at the surface of the maximally respiring region of tissue slice

T   thickness of tissue slice

α Volume fraction of extracellular fluid in tissue

γ Extracellular fraction of the cross-sectional area of the tissue

μ Shape factor of cell defined as Ab/v where A is the surface area of cell and v is volume of cell

MATHEMATICAL MODELS

A. Layered Models

In the layered models cellular and extracellular materials are represented by alternating layers. Two arrangements can be made for these layers, namely in series (Fig. 1B) and in parallel (Fig. 1C).

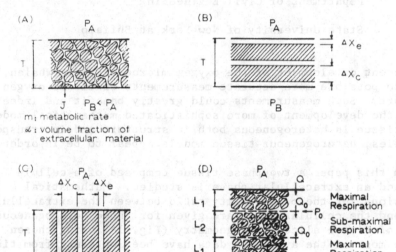

Fig. 1 Definition sketch of the heterogeneous tissue models.
(A) A tissue slice, (B) Series Model
(C) Parallel Model, (D) Non-Layered Model

1. <u>Series Model</u>. Applying the principle of mass balance to each layer and noting that the partial pressure drop across the tissue slice is equal to the sum of the pressure drops across each layer, the total out-flux from the tissue surface may be expressed as

$$J = \frac{K_e \cdot \Delta P}{(1 + \frac{1-\alpha}{\alpha} \cdot \frac{K_e}{K_c})\alpha T} - \frac{1}{2}(1-\alpha)mT. \qquad (1)$$

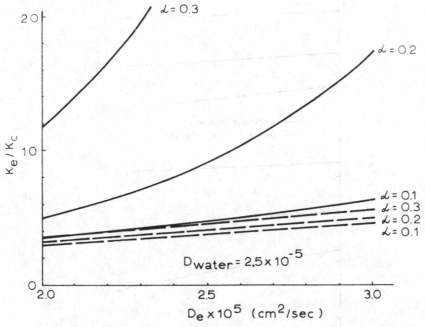

Fig. 2 Application of layered models to data of Gore and Whalen
(1968). Solid lines are based on parallel model; broken
lines are based on series model.

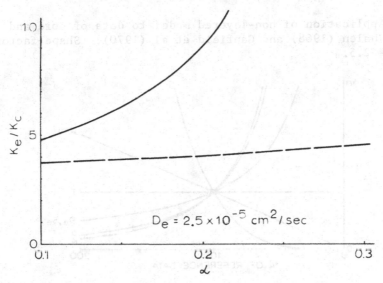

Fig. 3 Effect of α on the prediction of $\dfrac{K_e}{K_c}$. Data used are
those of Gore and Whalen (1968). Solid line is
based on parallel model; broken line is based on series
model.

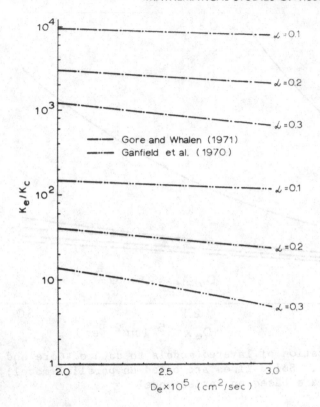

Fig. 4 Application of non-layered model to data of Gore and
        Whalen (1968) and Ganfield et al (1970). Shape factor μ
        = 2.5.

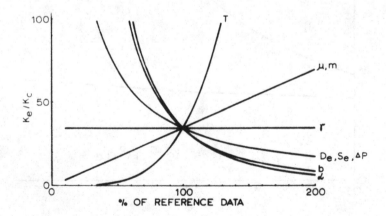

Fig. 5 Parametric variation of $\dfrac{K_e}{K_c}$ in the non-layered model (sub-
        maximal respiration only). Data used are those
        of Ganfield et al (1970).

From Eq.(1) the permeability ratio K_e/K_c may be obtained,

$$\frac{K_e}{K_c} = \frac{\frac{K_e \Delta P}{(1-\alpha)T}}{J + \frac{1}{2}(1-\alpha)mT} - \frac{\alpha}{1-\alpha} \tag{2}$$

2. __Parallel Model.__ Applying the mass balance principle,

$$J = \alpha K_e \frac{\Delta P}{T} + (1-\alpha)(K_c \frac{\Delta P}{T} - \frac{1}{2} mT) \tag{3}$$

The permeability ratio is then

$$\frac{K_e}{K_c} = \frac{1}{\frac{JT}{(1-\alpha)K_e \Delta P} - \frac{\alpha}{1-\alpha} + \frac{1}{2} \frac{mT^2}{K_e \Delta P}} \tag{4}$$

B. Non-Layered Model

A model proposed by Hills (1971) is modified to suit the geometry of the tissue slice (Fig. 1D). The tissue slice is divided into maximally respiring regions in which no anoxic core exists for any cell, and a sub-maximally respiring region in which each cell has a anoxic core.

For a tissue slice consisting of both maximally respiring and sub-maximally respiring regions,

$$\frac{K_e}{K_c} = \frac{(1-\alpha)\mu \lambda^2}{6\alpha b^2} \tag{5}$$

where

$$\lambda = \frac{T(W+Y) \pm \sqrt{(W+Y)^2 T^2 - (W+3Y)\left[(W-Y)T^2 - \frac{P_A K_e}{(1-\alpha)m}\right]}}{W + 3Y} \tag{6}$$

and

$$W = \frac{1}{2}\alpha - \frac{1}{\gamma}; \quad Y = \frac{1}{12\alpha} \tag{7}$$

For a tissue slice respiring sub-maximally only,

$$\frac{K_e}{K_c} = \left(\frac{2\gamma}{Q}\right)^4 P_A^3 \left(\frac{1-\alpha}{3\alpha b}\right)^2 2 m \mu K_e^3 \tag{8}$$

Eqs. (2), (4), and (8) may be used to compute the important parameter in tissue respiration, K_e/K_c, if the other physical and physiological parameters are known. A detailed derivation of these equations are given by Tai and Chang (1973) in a separate paper.

APPLICATION OF MODELS TO EXPERIMENTAL DATA

Experimental data of oxygen transfer in frog sartorius muscle (Gore and Whalen,1968) and in cat cerebral cortex (Ganfield et al, 1970) were used to compute K_e/K_c.

Fig. 2 shows the apparent heterogeneity of the frog sartorius muscle as manifested by the parameter K_e/K_c. The variation of K_e/K_c with α for both layered models is plotted in Fig. 3. It is very likely that the series model gives more resistance to oxygen transport through the tissue slice than does the real tissue, whereas the parallel model gives less. Thus, for a given K_e (which is assumed to be the oxygen permeability in water at 37°C) and a known J, K_e/K_c computed from the series model may be smaller than the real value and K_e/K_c computed from the parallel model may be greater than the real value. This reasoning agrees with the results in Fig. 2. If $\alpha = 0.2$ is assumed, K_e/K_c for frog sartorius muscle should be in the neighborhood of 6.

The application of the non-layered model to the same set of experimental data (Fig. 4) yields a K_e/K_c value several orders of magnitude higher than those computed from the layered models. In addition the data of Ganfield et al (1970) gives values of K_e/K_c nearly two orders of magnitude lower than those from Gore and Whalen (1968) when the same formula is used. These large discrepancies suggest that at least one of the predictive models is unreliable. A careful examination of the mathematical derivation as well as the numerical relationships of the various parameters involved indicates that the non-layered model seems likely to produce exaggerated results.

As an aid to estimate the errors of computed K_e/K_c caused by inaccurate data, a parametric study based on Eq.(8) and the data of Ganfield et al is shown in Fig. 5.

References

Ganfield, R.A., Nair, P. and Whalen, W.J. (1970) Mass transfer storage, and utilization of O_2 in cat cerebral cortex. Am.J. Physiol. 219: 814.

Gore, R.W. and Whalen, W.J. (1968) Relations among tissue P_{O2}, Q_{O2}, and resting heat production of frog sartorius muscle. Am. J. Physiol. 214: 277.

Hills, B.A. (1970) Respiration of tissue as a medium of heterogeneous permeability. Bull. Math. Biophys. 32:219.

Tai, R.C. and Chang,H.K. (1973) Oxygen transport in heterogeneous tissue. J. Theoret. Biol. in press.

Whalen, W.J., Riley, J. and Nair, P. (1967). A microelectrode for measuring intracellular P_{O2}. J. Appl. Physiol. 23: 798.

MATHEMATICAL MODELLING OF THE CORNEA AND AN EXPERIMENTAL APPLICATION

Ernest L. Roetman and Ronald E. Barr

Departments of Mathematical Sciences and
Ophthalmology, University of Missouri, Columbia,
Missouri, U.S.A.

Adequate oxygen pressure is important for proper respiration of the cornea. Should the anterior surface oxygen pressure environment decrease below about 20 torr for as little as a few hours, the cornea becomes edematous and vision is impaired. A common cause for concern on this subject is in the use of contact lenses, both hard methylmethacrylate and soft hydrophilic lenses.

In our laboratories we have given attention to the problem of oxygen properties of the cornea and the effects of various environmental conditions. After certain preliminary experiments for feasibility purposes, a first attempt has been made to model the cornea and its oxygenation properties.

The model envisions gas diffusion into a fluid filled porous media, which is multilayered and consumes oxygen. For the central portion of the cornea a one dimensional gas flow model, anterior-posterior, across a plane membrane infinitely extended in the plane normal to the gas flow has been assumed.

The model is based on Fick's diffusion laws and boundary conditions describing the oxygen pressures or pressure gradients at the anterior and posterior surfaces of the cornea.

Notation:

P(x,t) oxygen tension at $x = (x_1, x_2, x_3)$ and time t.

Q(x,t) rate of consumption per unit volume per unit time.

k   oxygen solubility in the tissue or aqueous.

897

D   oxygen diffusion constant in the tissue or aqueous.

Start with the oxygen diffusion equation

(1) $k \frac{\partial P}{\partial t} = kD\nabla^2 P - Q,$

and the boundary condition for an interface between two
physical regimes

$$P^+ = P^-$$

(2)

$$k^+ D^+ \frac{\partial P^+}{\partial n} = k^- D^- \frac{\partial P^-}{\partial n} \qquad\qquad - \Big|\begin{matrix} + \\ \to \\ n \end{matrix}$$

Then restrict the problem to one space dimension:

$$\nabla^2 P = \frac{\partial^2 P}{\partial x^2}$$

(3)

$$\frac{\partial P}{\partial n} = \frac{\partial P}{\partial x}$$

and assume steady state conditions:

(4) $\frac{\partial P}{\partial t} = 0$

If one applies this system to a two layer model the following
equations and solutions obtain.

$$k_1 D_1 \frac{d^2 P_1}{dx^2} = Q_1 , \quad \ell_0 < x < \ell_0 + \ell_1$$

(5)

$$k_2 D_2 \frac{d^2 P_2}{dx^2} = Q_2 , \quad \ell_0 + \ell_1 < x < \ell_0 + \ell_1 + \ell_2$$

$$P_1(\ell_0) = a_0$$

$$P_1(\ell_0 + \ell_1) = P_2(\ell_0 + \ell_1)$$

(6)
$$D_1 \frac{dP_1}{dx} (\ell_0 + \ell_1) = D_2 \frac{dP_2}{dx} (\ell_0 + \ell_1)$$

$$P_2(\ell_0 + \ell_1 + \ell_2) = a_2$$

where a_0 and a_2 are known or measurable oxygen tensions.
The solution is:

$$P_1(x) = \frac{Q_1}{2k_1 D_1} (x-\ell_0)(x-\ell_0-\ell_1) + a_0 + A_1(x-\ell_0)$$
(7)

$$P_2(x) = \frac{Q_2}{2k_2 D_2} (x-\ell_0-\ell_1)(x-\ell_0-\ell_1-\ell_2) + a_2 + A_2(x-\ell_0-\ell_1-\ell_2)$$

where A_1 and A_2 satisfy

$$\ell_1 A_1 + \ell_2 A_2 = -(a_0 - a_2)$$
(8)
$$D_1 A_1 - D_2 A_2 = -\frac{Q_1}{2k_1}\ell_1 - \frac{Q_2}{2k_2}\ell_2 .$$

Letting a_1 = oxygen tension at $\ell_0 + \ell_1$, substituting in
either of equations (7) and performing substitutions for A_1,
or A_2 from equation (8), one obtains

(9)
$$a_0 - a_1 = m(a_0 - a_2) + n$$

$$m = \ell_1 k_2 D_2 (\ell_1 k_2 D_2 + \ell_2 k_1 D_1)^{-1}$$
(10)
$$n = (\ell_1 k_2 D_2 + \ell_2 k_1 D_1)^{-1} [\frac{1}{2} Q_1 \ell_1^2 \ell_2 + \frac{1}{2} Q_2 \ell_1 \ell_2^2]$$

(11)
$$k_1 D_1 = \frac{1-m}{m} (\ell_1/\ell_2) k_2 D_2$$

$$Q_1 = \frac{n}{m} \frac{2k_2 D_2}{\ell_1 \ell_2} - \frac{\ell_2}{\ell_1} Q_2$$

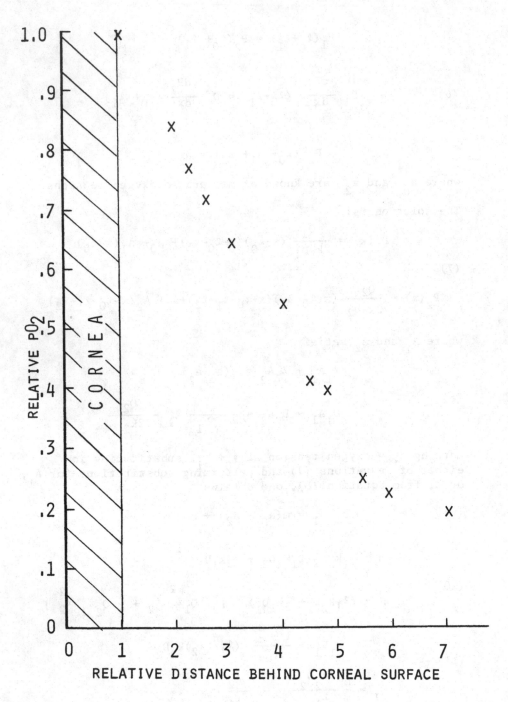

Fig. 1.   Relative change in oxygen tension with depth into the
          anterior chamber.

By obtaining a series of values for a , a_1 and a_2, and substituting into equation (9), one can determine the slope and intercept, m and n, substitute these into equation (10) and obtain values for $K_1 D_1$ and Q_1 equation (11).

On the experimental aspect of this presentation we have obtained preliminary information which has indicated the feasibility of working with the above model.

Male New Zealand white rabbits, weighing between 3 and 4 Kg. were anesthetized with intravenous injections of sodium pentabarbital and were mounted in a headholder. Under microscopic observation,an oxygen electrode, platinum-iridium insulated in glass with tip diameter of about two micrometers, was inserted through the limbus into the anterior chamber such that the tip was near the center of the pupil. The gases oxygen and nitrogen were then played on the corneal surface, at flow rates of 10-20 ml/minute, and the changes in oxygen tension in the aqueous was recorded. (The reference electrode was a Ag-AgCl wire inserted subcutaneously along the midline of the back.) The changes obtained were as follows: The mean pO_2 of the aqueous when the corneal surface was exposed to air was 32 torr. The aqueous pO_2 rose to 138 torr when oxygen was played on the cornea and decreased to 9 torr when nitrogen was played on the cornea.

Further experimentation has shown that an oxygen tension gradient exists, under the conditions of our experiments, along the visual axis in the anterior chamber. The gradient was as shown in Fig. 1.

These experiments showed that there would indeed be changes in a_1, and a_2 when a_0 is changed and that accurate measurements of ℓ_1 and ℓ_2 would be necessary, not only for its own sake but also to establish more accurate oxygen tension measurement statistics. From these data it appears that the model is amenable to experimental application.

INSTRUMENTATION AND CONTROL TECHNIQUES FOR DYNAMIC RESPONSE
EXPERIMENTS IN LIVING TISSUE

B. A. Bogue and W. J. Dorson, Jr.

Engineering Center, Arizona State University

Tempe, Arizona

SYSTEM BACKGROUND AND DESCRIPTION

The development of polarographic micro oxygen electrodes with tips less than 10μ has allowed the study of cellular and tissue oxygen levels in response to dynamic changes. Several investigators have recorded the in vivo behavior of pO_2 in the cerebral cortex tissue when arterial pO_2 was made to vary by a step change in the respired O_2 concentration. Of particular note are the reports by Bicher (1) in the cat cerebral cortex and Metzger in the cortex of rats (2). A transient overshoot in tissue pO_2 was observed during recovery from a one-minute induced hypoxia. This behavior is not predicted directly from first order models with constant metabolism (3). Subsequent analysis ascribed this behavior to delayed brain blood flow transients (4), but instantaneous flow rates were not recorded during the cited experiments.

The experimental system described herein is capable of repeating these earlier experiments with more complete measurements of variable physiologic phenomena. In addition, several different methods of achieving transient perfusion flows are being investigated. The arterial pO_2 can be made to oscillate by periodic variation of the inspired O_2 level. Of more direct concern to this report is a computer controlled blood exchange method to achieve sinusoidal arterial pO_2 delivery to brain tissue.

In order to obtain true dynamic data for the study of oxygen transport through tissue, it is necessary to go beyond step changes in respiratory gases and obtain a true dynamic input of arterial oxygen. To achieve this goal, a system has been developed to

903

produce a sinusoidal arterial pO_2. The concept applied was the exchange of venous blood for arterial, with the arterial pO_2 being a function of the rate of exchange. The exchange rate was controlled to produce the desired arterial pO_2 transients. A sinusoidal function was chosen due to its ease of production and potential for subsequent correlation.

The chamber arrangement through which the exchange was accomplished is shown in Figure 1. The total priming volume of the system was 15 ml with less than an additional 10 ml for the specific connectors and tubing, depending on their length. For normal flow rates the pressure drop through the arterial side was $\sim$ 10 mm Hg and the venous side $\sim$ 3 mm Hg (30 ml/min blood flow rate).

Oxygen tension was measured as it left the arterial side and as it entered the venous side with polarographic micro oxygen electrodes. (Transidyne #721, glass insulated pt wire electrodes.) The blood flow rate exiting from the arterial chamber was measured

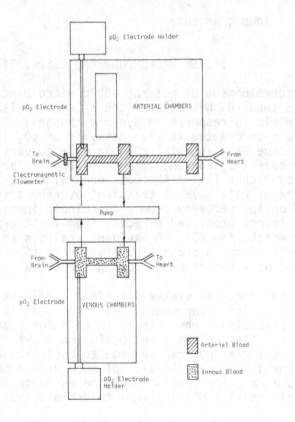

Figure 1. Schematic of Blood Exchange System

with a flow through electromagnetic probe system (IVM Model
EMF-130). The pump shown between the arterial and venous chambers
was a dual head totally occlusive roller pump, which had been modified to allow for speed control with D.C. voltage.

In order to determine the pump flow rate, it was necessary to
make a mass balance on the oxygen which can be stated as

$$\text{Accumulation} = \text{In} - \text{Out} \qquad (1)$$

A schematic of the chamber arrangement is shown on Figure 2. Since
arterial pO_2 was to be controlled, it was only necessary to consider chamber 4. By assuming chamber 4 to be perfectly mixed, the
following equation holds.

$$V_4 \frac{dC_{a,o}}{dt} = (Q_1 - Q_2) C_{a,i} + Q_3 C_{v,i} - Q_4 C_{a,o} \qquad (2)$$

From a total mass balance on all chambers it follows that

$$Q_1 - Q_2 = Q_4 - Q_3 \qquad (3)$$

where $Q_3 \approx Q_2$ by experimental means. Substituting Equation 3 into
Equation 2 and rearranging yields

$$V_4 \frac{dC_{a,o}}{dt} = Q_4(C_{a,i} - C_{a,o}) + Q_3(C_{v,i} - C_{a,i}) \qquad (4)$$

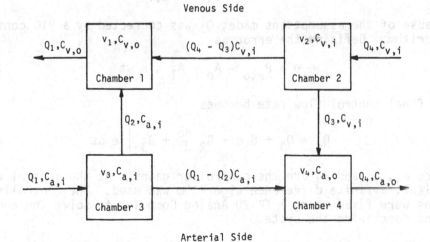

Figure 2. Arterial and Venous Mixing Chambers

In order to set up the control equations, the following assumptions were made and then later corrected. As pO_2 was measured directly, and not concentration, it was assumed that a linear approximation of the Bohr curve could be utilized over limited oscillation ranges as

$$C_j = Ap_j + B \tag{5}$$

where p is the pO_2. Substituting Equation 5 into Equation 4 results in

$$v_4 \frac{dP_{a,o}}{dt} = Q_4(P_{a,i} - P_{a,o}) + Q_3(P_{v,i} - P_{a,i}) \tag{6}$$

Since the desired form of $P_{a,o}$ is

$$P_{a,o} = A_o + A_1 \sin wt \tag{7}$$

it follows that

$$\frac{dP_{a,o}}{dt} = A_1 w \cos wt \tag{8}$$

By using Equations 8 and 7 in Equation 6, the required pump flow rate can be obtained as

$$Q_3 = \frac{v_4 A_1 w \cos wt - Q_4(P_{a,i} - A_o - A_1 \sin wt)}{P_{v,i} - P_{a,i}} \tag{9}$$

Because of the assumptions made, Q_3 was corrected by a PID control algorithm. Defining the error as

$$e = P_{a,o} - A_o - A_1 \sin wt \tag{10}$$

the final control flow rate becomes

$$Q_c = Q_3 + G_1 e + G_2 \frac{de}{dt} + G_3 \int e \, dt \tag{11}$$

where G_1, G_2 and G_3 are the controller gains. As the control was digital, a finite difference algorithm was used. The controller gains were fixed using a TR-20 Analog Computer to solve the equations describing the system.

The calculation of Q_3 and Q_c was done with a PDP-11 digital process control mini computer which also developed the appropriate

D.C. signal output to the pump. The values for Q_4, $P_{a,o}$ and $P_{v,i}$ were measured and transmitted to the computer through A-D inputs. The time required to calculate a sine or cosine function within the PDP-11 was $\sim$ 10 m sec as opposed to 20 μ sec to read a voltage. A separate frequency generator was employed to produce the sine and cosine function, allowing the sample and output frequencies in the PDP-11 to attain 20/second.

The format for the control program was established to allow changing of experimental conditions in minimum time. To accomplish this, the initial input contained the location and gains of each analog signal along with their calibration factors, the sample frequency, and the steady state arterial pO_2. The dynamic portion of the program allowed for changes in the variables A_0, A_1 and w for the desired arterial pO_2 and the frequency dependent controller gains to prepare for subsequent experiments.

Included in the control program was a check so that $Q_C < Q_4$ at all times. This prevented an internal loop flow which would have made the control system fail and also placed a limit on allowed combinations of frequency and amplitude.

The frequency limit for the system was due to the mechanical pump inertia which responded to a changing voltage input with a ± 3 db limit from D.C. to 1 Hz.

EXPERIMENT

The above system was used in experimentation with cats to obtain dynamic tissue O_2 data. The two carotid arteries and external jugular veins of the cats were cannulated to channel the blood through the exchange chambers. The remaining veins and arteries were clamped to insure that the total arterial blood supply was measured and that venous return was sufficient for the control pump exchange process.

After assembly, the chambers were first flushed with CO_2 and then primed with heparinized normal saline. The animals were anesthetized with nembutal and maintained on heparin to prevent clotting. The pleural cavity of the animal was vented to eliminate both spontaneous breathing and excessive motion of the animal due to breathing. Ventilation was maintained with a positive pressure respirator through an L-shaped tracheal cannula.

The cat was then mounted in a stereotaxic frame bolted to a 60 lb aluminum plate which rested on 8 inner tubes, arranged in a double stack at each of the four corners. Their purpose was to eliminate any external vibrations from affecting the experimental measurements. Above the cat, mounted to the plate, was a micro-drive which provided tissue electrode movement in 2.5 μ steps. This tissue oxygen electrode was also connected to a high input impedance unity gain differential amplifier, and then to a series of A.C. amplifiers and filters. Thus, it was possible to simulta-neously record action potentials with the O_2 electrode.

Arterial and tissue pO_2's were analyzed with a correlator for on-line cross correlation to give an immediate indication of the tissue response dynamics. All signals, including arterial, venous and tissue pO_2's, blood flow rate and the sine input were recorded on FM channels of a magnetic tape recorder with action potentials recorded directly. The animal's EEG and EKG were also monitored.

Oscillations in arterial pO_2 were also produced by changing inlet gases to the respirator. Tissue pO_2 was measured as before, and arterial pO_2 was determined by piercing one of the carotid arteries with a micro electrode. Qualitative arterial flow changes were followed with a wrap-around electromagnetic probe. The high-est possible frequency was a decade below that possible with the exchange system.

Several methods are available for data analysis which give rapid results for the large quantity of data this system produces. Auto- and cross-correlations are obtained directly and the PDP-11 provides an interface with larger computation facilities for comparison with theoretical models.

Results have been obtained with each described method of changing the arterial pO_2. The computer controlled exchange sys-tem yields well-defined oscillations and experiments have covered the range of 0.01 to 1 Hz. The ratio of tissue to arterial pO_2 varied from 0.45 to zero while the phase lag changed from $\sim 60°$ to $\sim 32\pi + 290°$ respectively. With the exchange bypass system, numerous physiological variations can be investigated and subjected to dynamic analyses.

The goal of this phase of the investigation is to develop and test theoretical models which include both autoregulatory and meta-bolic mechanisms. With the capability for simultaneous measure-ments during phasic changes, a detailed comparison of theoretical models is possible. In addition, the use of correlation and sig-nal averaging techniques allows low level responses to be separated from both noise and respirator induced artifacts. The use of the described system may make it possible to deduce rate-limiting meta-bolic contributions to the dynamic behavior of tissue pO_2 in addi-tion to quantifying autoregulatory responses.

NOMENCLATURE

A	intercept of Bohr curve approximation
A_0	mean desired arterial pO_2
A_1	amplitude of desired arterial pO_2 oscillation
B	slope of Bohr curve approximation
$C_{a,i}$	oxygen concentration of arterial blood entering chambers
$C_{a,o}$	oxygen concentration of arterial blood leaving chambers
$C_{v,i}$	oxygen concentration of venous blood entering chambers
$C_{v,o}$	oxygen concentration of venous blood leaving chambers
e	error
G_1	proportional control gain
G_2	derivative control gain
G_3	integral control gain
$P_{a,i}$	oxygen partial pressure of arterial blood entering chambers
$P_{a,o}$	oxygen partial pressure of arterial blood leaving chambers
$P_{v,i}$	oxygen partial pressure of venous blood entering chambers
Q_c	control pump flow rate
Q_1	blood flow rate entering arterial chambers
Q_2	arterial to venous exchange flow rate
Q_3	venous to arterial exchange flow rate
Q_4	blood flow rate leaving arterial chambers

REFERENCES

1. Bicher, H. I., and M. H. Knisely, "Brain tissue reoxygenation time, demonstrated with a new ultramicro oxygen electrode," *J. Appl. Physiol.* 28:387-390, 1970.

2. Metzger, H., W. Erdmann and G. Thews, "Effect of short periods of hypoxia, hyperoxia and hypercapnia on brain O_2 supply," *J. Appl. Physiol.* 31:751-759, 1971.

3. Reneau, D. D., Jr., D. F. Bruley and M. H. Knisely, "A mathematical simulation of oxygen release, diffusion, and consumption in the capillaries and tissue of the human brain," in *Chemical Engineering in Medicine and Biology*, ed. by D. Hershey, p. 135-241, Plenum Press, New York, 1967.

4. Reneau, D. D., Jr., H. I. Bicher, D. F. Bruley and M. H. Knisely, "A mathematical analysis predicting cerebral tissue reoxygenation time as a function of the rate of change of effective cerebral blood flow," *Blood Oxygenation*, ed. by D. Hershey, p. 175-200, Plenum Press, New York, 1970.

OXYGEN TRANSPORT IN SKELETAL MUSCLE: HOW MANY BLOOD CAPILLARIES

SURROUND EACH FIBRE?

A.C. Groom & M. J. Plyley, Department of Biophysics,

University of Western Ontario, London, Ontario,

Canada. N6A 3K7

Implicit in a number of recent studies on the capillary bed
in skeletal muscle is the idea that exercise training (1,2,3) or
adaptation to the chronic hypoxia of high altitude (4,5) might
lead to a compensatory increase in capillary density per mm^2 of
cross-section. Although such an increase might occur without an
actual change in the ratio of capillaries to fibres (C/F ratio)
simply as a result of the fibres growing smaller in diameter, the
underlying assumption is that the C/F ratio might itself increase
and that in some way this would lead to improved oxygenation of
the tissue. This is a difficult concept to accept since blood
capillaries, which run parallel to the fibres, are never found
within the fibres themselves but are located around the perimeter
of each fibre. The question we have to ask ourselves is this:
how many capillaries are needed around the perimeter of a muscle
fibre in order to provide uniform O_2 transport to that fibre?

An elementary consideration suggests that a fibre of
circular cross-section whose O_2 consumption is spread uniformly
throughout its mass would require only three vessels (Fig. 1);
however, if the fibre cross-section were somewhat elongated, then
four or more vessels might be required. In the array of fibres
within a muscle bundle each capillary will, because of its
location, be shared by two or more fibres. Therefore the C/F
ratio will always be less than the mean number of capillaries
surrounding a fibre, and even if the latter were a constant for
all muscles the C/F ratio could vary depending on whether the
array formed by the fibres was square (6) or hexagonal (7). It
seemed to us that the basic parameter in ensuring uniformity of
O_2 delivery to striated muscle fibres was not the mean radius of
the 'Krogh' cylinder, nor the C/F ratio, but the mean number of

911

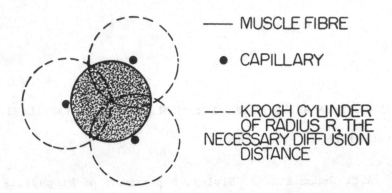

ONLY 3 CAPILLARIES ARE NECESSARY TO SUPPLY
O_2 UNIFORMLY TO A CIRCULAR MUSCLE FIBRE

Fig. 1 - Illustrates the basic hypothesis (see text)

capillaries surrounding each fibre. We therefore began to search
the literature for values of capillary and fibre densities per
mm^2 from which, on the assumption that each vessel supplies (say)
three fibres approximately, we might compute numbers of vessels
surrounding a fibre.

Examination of the published values for capillary density in
mammalian skeletal muscle proved very disheartening. Even for the
same muscle in the same species the values reported vary by a
factor of 3 to 6. For various muscles of cat, dog, rabbit, rat
and guinea-pig, values reported prior to 1950 lie within the range
790 to 5900 vessels/mm^2 (mean 2584 $\pm$ 299 S.E.; N = 16) whereas for
these same species values reported in papers appearing since 1950
range from 219 to 1990 vessels/mm^2 (mean 738 $\pm$ 152 S.E.; N = 15).
The means are in the ratio 3.5:1 and some of the earlier values
cited imply the presence of 10 to 20 capillaries around each
muscle fibre, a result which appears seriously to challenge our
hypothesis. Also disquieting are several reports that the C/F
ratio for red fibres is approximately twice that for white fibres
(7,8). Whilst this could be due merely to a difference in the
number of fibres sharing one capillary in these two situations it
could, alternatively, indicate a difference in the actual number
of capillaries surrounding each fibre.

We have therefore undertaken a series of experiments to
determine, in several muscles from several different species, the

numbers of vessels surrounding individual fibres and here present
a preliminary report of our results.

MATERIALS AND METHODS

These studies were performed on dogs, cats, rabbits, rats
and guinea-pigs. The animals were anesthetized and given heparin
(500 i.u./Kg as an anticoagulant. The gastrocnemius and soleus
muscles were perfused via the femoral arteries (dogs, cats,
rabbits) or the absominal aorta (rats, guinea-pigs) with a
silicone elastomer of viscosity 20 centipoise (Microfil: Canton
Biomedical Products, Inc.). This material fills the capillary
network and passes on into the venous outflow; it remains fluid
for about 15 min. after which it begins to set. No attempt was
made to keep the perfusion pressure within physiological limits
since we were concerned only to fill, as far as possible, all the
capillaries in the vascular bed. One hour after perfusion the
muscles were dissected out and fixed in 10% buffered formalin for
24 hours. Small pieces of tissue were then cut from each muscle
and used for the preparation of histological sections. Transverse
sections 12-20 μm thick were stained with haematoxylin and eosin
(H & E), methylene blue, or Gomori trichrome and examined under
the microscope. Photomicrographs were prepared and the number of
capillaries (filled with Microfil) around each of 500 to 600 fibres
were counted for each muscle preparation. Four to five such
preparations of both lateral gastrocnemius and soleus were examined
in each species.

RESULTS AND DISCUSSION

We have found that haematoxylin stains darkly the nuclei
located peripherally around each fibre and such nuclei could
easily be confused with capillaries, giving falsely high counts.
Examination of the staining procedures used by the earlier group
of workers (see Introduction) shows that the use of haematoxylin
was almost certainly responsible for the high values of capillary
density they reported. Thus the mean of all the published values
for the above species using haematoxylin is 2567 ± 370 S.E.
vessels/mm^2 (N = 13) compared with 697 ± 158 S.E. (N = 14) using
other methods. Neither methylene blue nor Gomori trichrome stain
the nuclei, but Gomori trichrome has the advantage that it stains
the erythrocytes a brilliant red. Using this stain, therefore,
capillaries unperfused by Microfil but containing red cells may
be detected and included in the vessel counts. The results which
follow have been obtained using this method.

In all muscles examined the number of vessels surrounding
each fibre lay between 0 and 9. The percentages of fibres having

Table I

PERCENTAGE OF FIBRES SURROUNDED BY N VESSELS IN GASTROCNEMIUS ('G') AND SOLEUS ('S') OF FIVE MAMMALIAN SPECIES.

N	DOG		CAT		RABBIT		RAT		GUINEA PIG	
	G	S	G	S	G	S	G	S	G	S
0	0.1	-	0.7	0.6	0.8	0.6	0.7	0.4	0.8	0.5
1	2.6	1.1	3.0	4.3	6.8	5.7	4.0	2.9	6.6	6.0
2	14.0	8.6	11.5	12.5	17.2	16.6	14.1	13.0	17.7	20.3
3	32.6	26.3	24.5	21.9	28.6	29.9	23.3	20.2	30.3	32.5
4	33.3	35.0	33.9	33.1	25.3	29.3	25.9	28.5	28.3	27.9
5	14.2	22.3	24.6	19.1	14.8	13.2	19.5	20.3	12.4	10.3
6	2.4	5.6	5.4	6.6	5.1	3.9	9.0	9.0	2.9	2.0
7	0.6	1.0	1.1	1.7	1.1	0.6	2.7	4.1	0.7	0.5
8	0.1	0.1	-	0.2	0.4	0.3	0.8	1.3	0.1	-
9	-	-	-	-	-	-	0.2	0.2	-	-

various numbers of vessels between these limits are shown in
Table I, and a marked similarity of these values for all five
species is immediately obvious. In every case the mean number of
vessels around a fibre lies between 3.22 (Guinea-Pig soleus) and
4.01 (Rat soleus). In view of these close similarities we have
lumped together the data from all ten muscles and have computed
overall mean values ± S.E. for the distribution of the number of
vessels surrounding individual fibres (Fig. 2). This distribution
is symmetrical with its mean at 3.62. The same results have been
obtained from gracilis and tibialis anterior of guinea-pig, from
masseter and striated fibres in tongue of cats.

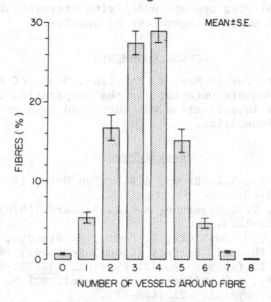

Fig. 2 - Frequency distribution for number of vessels
surrounding a fibre (grouped data of Table I)

The finding that in all muscles examined the mean number of
capillaries surrounding a fibre lay between three and four
suggests that nature attempts to achieve a uniform supply of O_2
to each fibre with the greatest economy in blood vessels. A
greater rate of O_2 supply than can be obtained with this number
of vessels appears to be achieved by having smaller fibres rather
than by increasing the number of capillaries. Our mean value of
3.62 agrees well with the value 3.6 determined for tenuissimus
muscle of cat (9) but differs significantly from the ranges 3.9
to 5.3 reported for gastrocnemius of guinea-pig (2) and 6.8 to
9.1 reported for various leg muscles of guinea-pig (4).

It has been shown recently that in cat, rat and guinea-pig,
the soleus consists entirely of red fibres whereas the lateral
gastrocnemius contains at least 60 percent white fibres (10).

Nevertheless we found little difference in the numbers of vessels around a fibre for any of these muscles, i.e. no difference between red and white fibres, agreeing with recent reports (2,9).

We made similar studies in gastrocnemius of frogs and found essentially the same number of vessels around each fibre (2.95 $\pm$ 0.1 S.E.) as in mammals. The cross-sectional area of these frog fibres was 16 times greater than that of striated muscle fibres in cat tongue, yet the number of vessels around each fibre was not significantly different. These results suggest that a mean value of 3 to 4 capillaries surrounding each fibre may constitute a unifying concept underlying microvascular architecture in all striated muscle, regardless of species.

ACKNOWLEDGEMENTS

Thanks are due to Mrs. Z. Pattison, Dept. of Pathology, University of Western Ontario, for the preparation of histological sections. This investigation was supported by a grant from the Ontario Heart Foundation.

REFERENCES

1. Carrow, R.E., R.E. Brown, & W.D. Van Huss (1967). Anat. Rec. 159:33-40.
2. Mai, J.V., V.R. Edgerton, & R.J. Barnard (1970). Experientia 26:1222-1223.
3. Cotter, M., O. Hudlicka, D. Pette, H. Staudte, & G. Vrbova. Proc. Physiol. Society (Lond.) December, 1972.
4. Valdivia, E. (1958). Am. J. Physiol. 194:585-589.
5. Cassian, S., R.D. Gilbert, C.E. Bunnel, & E.M. Johnson (1971). Am. J. Physiol. 220:448-451.
6. Krogh, A. (1930). The Anatomy and Physiology of Capillaries. New Haven: Yale Univ. Press; reprinted by Hafner (New York) 1959, p61-65.
7. Schmidt-Nielsen, K. & P. Pennycuik, (1961). Am. J. Physiol. 200:746-750.
8. Smith, R.D. & R.P. Giovacchini (1956). Acta. Anat. 28:342-358.
9. Eriksson, E., and R. Myrhage (1972). Acta. Physiol. Scand. 86:211-222.
10. Aviano, M.A., R.B. Armstrong, & V.R. Edgerton (1973). J. Histochem. Cytochem. 21:51-55.

FILTERING AND PREDICTION OF BLOOD FLOW AND OXYGEN

CONSUMPTION FOR PATIENT MONITORING

P. David Wilson

Institute for Emergency Medicine

University of Maryland Hospital, Baltimore, Maryland

Suppose data is to be electronically acquired at discrete times Δ, 2Δ, 3Δ,..., $t\Delta$,... (starting from an arbitrary origin) on arterial and venous oxygen concentration as well as an <u>independent variable</u> which is either blood flow rate or oxygen consumption, and that the remaining <u>dependent variable</u> (O_2 consumption or blood flow) is to be <u>predicted from the data</u>. We refer to "time $t\Delta$" merely as "<u>time t</u>". Let y_{1t} be the observed value of the <u>independent variable</u> at time t and let y_{2t}, y_{3t} be the observed values of the arterial and venous O_2 concentrations respectively at time t. The observations are <u>physiological state</u> values corrupted by <u>noise or observation error</u>. For j = 1,2,3 corresponding to the observation subscripts, let $\tilde{x}_j(t)$ be the jth physiological state (existing in continuous time) and let v_{jt} be the noise or observation error of the jth observation at time t.

The observation equation is thus

$$\underline{y}_t = \underline{\tilde{x}}_t + \underline{v}_t \tag{1}$$

in which the underbar denotes a column vector of (in this case) three elements. The notation $\underline{\tilde{x}}_t$ is used to denote the vector of state values at the discrete time t, while $\underline{\tilde{x}}(t)$ denotes the vector of continuous time functions. In (1), the sequence $\{\underline{v}_t\}$ is modeled as a sequence of independent random variables with mean vector equal to zero

917

and variance-covariance matrix

$$\text{var}(\underline{v}_t) = R \tag{2}$$

for all t. We assume the existence of a long time or ensemble average $\underline{c}$ defined by

$$\underline{c} = \lim_{n\to\infty} \frac{1}{n} \sum_{t=1}^{n} \underline{y}_t = \lim_{T\to\infty} \frac{1}{T} \int_0^T \underline{\tilde{x}}(t)\,dt \tag{3}$$

The filtering and prediction algorithms to be presented here assume the knowledge of $\underline{c}$ and so we require the existence of a previously obtained set of data $\{\underline{y}_n\}$ sufficiently large for identification of $\underline{c}$ by averaging $\{\underline{y}_n\}$ until convergence has occurred. Define the working observation equation by

$$\underline{z}_t = \underline{x}_t + \underline{v}_t \tag{4}$$

in which $\underline{z}_t = \underline{y}_t - \underline{c}$ and $\underline{x}_t = \underline{\tilde{x}}_t - \underline{c}$, are the working observation and state respectively.

The dynamics of the state are assumed to be represented by the model

$$\underline{\dot{x}}(t) = F\,\underline{x}(t) + \underline{w}(t) \tag{5}$$

in which $\underline{\dot{x}}(t)$ represents the vector of derivatives of $x(t)$ with respect to time, $\underline{w}(t)$ is a vector of unknown forcing functions which disturb the steady state, and F is a matrix of unknown coefficients. The solution of (5) is of the form

$$\underline{x}_t = T\,\underline{x}_{t-1} + \underline{u}_t \tag{6}$$

in which

$$T \equiv T_{t,t-1} = e^{F\Delta} \equiv \sum_{n=0}^{\infty} \frac{1}{n!}(F\Delta)^n \tag{7}$$

is called the transition matrix, and

$$\underline{u}_t \equiv \underline{u}_{t,t-1} = \int_{t-1}^{t} T(t,s)\,\underline{w}(s)\,ds \tag{8}$$

are called <u>state disturbances</u>.

We model the state disturbances as random variables with mean zero, but whose actual values may be (implicitly) estimated from the data. Explicitly we assume the sequence $\{\underline{u}_t\}$ to be a sequence of independent, zero mean random variables, independent of the sequence $\{\underline{v}_t\}$, and having variance-covariance matrix

$$\text{var}(\underline{u}_t) = Q \tag{9}$$

for all t. We also assume $\underline{x}_0$ is random, with mean $\hat{\underline{x}}_0$ and variance S_0, and that it is independent of $\{\underline{v}_t\}$ and $\{\underline{u}_t\}$.

Optimal estimation theory may be applied to the above model to obtain optimal estimates of $\underline{x}_t$ based on $Z_\tau = \{\underline{z}_1, \underline{z}_2, \ldots, \underline{z}_\tau\}$. Denote the optimal estimate by $\hat{\underline{x}}_{t|\tau}$. If "optimal" is taken to mean minimum-mean-square-error or Bayes (with quadratic loss), then

$$\hat{\underline{x}}_{t|\tau} = E(\underline{x}_t|Z_\tau) \tag{10}$$

in which E denotes the expectation operator in the conditional probability density function (pdf) of $\underline{x}_t$ conditioned on (or given) the values of Z_τ. For $\tau = t$ the estimation is called <u>filtering</u> and for $\tau < t$ it is called <u>prediction</u>.

Under the assumption that the probability law governing $\{\underline{v}_t\}$, $\{\underline{u}_t\}$, and $\underline{x}_0$ is Gaussian, the algorithm to be given for computing $\hat{\underline{x}}_{t|\tau}$, $\tau \leq t$, is optimal; for other probability laws, the algorithm is optimal among the class of estimators which are <u>linear</u> functions of the data.

Denote the estimation error variance by

$$S_{t|\tau} = \text{var}(\underline{x}_t - \hat{\underline{x}}_{t|\tau}) \tag{11}$$

and let $\hat{\underline{x}}_t \equiv \hat{\underline{x}}_{t|t}$, $S_t \equiv S_{t|t}$. Suppose S_{t-1} and $\hat{\underline{x}}_{t-1}$ have been computed. (The starting procedure will be given below). The one step prediction equations are

$$\hat{\underline{x}}_{t|t-1} = T \, \hat{\underline{x}}_{t-1} \tag{12a}$$

$$\tilde{S}_t \equiv S_{t|t-1} = T S_{t-1} T' + Q \tag{12b}$$

in which the prime denotes transposition. The **filter equations** are

$$\hat{\underline{x}}_t = \hat{\underline{x}}_{t|t-1} + B_t (\underline{z}_t - \hat{\underline{z}}_{t|t-1}) \tag{13}$$
$$S_t = \tilde{S}_t - B_t \tilde{S}_t$$

in which $B_t = \tilde{S}_t [\tilde{S}_t + R]^{-1}$ and $\hat{\underline{z}}_{t|t-1} = \hat{\underline{x}}_{t|t-1}$. The **n-step prediction equations** are

$$\hat{\underline{x}}_{t+n|t} = T^n \hat{\underline{x}}_t$$
$$S_{t+n|t} = T^n S_t (T')^n + \sum_{k=1}^{n} T^{n-k} Q (T')^{n-k}. \tag{14}$$

For starting the algorithm, $\underline{x}_0$ and S_0 are generally unavailable. It can be shown that, from the viewpoint of Bayes estimation theory, an informationless aprori distribution is achieved by setting $S_1 = R$ and $\hat{\underline{x}}_1 = \underline{z}_1$.

Knowledge of the matrices T, Q, and R is required before computation can proceed. These may be obtained from the same long run of data required for computation of $\underline{c}$ as follows: Let $V_t = \text{var}(\underline{x}_t)$ and assume that $\{\underline{x}_t\}$ is a wide-sense stationary stochastic process, which implies that T has some norm less than unity. Under this condition $V_t = V$ for all t and

$$V = TVT' + Q . \tag{15}$$

Define C_j to be the expected outer (Kroneker) product of $\underline{z}_t$ and $\underline{z}_{t-j}$:

$$C_j = E(\underline{z}_t \underline{z}'_{t-j}), \quad j = 0,1,2. \tag{16}$$

The matrices T, Q, and R may be identified from the relations

$$C_0 = V + R$$
$$C_1 = TV \tag{17}$$
$$C_2 = T^2 V$$

along with (15). The C_j matrices may be identified as

long run converged averages:

$$C_j = \lim_{n\to\infty} \frac{1}{n} \sum_{t=1}^{n} \begin{pmatrix} z_{1t}z_{1,t-j} & z_{1t}z_{2,t-j} & z_{1t}z_{3,t-j} \\ z_{2t}z_{1,t-j} & z_{2t}z_{2,t-j} & z_{2t}z_{3,t-j} \\ z_{3t}z_{1,t-j} & z_{3t}z_{2,t-j} & z_{3t}z_{3,t-j} \end{pmatrix} \quad (18)$$

From the C_j matrices compute

$$V = C_1 C_2^{-1} C_1$$

$$T = C_2 C_1^{-1}$$

$$R = C_0 - V \quad (19)$$

$$Q = V - TVT'.$$

The matrix V should be symmetric. If slight asymmetry exists due to uneven round-off error, this should be corrected by averaging the off-diagonal elements before proceeding to compute R and Q. The matrix T should have a norm less than unity. The square root of the modulus of the maximum eigenvalue of T'T is a norm for this check. If this norm is not less than unity, the assumption of stationarity of $\{\underline{x}_t\}$ may not be valid.

Using the output of the algorithms (12), (13), (14), and assuming the total body O_2 volume to be constant, the dependent variable d_t (O_2 consumption or blood flow) may be estimated from the well-known relation

$$0 = (\tilde{x}_2(t) - \tilde{x}_3(t)) \, f(t) - g(t) \quad (20)$$

where $f(t)$ is blood flow rate, and $g(t)$ is O_2 consumption rate. Its approximate error variance is the quadratic form

$$\mathrm{var} \, (d_{t+n} - \hat{d}_{t+n|t}) \doteq \underline{\ell}' S_{t+n|t} \, \underline{\ell}$$

$$\underline{\ell}' = \begin{cases} [(\hat{x}_2 - \hat{x}_3), -\hat{x}_1, \hat{x}_1]/(\hat{x}_2 - \hat{x}_3)^2, & \text{if } d_t = f_t \\ [(\hat{x}_2 - \hat{x}_3), \hat{x}_1, -\hat{x}_1], & \text{if } d_t = g_t. \end{cases} \quad (21)$$

In $\underline{\ell}$ in (21), $\hat{x}_i = \hat{\tilde{x}}_{i,t+n|t}$ for $i = 1,2,3$.

REFERENCES

1.   Wilson, P. David: Adaptive Smoothing and Prediction
      of a Nonstationary, Multivariate Time Series; An
      Approach to Computer Monitoring of Patients in an
      Intensive Care Unit. Doctoral Dissertation, Johns
      Hopkins University, 1970.

2.   Leondes, C.T. (ed.): Theory and Application of
      Kalman Filtering, North Atlantic Treaty Organization
      Advisory Group for Aerospace Research and Develop-
      ment, AGARDograph No. 139.

3.   Leibelt, P. B.: An Introduction to Optimal Estim-
      ation, Addison-Wesley Publishing Co., 1967.

4.   Wilson, P. David: Optimal Estimation Theory and
      Method for Patient Monitoring. In preparation for
      publication.

COMPUTER ANALYSIS OF AUTOMATICALLY RECORDED OXYGEN DISSOCIATION CURVES

Bruce F. Cameron

Papanicolaou Cancer Research Institute, 1155 N.W. 14th

Street, Miami, Fla. 33136, and Division of Hematology,

Dept. of Medicine, University of Miami School of Medicine

The physiologic function of the erythrocyte is the transport of oxygen from lungs to tissue. The chemistry of this process is represented by a non-linear functional relationship between oxygen tension and oxygen saturation of the contained hemoglobin. This curve is usually characterized by two parameters, P_{50} (the oxygen tension at half-saturation) and the Hill constant, n (a measure of sigmoidicity).

Since this function is non-linear, it would be preferable for complete characterization of oxygen dissociation that the entire curve be measured and analyzed rather than P_{50} alone or a discrete number of points on the curve around 50% saturation, such as is required to calculate n.

This communication describes a computer program for analysis of continuously recorded oxygen dissociation curves. Routine calculation of P_{50} and n is carried out automatically, and the complete curve expressed as % saturation vs. pO_2 may be examined. The analysis was devised for curves obtained by the method of Colman and Longmuir (1), modified by Cameron (2). Recently a commercial instrument (Radiometer DCA-1 Dissociation Curve Analyzer) has become available and the same program may be used to analyze dissociation curves produced by this instrument.

923

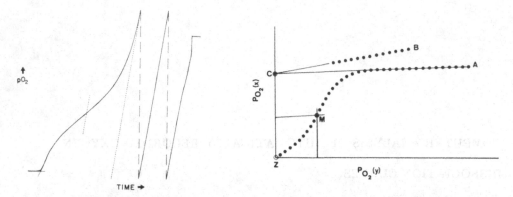

Fig. 1: Normal erythrocyte oxygen dissociation curve by metabolic
deoxygenation. The Y-axis shows automatic scale changes for
expanded accuracy. Note extrapolation at zero point.

Fig. 2: Dissociation curve as recorded from the Radiometer DCA-1.
Figure from product literature (10), reproduced by permission.
The axes are rotated to conform to variable notation used in the
computer analysis.

Both of these techniques are based on measurement of the pO_2 of a
suspension of erythrocytes or a hemoglobin solution. A typical
dissociation curve of normal erythrocytes in pH 7.4 phosphate buffer
at 37° is given in Fig. 1. The principle of the method is described
in detail by Cameron et al. (3). Fig. 2 is a representation of the
experimental data presentation from the DCA Dissociation Curve
Analyzer (4).

For both methods, the experimental trace in the absence of hemoglobin is a straight line with slope proportional to the partition coefficient
for oxygen in the liquid phase. The experimental trace deviates from
linear when the pO_2 falls sufficiently for oxygen to be released from
its combined state as oxyhemoglobin. The dissociation curve is then the
deviation of the experimental trace from the (linear) control trace.

The assumption that the deviation of the experimental from the control
trace is a linear function of % saturation of the hemoglobin with oxygen
presumes only that there is no other oxygen-binding material, and that
the hemoglobin is not heterogeneous. If it is heterogeneous, the observed
dissociation curve will be a linear combination of the individual dissociation curves.

SUBROUTINES

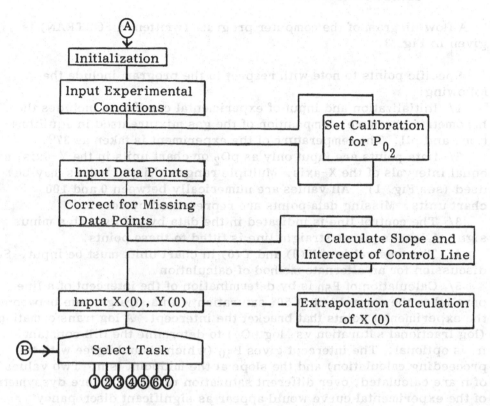

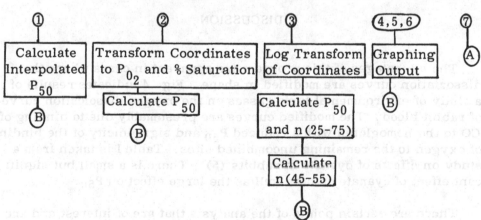

Figure 3: Flow Diagram of the Analysis

COMPUTER ANALYSIS

A flow diagram of the computer program (written in FORTRAN) is given in Fig. 3.

Specific points to note with respect to the program include the following:

1/ Initialization and input of experimental conditions includes the barometric pressure, composition of the gas mixture used in equilibration, and pH. The temperature of the experiment is taken as 37°.

2/ Data points are input only as pO_2 or chart units in the Y-axis, at equal intervals of the X-axis. Multiple ranges of input values may be used (see Fig. 1). All values are numerically between 0 and 100 chart units. Missing data points are represented by zero.

3/ The control line is indicated in the data by points with a minus sign. A least squares straight line is fitted to these points.

4/ The true value of $X(0)$ and $Y(0)$ in chart units must be input. See discussion for an alternate method of calculation.

5/ Calculation of P_{50} is by determination of the intercept of a line parallel to the control but at 50% saturation with a straight line between the experimental points that bracket the intercept. A log transformation (log fractional saturation vs. log pO_2) to determine the Hill constant, n, is optional. The intercept gives P_{50} (which should agree with the preceeding calculation) and the slope at the midpoint is n. Two values of n are calculated, over different saturation ranges. Severe dyssymetry of the experimental curve would appear as significant discrepancy between these two calculations.

DISCUSSION

The advantages of this program are apparent in cases where the dissociation curves are modified in shape. Fig. 4 indicates results of a study of environmental toxic gasses on the oxygen dissoication curve of rabbit blood. The modified curves are presumably due to binding of CO to the hemoglobin with a reduced P_{50} and sigmoidicity of the binding of oxygen to the remaining uncombined sites. Table 1 is taken from a study on effects of cyanate in rabbits (5). There is a small but significant effect of cyanate on n as well as the large effect on P_{50}.

There are certain points of the analysis that are of interest and are under current investigation in our laboratory.

Table I

Representative data from rabbits treated with cyanate (NaNCO)

I.P. daily 1 month dose mg.	ΔP_{50} mm Hg	Δn
25	-9.6	-0.52
50	-15.6	-0.71
100	-19.4	-0.94

One problem with automatic analysis is that for both experimental methods used in this study, the zero point is uncertain. In one (3)
this is because the control may be linear down to very low pO_2 but not
zero (see Fig. 1); in the other (4) because data at zero pO_2 requires long
equilibration with an oxygen-free gas mixture. It would be useful if
an adequate extrapolation function could be devised to calculate X(0).
In Fig. 5 a set of curves is given in which extrapolation to zero was
done by least squares, curve fitting the lower end of the experimental
data to Chebyshev polynomials of varying degree(6). It appears
that such an extrapolation can be used, and that a polynomial of quadratic
degree is necessary and sufficient, but this result is still preliminary.

The Adair model of sequential cooperative binding could be used
for fitting the data. However, where the Adair equations have been fit
to experimental data, the significance of the calculated values of binding
constants is low (7) or certain of them may even be negative (8,9).

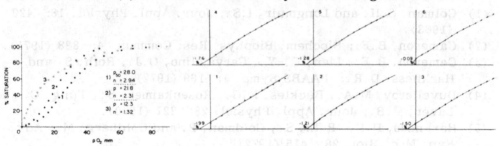

Fig. 4: Oxygen dissociation curves of rabbit erythrocytes.
         1) Normal. 2) Cigarette smoke. 3) Automobile exhaust.

Fig. 5: Chebyshev polynomial curve fit to the lower 10% of a normal
         dissociation curve, degree 1-5. The "correct" extrapolated
         X-intercept is ca. (-1.2).

For these reasons an arbitrary fit to the experimental curve was chosen. The fit to Chebyshev polynomials is a convenient one, since it converges when a close fit is obtained, and higher order terms in the expansion may be dropped when their magnitude is small compared to leading terms.

The problem of graphing the output of the program has also been approached by Chebyshev polynomial fitting. As in the case of extrapolation, the results are only preliminary. For a "normal" curve, a fit of degree 5 is adequate for graphic output; at this level, the maximum absolute error is under 1%.

ACKNOWLEDGEMENTS

This work has taken place over a number of years, supported at various times by Grant No. 661 from the Council for Tobacco Research, Grant No. SO1-RR05690 from the National Institutes of Health, and Contract NHL1-72-2926B from the National Institutes of Health. A listing of the FORTRAN program may be obtained from the author. The curves in Fig. 4 are from unpublished studies by S. Peron and B.F. Cameron.

REFERENCES

(1) Colman, C.H. and Longmuir, I.S.; Jour. Appl. Physiol. 18: 420 (1963).
(2) Cameron, B.F.; Biochem. Biophys. Res. Commun. 4: 888 (1971).
(3) Cameron, B.F., Lian, C.-Y., Carvajalino, O.J., Roth, S. and Harkness, D.R.; PAABS Symp. 1: 169 (1972).
(4) Duvelleroy, M.A., Buckles, R.G., Rosenkaimer, S., Tung, C. and Laver, M.B.; Jour. Appl. Physiol. 28: 227 (1970).
(5) Harkness, D.R., Roth, S., Goldman, P. and Goldberg, M.; Adv. Exp. Med. Biol. 28: 415 (1972).
(6) IBM Scientific Subroutines Package.
(7) Roughton, F.J.W., Otis, A.B. and Lyster, R.L.J.; Proc. Roy. Soc. B144: 29 (1955).
(8) Yoder, R.D., Seidenfield, A., Lopez, W.M. and Suwa, K.; Computers Biomed. Res. 6: 14 (1973).
(9) Margaria, R.; Clin. Chem. 9: 745 (1963).
(10) Radiometer A/S, "Provisional Instruction Manual for the Dissociation Curve Analyzer Type DCA 1", p. 19 (1972).

A POLYGONAL APPROXIMATION FOR UNSTEADY STATE DIFFUSION OF OXYGEN INTO HEMOGLOBIN SOLUTIONS

R. L. Curl and J. S. Schultz

Dept. of Chemical Engineering, The University of
Michigan

Ann Arbor, Michigan 48104

Several authors have grappled with the problem of developing methods for conveniently calculating the rate of oxygen transfer into a film of whole blood. The problem is very difficult because blood is a heterogeneous fluid and also the hemoglobin-oxygen reaction is so complex. In order to obtain manageable analytical equations one is forced to assume that blood is a homogeneous fluid and that the saturation of hemoglobin is determined by equilibrium with the local dissolved oxygen concentration. One might question whether the latter assumption is valid, especially in situations where the blood film is exposed to a new gas environment for very short periods of time as in the disc oxygenator. Even if changes in oxygen concentration are occurring slowly enough so that one might assume that hemoglobin is in chemical equilibrium with the local oxygen concentration, there is still a problem of obtaining an approximation for the non-linear oxygen hemoglobin saturation curve that can be used to generate valid analytical expressions for the unsteady state oxygenation process.

A typical hemoglobin oxygenation saturation curve is shown in Figure 1. A number of different approaches using straight lines have been tried to obtain approximate analytical estimates of oxygen diffusion rates and oxygen concentration profiles for the transient diffusion of oxygen into semi-infinite hemoglobin solutions (Roughton, 1959).

Here we have generalized the approach to the use of several line segments of arbitrary length and slope (polygonal) to approximate the saturation curve as shown in Figure 1. The placement of two lines, or digonal approximation, is somewhat arbitrary, but one may expect that a reasonable procedure is to qualitatively minimize the area between saturation curve and the straight line segments. This approximation is more general than those pre-

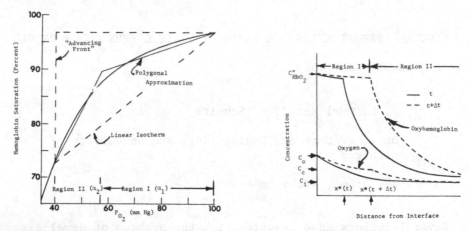

Figure 1. Typical non-linear saturation curve for hemoglobin solutions and some methods of linearization.

Figure 2. Schematic unsteady-state oxygen profiles for diffusion into a semi-infinite hemoglobin solution. Two-line polygonal approximation to saturation curve.

viously developed in the literature and can be shown to reduce to other models, i.e., advancing front and linear isotherm cases, when the slopes of the two lines are varied.

Although the analysis is presented here in the context of O_2 diffusion into hemoglobin solutions, the equations and results are valid to the extent of the linearized binding curve, for any solute which reversibly and instantly binds with a non-permeable solute.

The form of the linearized "oxygenation curve" or isotherm is given mathematically as

Region I: $C'_{HbO_2} = \alpha_1 C' + \Delta_1$ where $C_c \leq C \leq C_o$

Region II: $C''_{HbO_2} = \alpha_2 C' + \Delta_2$ where $C_i < C < C_c$

where C_{HbO_2} (moles/cm^3) is the concentration of bound oxygen (oxyhemoglobin), C is the concentration of free oxygen, C_c is the critical oxygen concentration where the linearized saturation lines intersect. α_1 and α_2 are the slopes of the lines above and below the critical O_2 concentration, as indicated in the figure, and the other constants Δ_1 and Δ_2 are given by the intersection of the lines with the coordinate axes.

Using this model, we have obtained a solution to the problem of transient oxygen diffusion into a semi-infinite film of blood. The mathematical development starts with the recognition that two zones (',") will exist in the film, meeting at distance coordinate x*, corresponding to the critical oxygen concentration C_c as shown in Figure 2. The dashed line indicates the temporal change in the oxygen and oxyhemoglobin concentrations within a short period of time. There is a discontinuity in the slope of the curves at x=x* corresponding to the change from region I to region II. The position of x* moves further away from the interface with time.

If the assumption is used that all forms of hemoglobin have the same diffusivity, then it can be shown that everywhere C_{HbO_2} + C_{Hb} = C_{Hbt}, that is, the total hemoglobin concentration is constant at all positions with time.

Because of the differences between the regions, it is necessary to write a set of equations for each region, based on a material balance for total O_2 in all forms:

Region I   Region II

$$(D + \alpha_1 D_{Hb}) \frac{\partial^2 C'}{\partial x^2} = (1+\alpha_1) \frac{\partial C'}{\partial t} \; ; \quad (D + \alpha_2 D_{Hb}) \frac{\partial^2 C''}{\partial x^2} = (1+\alpha_2) \frac{\partial C''}{\partial t}$$

@ x=0; $C' = C_o$ @ x = ∞ $C'' = C_c'$

@ x=x* ; $C' = C''$, and

$$(D + \alpha_1 D_{Hb}) \frac{\partial C'}{\partial x}\Big|_{x=x^*} = (D + \alpha_2 D_{Hb}) \frac{\partial C''}{\partial x}\Big|_{x=x^*}$$

@ t=0; $C' = C_i$ (x*=0) $C'' = C_i$

where D is oxygen diffusivity. In order to solve these equations simultaneously some non-dimensional notation and variables are introduced.

$$\phi = \frac{C-C_i}{C_o-C_i} \; ; \quad \eta = \frac{x-x^*}{\sqrt{4Dt}} \; ; \quad x^* = \beta \sqrt{4Dt}$$

$$R = D_{Hb}/D$$

Rewriting the differential equations and boundary conditions in terms of the new parameters gives:

$$\frac{d^2\phi'}{d\eta^2} + 2\left(\frac{1+\alpha_1}{1+R\alpha_1}\right)(\eta+\beta)\frac{d\phi'}{d\eta} = 0 \qquad \frac{d^2\phi''}{d\eta^2} + 2\left(\frac{1+\alpha_2}{1+R\alpha_2}\right)(\eta+\beta)\frac{d\phi''}{d\eta} = 0$$

@ x=0 $\eta=-\beta$; $\phi'=1$ @ x=∞ $\eta\to\infty$ $\phi''=0$

@ x=x* $\eta = 0$

$$(1+\alpha_1 R)\frac{d\phi'}{d\eta}\Big|_{\eta=0} = (1+\alpha_2 R)\frac{d\phi''}{d\eta}\Big|_{\eta=0} \;\; ; \;\; \phi' = \phi'' = \phi_c = \frac{C_c-C_i}{C_o-C_i}$$

These equations can be solved formally by letting p = dϕ/dη resulting in equations of the form

$$\frac{dp'}{d\eta} + 2\left[\frac{1+\alpha_1}{1+R\alpha_1}\right](\eta+\beta)\;p' = 0 \qquad \frac{dp''}{d\eta} + 2\left[\frac{1+\alpha_2}{1+R\alpha_2}\right](\eta+\beta)\;p'' = 0$$

which can be solved, together with the boundary conditions, to obtain an equation for the flux of oxygen into the surface at any time t

$$N = \frac{(1-\phi_c)}{\text{erfg}_3}\;\sqrt{(1+\alpha_1)(1+R\alpha_1)}\;(C_o-C_i)\;\sqrt{D/\pi t}$$

where the units of N are mole/cm^2-sec. The constant g$_3$ defined as g$_3$=$\beta\sqrt{(1+\alpha_1)/(1+R\alpha_1)}$ is given implicitly by

$$\left[\frac{1+R\alpha_2}{1+R\alpha_1}\right]\left[\frac{\phi_c}{1-\phi_c}\right] = \frac{\exp[\bar{\alpha}g_3^2][1-\text{erf}(\sqrt{(1+\bar{\alpha})}g_3)]}{\sqrt{(1+\bar{\alpha})}\;\text{erfg}_3}$$

where 1+$\bar{\alpha}$ = [(1+α_2)(1+Rα_1)]/[(1+α_1)(1+Rα_2)].

The total oxygen transferred into the film from time zero to time t is obtained by

$$V = \int_o^t N dt = \frac{(1-\phi_c)}{\text{erfg}_3}\;\sqrt{(1+\alpha_1)(1+R\alpha_1)}\;(C_o-C_i)\;\sqrt{4Dt/\pi}$$

The concentration profile of dissolved oxygen in regions I and II are given by the following two expressions

$$0 < x < x* \quad \phi' = 1 + (\phi_c-1)\frac{\text{erf}\left[\sqrt{(1+\alpha_1)(1+R\alpha_1)}\;\dfrac{x}{\sqrt{4Dt}}\right]}{\text{erfg}_3}$$

$$x* < x < \infty \quad \phi'' = \phi_c\frac{1 - \text{erf}\left[\sqrt{(1+\alpha_2)(1+R\alpha_2)}\;\dfrac{x}{\sqrt{4Dt}}\right]}{1 - \text{erf}\left[g_3\sqrt{1+\bar{\alpha}}\right]}$$

Oxyhemoglobin concentration profiles may then be obtained directly from the saturation curve. For desorption, these same relations apply except with α_1 and α_2 interchanged.

It can be seen that besides some constants both of these expressions are functions of the group $x/\sqrt{4Dt}$ and therefore one can plot a "universal" concentration profile using the single parameter $x/\sqrt{4Dt}$ as the absicissa for given values of the constants α_1, α_2, R, ϕ_c, etc.

We are now in position to compare various approximate models for unsteady-state equilibrium diffusion into semi-infinite hemoglobin solutions. We can see the relationship of previous models to the one presented here in terms of extreme values for the constants α_1, α_2, and R.

	Hemoglobin Fixed R = 0	Hemoglobin Mobile R > 0
"Advancing Front"	$\alpha_2=\infty$, $\alpha_1=0$, $\phi_c=0$ Hill (1929)	$\alpha_2=\infty$, $\alpha_1=0$, $\phi_c=0$ Roughton (1960)
"Linear Isotherm"	$\alpha_1>0$; $\phi_c = 1.0$, or $\alpha_1=\alpha_2$ Crank (1956)	$\alpha_1>0$, $\phi_c=1.0$, or $\alpha_1 = \alpha_2$

The equations for total oxygen transport into the film after a time t are generally of the form

$$V = k \, (C_o - C_i) \, \sqrt{4Dt/\pi}$$

where the factor k for our model is $(1-\phi)\sqrt{(1+\alpha_1)(1+R\alpha_1)}/\mathrm{erfg}_3$. The k factors for other models are summarized below.

	Hemoglobin Fixed	Hemoglobin Mobile
"Advancing Front"	$k = \dfrac{1}{\mathrm{erfg}_1}$ $g_1 e^{g_1^2} \mathrm{erfg}_1 = \dfrac{1}{M\sqrt{\pi}}$	$k = \dfrac{1}{\mathrm{erfg}_2}$ $\dfrac{\mathrm{erfg}_2 \exp[-g_2^2 \frac{(1-R)}{R}]}{1-\mathrm{erf}(g_2 \sqrt{R})} = \dfrac{1}{M\sqrt{R}}$
"Linear Isotherm"	$k = \sqrt{1+M}$	$k = \sqrt{(1+M)(1+RM)}$

where $M = \Delta C_{HbO_2}/(C_o - C_i)$ and ΔC_{HBO_2} is the difference in the concentrations of oxyhemoglobin in equilibrium with the C_o and C_i respectively. M is related to the other constants by the equation $M = \phi_c \alpha_2 + (1 - \phi_c) \alpha_1$. When M is large, the advancing front equation, for R = 0, simplifies to

$$V = 2(C_o - C_i)\sqrt{MDt}$$

An interesting comparison of the models can be made using some universal concentration profiles given by Diendorf et al.

They have calculated the universal oxygen concentration profiles for the diffusion of oxygen into a non-mobile hemoglobin solution for the exact saturation curve, and the limiting cases of the "advancing front" and "linear isotherm" models. These profiles are shown in Figure 3, along with the curve determined from the 2-segment polygonal approximation presented in this paper. There is not a very great difference in the profiles between the various approaches but our approximation lies on the exact solution within the accuracy of the lines on the figure.

A comparison of oxyhemoglobin concentration profiles is more sensitive for differences in the models; these are presented in Figure 4.

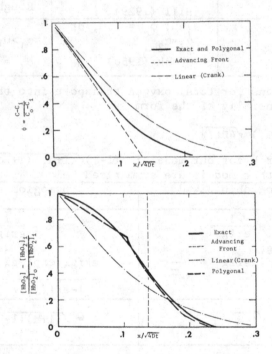

Figure 3. Comparison of dedimensionalized oxygen concentration profiles for various approximations and exact solution. Some curves reproduced from Diendorf, et al. Initial P_{O_2} = 40 mm Hg; surface P_{O_2} = 100 mm Hg.

Figure 4. Comparison of dedimensionalized oxyhemoglobin profiles for various approximations and exact solution.

The deficiencies of the advancing front and linear models in predicting the oxyhemoglobin concentration field show up clearly in this figure. However, the 2-segment approximation coincides remarkably well with the exact solution.

In these examples, where M is large, an estimate of the total absorbed O_2 according to each model, is given roughly by the area under the respective curve in Figure 4. It is apparent that all the models give about the same result for total O_2 uptake as the exact solution, indicating that the value of the method introduced here is primarily in obtaining an analytical approximation to the concentration profiles.

Higher order polygonal approximations are possible, but are considerably more complicated as additional simultaneous implicit transcendental equations are generated. These more exact approximations do not seem warranted especially when one considers the initial simplifications of homogeneous solutions and chemical equilibrium.

Acknowledgment: This work was partially supported by PHS Grant 15152 and a Research Career Development Award to J.S. Schultz, No. 1-K4-GM-8271.

References

1. Roughton, A.W., Progr. Biophys. Chem. <u>9</u>: 55-105 (1959).

2. Hill, A.V., Proc. Roy. Soc., London, Ser. <u>B104</u>: 39-96 (1929).

3. Crank, J. The Mathematics of Diffusion, Oxford (1956).

4. Diendorf, J.A., Lightfoot, E.H., and Solen, K.A., Chem. Eng. Symp. Ser., Vol. 67, No. 114: 75-87 (1971).

GASEOUS TRANSPORT IN HEMOGLOBIN SOLUTIONS

George C. Frazier and S. E. Shumate II

The University of Tennessee

INTRODUCTION

Gas exchange between blood and tissue is a complicated process because of factors such as the multiphase nature of the system (red cells, plasma, blood vessel walls, tissue), diffusion-chemical reaction interaction and associated non-linearities, and the irregular and sometimes poorly defined geometry of the system. In view of the complexity of such a system, one therefore seeks a reasonably simple model, or models, which provide a relatively good description of the gas exchange process.

A successful model must contain a number of elements which, when acting together, produce effects consistent with physical reality, both qualitatively and quantatively, at least in some measure. This work is directed toward the development of an understanding of one of the elements, the diffusion-reaction interaction process, which must be contained in the larger model. The approach taken is to develop a model of the diffusive-multireaction process in a single, well defined phase. One of the first tasks is the experimental verification of the model. Data concerning certain of the diffusion and reaction rate coefficients are required in this development, and the experimental method described here can be used to evaluate some of these parameters which are not currently available in the literature.

The purpose of this communication is to describe the present state of development of the model, the experimental approach, and to present some preliminary results for the diffusion of carbon dioxide through hemoglobin solutions.

MODEL AND MATHEMATICAL DESCRIPTION

The association reactions of the gaseous species of interest in living systems, such as oxygen and carbon dioxide, with carriers, such as hemoglobin and myoglobin, are second order and therefore nonlinear in general. An added complication is that several of these nonlinear reactions may be occurring simultaneously. For example, in the simultaneous transfer of carbon dioxide and oxygen through a hemoglobin solution, the minimum number of reactions which must be taken into account is five (Ulanowicz and Frazier, 1970) and if the full Adair scheme is taken into account, this number is increased to eight. An approach to this problem is to linearize the system-governing equations, and then to establish the range of conditions over which the subsequent solution is a reasonably good approximation of the physical process.

A linearization technique which is suitable to problems of this type was developed by Friedlander and Keller (1965) and applied to a single reaction, single phase system. This method was later extended to the multireaction, single phase and two phase systems (Ulanowicz and Frazier, 1968, 1970).

The linearization technique is by way of the chemical affinity function (Prigogine, 1965) and for a steady, planar, one dimensional, diffusional transport system, the result for the normalized flux, W_A, of the gaseous species through the system is:

$$W_A = [1 - \gamma^2(1 - \tanh \lambda\ell/\lambda\ell)]^{-1} \tag{1}$$

W_A is the augmentation, or facilitation, coefficient, ℓ is the film thickness, and for a reaction of the type

$$A + B \underset{k_2}{\overset{k_1}{\rightleftharpoons}} C \tag{2}$$

the parameters λ and γ are defined as follows:

$$\lambda^2 = k_1 \left(\frac{\bar{C}_B}{D_B}\right) \left[\frac{D_B}{D_A} + \frac{\bar{C}_A}{\bar{C}_B} + \frac{\bar{C}_A}{\bar{C}_C} \frac{D_B}{D_C}\right] \tag{3}$$

$$\gamma^2 = \bar{C}_B k_1/(D_A \lambda^2) \tag{4}$$

where the $\bar{C}_i$ are the mean concentrations, the D_i are the diffusion coefficients, and k_1 the forward rate coefficient. Refer to Frazier and Ulanowicz (1968) for details of the solution and the assumptions and restrictions upon which it is based.

Certain of the physical and chemical parameters of the system can be evaluated by interpreting appropriate experimental data in terms of eq. (1).

First, consider a series of diffusion experiments conducted on "thick" films, $\ell \rightarrow$ "large," in which the hemoglobin concentration is varied. For this case, eq. (1) reduces to:

$$W_{A|\ell \rightarrow \infty} = (1 - \gamma^2)^{-1} \tag{5}$$

After substituting for γ^2 from eqs. (3) and (4), one obtains:

$$W_{A|\ell \rightarrow \infty} = 1 + \left(\frac{D_A}{D_B} \bar{C}_A + \frac{D_A}{KD_C} \right)^{-1} \bar{C}_B \tag{6}$$

K is the equilibrium constant for reaction (2). Therefore, a plot of $W_{A|\ell \rightarrow \infty}$ vs. $\bar{C}_B$ should yield a straight line whose slope, S, is:

$$S = \left(\frac{D_A}{D_B} \bar{C}_A + \frac{D_A}{KD_C} \right)^{-1} \tag{7}$$

Thus, by knowing D_A and K, we can conduct two series of experiments, each with a different, known concentration of A, and calculate D_B and D_C from the two S's. Alternately, if the diffusion coefficients are known, one can calculate the equilibrium constant K.

It should be pointed out that one should not expect eq. (6) to hold over a wide range in the mean concentration of B, $\bar{C}_B$, because of the assumptions and restrictions upon which it is based. However, data of Wittenberg (1966) indicate that it is applicable for moderately low values of $\bar{C}_B$, for the oxygen-hemoglobin and oxygen-myoglobin systems.

One can also evaluate the forward rate coefficient k_1 in another series of experiments. But first, the parameter γ^2 must be determined. This is done by use of eq. (5), where $W_{A|\ell \rightarrow \infty}$ is measured in a single experiment at large film thickness. To get the forward rate coefficient, one next conducts a series of experiments using film thicknesses of intermediate values. Based on what is known about the properties of the ligand-Hb systems, it appears this series of experiments could be conducted with Millipore filters whose thicknesses are in the range of 100 μ, which are available commercially. The analysis and data interpretation are as follows.

For values of film thickness, ℓ, such that $\lambda\ell > 10$, eq. (1) reduces to:

$$(1 + \gamma^2)W_A = \left[1 - \frac{\gamma^2}{(1 + \gamma^2)\lambda\ell} \right]^{-1} \tag{8}$$

The parameter γ^2 is always less than unity, so the second term in brackets on the right will be small relative to unity for ℓ sufficiently large. Hence, one may expand the right-hand side in binomial series, yielding:

$$(1 + \gamma^2)W_A \doteq 1 + \frac{\gamma^2}{(1 + \gamma^2)\lambda} \cdot \frac{1}{\ell} \qquad (9)$$

Therefore, a plot of $(1 + \gamma^2)W_A$ vs. $1/\ell$ should yield a straight line of slope, S,

$$S = \frac{\gamma^2}{(1 + \gamma^2)\lambda} \qquad (10)$$

As γ^2 was determined as described above, one may now evaluate λ, and compute the rate coefficient, k_1, from eq. (3).

Although the above linearized theory gives predictions that are in qualitative agreement with observations, a rigorous test of its validity has not been made. Nor has the range of conditions for which it may be applicable been established. For some puzzling results in this connection for the CO + Hb system, refer to Frazier (1973).

The above theory can be extended to the multireaction case, as was done by Ulanowicz and Frazier (1970) for the O_2-CO_2-Hb system, although the results are not as readily interpreted in terms of the system properties as for the single reaction case considered above. The matrices and partitioned matrices associated with the multireaction case are, however, amenable to computation by electronic means.

EXPERIMENTAL APPROACH

The experimental arrangement used is similar to that of Wittenberg (1966), with certain modifications. This experiment is of the steady state, one dimensional type. Wittenberg produced thin films of Hb solutions by soaking the solution into Millipore filters of various thicknesses. The membrane is mounted in a two compartment diffusion cell and thermostated. The ligand, or carbon dioxide, is supplied to one compartment of the cell with a predetermined amount of a diluent such as nitrogen at a steady rate, and a second carrier stream sweeps the diffusing species from the opposite side of the cell. This second gaseous stream is analyzed for the diffusing substance by means of standard gas chromatography. One can then compute the flux of the diffusing species through the Hb film from the carrier stream composition, its known flow rate, and the cell parameters. A sketch of the experimental arrangement is provided as Figure 1.

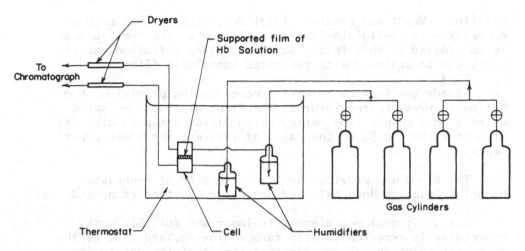

FIG. 1 SKETCH OF THE EXPERIMENTAL ARRANGEMENT

In order to test the linearized theory and to obtain some
of the system properties, it is necessary to conduct diffusion
measurements over a relatively wide range of hemoglobin solution
film thicknesses. Preliminary estimates indicate this range
should be from about 20 μ to 5 mm. It is necessary to cover this
range by two techniques. For the smaller part of this range, the
Millipore filters can be used. They are available in thicknesses
from about 25 μ up to several hundred microns. For the larger
part of the range, films of Hb solution can be supported on thin
membranes which are permeable to the gaseous diffusing species.
This technique was demonstrated to be feasible in our prototype
cell, where a General Electric MEM-213 film, one mil. thick,
was used to support Hb solutions up to about 5 mm in depth
(Shumate, 1973). Hence, measurements over a wide range in film
thicknesses are now possible. However, it is necessary to
calibrate the cell for both the Millipore filter and the supported
solution cases in order to arrive at the cell constants. We
calibrated the cell for the supported solution technique by diffusing carbon dioxide through a 2.5 mm layer of pure water. The
value of the diffusivity obtained for this system agreed within
2% of literature values, well within the experimental error.

Calibration of the chromatograph was done using certified
gas mixtures purchased from the Matheson Company.

Control of the thermostat is to ±0.01°C, using a Brookstat
regulator. Close control of the temperature is required in order
to prevent natural convection within the supported hemoglobin

solution. Wendt and Frazier (1973) demonstrated that natural
convection is negligible in liquid films when the temperature
is controlled at this level, so steady state diffusion experiments can be achieved with supported hemoglobin films.

Steady gas flow rates were produced using precision, Moore
Nullmatic pressure regulators. The flow rates were measured
using a soap bubble flow meter, which yields a reproducibility
of better than $\pm 0.5\%$ in the range of interest for these experiments.

The blood was provided by the University of Tennessee
Memorial Research Hospital, and processed by standard procedures.

One additional measurement is important for this work,
especially in experiments with carbon dioxide, and that is the
pH of the solution. A considerable amount of insight to the
diffusion-reaction process could be gained if the pH of the
surface layer of the solution could be measured during the
transport process. This is a problem we have not solved as
yet, but it may now be possible to make such measurements
with the microprobe of Bicher and Ohki (1972), and efforts
will be taken to incorporate a pH probe of this type in our
diffusion cell.

RESULTS FOR THE CARBON DIOXIDE-HEMOGLOBIN SYSTEM

One of the basic questions associated with gaseous exchange
between blood and tissue is that of whether or not the chemical
reactions can be considered to be proceeding at quasi-equilibrium
during the process. For the movement of oxygen to and from the
red cell, it appears that this may be a good assumption, as
estimates of the characteristic reaction-diffusion length, λ^{-1},
are in the range of the red cell dimension.

However, less is known about the carbon dioxide-hemoglobin
exchange process. One estimate (Ulanowicz and Frazier, 1970)
for the case of CO_2 diffusing through Hb solutions gave a value
of $\lambda^{-1} = \mathcal{O}(10^{-1})$ cm. This estimate was based on the assumption
that the rate coefficient for the carbon dioxide-hemoglobin
association is approximately the same as that for other carbon
dioxide-amide reactions, as the former rate coefficient apparently
has never been measured. If this estimate for λ^{-1} is at least
approximately correct, then it would appear that the CO_2 + Hb
reaction can be displaced appreciably from equilibrium during the
exchange process in the red cell.

In an effort to develop a better understanding of this situation, we have conducted a series of steady diffusion experiments,
with carbon dioxide passing through a 10 g/100 ml hemoglobin solution. The pH of our Hb solution after processing the blood was
6.3, and this was adjusted to 7.4 by the addition of 0.1M sodium
hydroxide solution. The temperature of the diffusion cell was
25°C, and the partial pressure of the carbon dioxide on the top
side of the cell was maintained at one atmosphere. The Hb solution film thickness was varied from nominal values of 1 to 5 mm.

The pH of the Hb solution was measured immediately after the
diffusion runs and was found to be 6.4, the reduction from the initial value of 7.4 being attributed to the presence of carbon dioxi-
dide in the solution. On standing in air, the pH of these solutions
gradually rose to a value of 8.1. As this value is higher than the
initial value of 7.4, it is apparent that some changes, probably
denaturation of the Hb, took place during the course of the experiments.

The carrier stream containing the carbon dioxide from the
botton of the cell was analyzed periodically using the gas chromatograph, and after steady state was achieved, the carbon dioxide
flux through the Hb solution was computed. The results are given
in the table below for two runs under similar conditions. These
fluxes were normalized to obtain the augmentation coefficients of
carbon dioxide through the film, computed by assuming the reaction
with Hb did not occur. The pH of 6.4 was used in order to estimate
the gas solubility under experimental conditions. The resulting
augmentation coefficients are shown in Figure 2.

As can be seen in Fig. 2, the scatter in the data is too great
in these prototype runs to draw firm conclusions at this time, but
the facilitation coefficient, $W_{CO_2} \equiv J_{CO_2}/J_{CO_2}$ inert appears almost
constant, rising only slightly in the film thickness range investigated. A direct comparison of the experimental and theoretically
calculated facilitation coefficients is not possible at this time,
as the stoichiometry of the carbon dioxide-hemoglobin reaction is
not well established. However, the lack of significant curvature
in the experimental data in Figure 2 indicates that the reactions

Nominal Film Thickness cm.	$J_{CO_2} \times 10^9$ g moles/(cm^{2}sec)	
	EXPT. 1	EXPT. 2
0.1	3.57	4.71
0.2	2.06	1.82
0.3	1.53	1.35
0.4	0.97	1.16
0.5	0.78	0.92

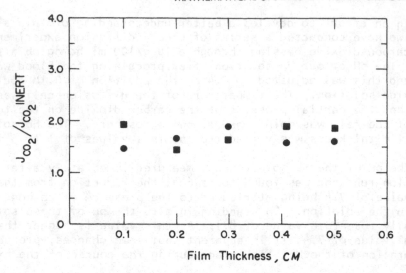

FIG. 2 Experimental Values of CO_2 Augmentation Coefficient

are essentially at equilibrium, which suggests that the overall
rate coefficient for these reactions is larger than that for
known carbon dioxide-amine reactions, which are in the range of
10^4 liter/(mole sec). Additional results and their interpretation will be provided in a later communication.

REFERENCES

Bicher, H. I., and Ohki, S. (1972), Biochim. Biophys. Acta., 255,
     900-904.

Frazier, G. C., and Ulanowicz, R. E. (1970), Chem. Eng. Sci., 25,
     549-577.

Frazier, G. C. (1973), "Some Inconsistencies in the Carbon Mono-
     xide-Human Hb System during Simultaneous Diffusion and
     Reaction," submitted for publication to J. Biol. Chem.

Friedlander, S. K., and Keller, K. H. (1965), Chem. Eng. Sci., 20,
     121-129.

Progogine, I. (1965), Thermodynamics of Irreversible Processes,
     John Wiley and Sons, New York.

Shumate, S. E. (1973), M. S. Thesis, The University of Tennessee.

Ulanowicz, R. E., and Frazier, G. C. (1970), Math. Biosci., 7,
     111-129.

Wendt, J.O.L., and Frazier, G. C. (May, 1973), "Measurement of
     Liquid Phase Diffusivities," scheduled to appear in I and EC
     Fundamentals.

Wittenberg, J. B. (1966), J. Biol. Chem., 241, 104-114.

MECHANOCHEMICAL PUMPING AS AN ADJUNCT TO DIFFUSION IN OXYGEN TRANSPORT

H. G. Clark and W. D. Smith

Department of Biomedical Engineering

Duke University, Durham, N. C. 27706

The interconversion of mechanical and chemical energy has been recognized for many years. Some examples of conversion of mechanical energy to chemical energy are the decrease in molecular weight of rubber due to milling (1) and the initiation of styrene polymerization by fracturing sodium chloride crystals. Muscle contraction is an example of transformation of chemical to mechanical energy.

Kuhn (2) and Katchalsky (3) proposed that the reversible swelling of gels in a chemically altered environment could be utilized as a source of useful work. When suitable constraints are applied to the gel geometry it can be demonstrated that ionizeable polymers both natural and synthetic can function as a muscle-like device which can undergo thousands of reversible expansion and contraction cycles. This occurs when the polymer is alternately exposed to acid and base (4), or for other systems when the muscle analog is alternately oxidized or reduced (5). Although the effect may be observed with most ionizeable polymers, both natural and synthetic, the significance of this mechanism to smooth muscle action or other physiological processes is not clear (6).

In blood there are ionizable proteins subjected to mechanical stress. The stress is related to the velocity gradient and therefore will be greatest for cells near the walls of large vessels, cells in flow around obstructions, and cells in flow through capillaries. In the first two cases the cells exposed to maximum stress are platelets and endothelial cells since red blood cells (RBC) tend to move away from the walls in large vessels. In the

latter case, particularly where RBC are moving through capillaries
which are smaller than the RBC diameter, a strain must be experienced by the cell membrane and/or the RBC contents. Is this
stress and its concommitant strain large enough to produce a
mechanochemical interaction and if so what are the physiological
consequences? It was to answer this question that the present
investigation was begun.

   If strain of RBCs in transit through capillaries produces a
chemical alternation what effect might this have? One possibility
is that a strain of the membrane would produce a pH shift. This
shift need not be symmetric on both sides of the membrane. A local
pH change external to the RBC would be rapidly dissipated by convection and diffusion (7). By contrast, a pH shift within the cell
would be controlled by cell membrane permeability. Thermodynamic
calculations have shown that during rest the hemodynamic power
is several orders of magnitude greater than the rate of free energy
change needed to produce a pH shift of 0.1 unit in the RBC interior.
During exercise this difference is still greater, so that even with
a low over-all efficiency, such a process is thermodynamically
feasible.

   Increased intracellular hydrogen ion would react with bicarbonate to produce CO_2 and water. This reaction is rapid due to
carbonic anhydrase. The extent of this reaction is, of course,
modified by concentrations of extracellular CO_2.

   Assuming that shear promotes ionization of intracellular acid
groups, the pH shift will be moderated in the afferent end of the
pulmonary capillaries due to bicarbonate conversion to CO_2. In
systemic capillaries lower bicarbonate and CO_2 in the afferent
end will allow a greater pH lowering and consequently have a
larger effect on hemoglobin capacity for oxygen.

   The effect of a shear dependency of intracellular pH would
alter oxygen affinity according to the well known Bohr shift.
"Pumping" of oxygen would then ensue as shown in Figure 1. A cell
entering the capillary would be strained and, assuming this
lowered pH, oxygen tension would be elevated. The RBC would then
lose oxygen to its environment as it traversed the capillary.
Upon exiting the capillary the strain would relax and a coupled pH
alteration would take place causing the oxygen tension to fall
back to a position on the equilibrium curve corresponding to the
new PCO_2. During this entire loop the driving force for oxygen
transfer from RBC to environment would be increased over what
would be expected without shear induced pH decreases. It is apparent that due to different morphology of pulmonary and systemic
capillary beds, different shearing rates, and different ambient
PCO_2 that this is not a symmetric system and that an exactly com-

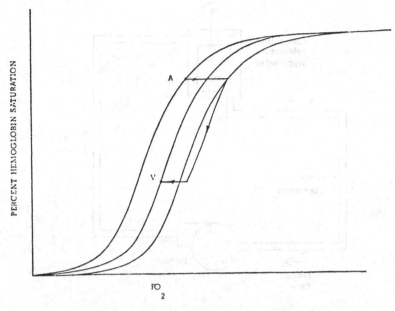

Figure 1. Scheme for Augmentation of Oxygen Transport between Arterial End (A) and Venous End (V) of Capillary

pensating decrease in driving force for oxygen transport would not be found between the systemic and the pulmonary capillary beds. The magnitude of such shifts and possible benefits to the energy budget of an animal are under study and will be reported later.

Direct experimental evidence for mechanochemical pumping has been sought in our laboratory by measuring oxygen solubility in blood during shearing. Two forms of appartus have been constructed. In the first of these blood is passed through a capillary slit (0.002 cm x 2.5 cm x 9 cm) at various flow rates. The blood is first saturated with a nitrogen, oxygen, CO_2 mixture and the entire system purged with the same gas. Oxygen electrodes connected to the shear zone and the unsheared zone measure oxygen tension. There is a small amount of oxygen consumed by the electrode, but otherwise there is no O_2 gradient. Due to the small volume of fluid in the shear zone a small increase in measured O_2 may be predicted with increasing fluid flow rates. When plasma, saline, or water is used as the fluid the small effect is indeed found. When fresh blood stabilized with sodium heparin or when a saline suspension of RBC is used, however, a substantially larger shift has been observed in the opposite direction. Although this could indicate a changing oxygen capacity of the hemoglobin in the RBC, the shift is not in the expected direction, since to be consistent with the data of Salzano et al (8) and others (9) an increased plasma

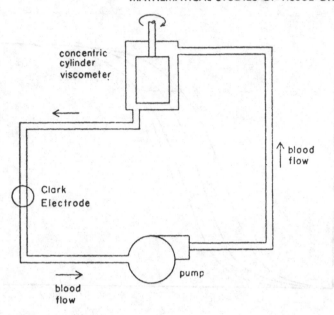

Figure 2. Apparatus for Measuring Oxygen Saturation under Shear

P_{O_2} was expected. The largest observed shift of oxygen tension
was above 20%. Similar observations were made on blood from
three donors. Measurements were made with wall shear stress in
the physiologic range. Although it produces a physiological flow,
the capillary slit system presents difficulties due to possible
flow and pressure artefacts, lack of temperature control, and difficulty in precisely defining the velocity gradient.

    In order to check for instrument artefact a second system is
being constructed. The Couette flow appartus, Fig. 2 holds a
volume of blood in an annulus between a stationary and a rotating
cylinder. A pump circuit produces secondary flow and allows the
blood to equilibrate with the sealed gas space. The whole system
is equilibrated with a gas mixture while both cylinders are
stationary. When shearing begins, if a mechanochemical effect
exists, a driving force for mass transfer will be created. A
rapid increase of oxygen tension in the blood would produce a
slower rise of oxygen pressure in the gas. As blood leaves the
shear zone through the pump circuit and the strain relaxes, a
decrease of oxygen tension should be measured which would correspond to the actual oxygen content of the blood as usually determined in low shear systems. A reservoir of osygen permits the
measurement of the entire equilibrium curve from a single sample.

Other evidence for possible mechanochemical pumping of blood gases exists in the literature. The <u>in vivo</u> data of Salzano et al (8) shows that a significant number of exercising subjects show alveolar PCO_2 higher than arterial PCO_2. Similar results have been obtained in other laboratories (9). This seems to imply active pumping of CO_2 under some conditions. Mechanochemical pumping would seem to be a simple mechanism for explaining these observations.

There are other phenomena which must be reexamined if the magnitude of a mechanochemical effect under physiological shear is sufficiently great to alter intracellular pH. These include increased oxygen transport rates, which have been attributed to elevated diffusion coefficients in shear (10) but may be due to increased driving force. Also, release of adenine nucleotides and other substances from red cells in subhemolytric shear fields (11) may be due to a mechanochemically altered Donnan potentials rather than altered pore geometry.

Investigations are continuing to determine the effect of rate of shear, temperature, partial pressure of gases, donor variability, anticoagulents, etc. on mechanochemically induced changes in blood cells.

Supported in part by NIH Grant 14228

References

1. Gallay, W.; Mastication and Plasticity p. 150-179 in "Chemistry and Techology of Rubber," C. C. Davis, ed., Reinhold (1937).

2. Kuhn, W.; Experientia, <u>5</u>, 318 (1949).

3. Katchalsky, A.; Experientia, <u>5</u>, 319 (1949).

4. Kuhn, W. and M. Thuerkauf; Kolloid Z., <u>184</u>, 114-17 (1962).

5. Kuhn, W.; Gazz. Chim. Ital., <u>92</u>, 951-67 (1962).

6. McClare, C. W. F.; J. Theor. Biol., <u>30</u>, 1 (1971).

7. Burton, A. C.; "Physiology and Biophysics of the Circulation," p. 56, Yearbook Medical Publisher, Chicago (1966).

8. Salzano, J. V., W. H. Bell, W. B. Weglicki, and H. A. Saltzman; "Proceedings of the Third Symposium on Underwater Physiology," C. J. Lambertson ed.; p. 351; The Williams and Wilkens Co., Baltimore (1966).

9.   Gurtner, G. H., S. H. Song and L. E. Farhi; Respir. Physiol.,
     7, 173-187 (1969).

10. Keller, K. H., Federation Proc., 30, 1591 (1971).

11. Blackshear, P. L. Jr., Federation Proc., 30, 1709 (1971).

AN ANALYSIS OF THE COMPETITIVE DIFFUSION OF O_2 AND CO THROUGH HEMOGLOBIN SOLUTIONS

S. R. Suchdeo, J. D. Goddard, and J. S. Schultz

Department of Chemical Engineering, University of

Michigan, Ann Arbor, Michigan 48104

1. INTRODUCTION

The subject of carrier-mediated or "facilitated" transport has received widespread attention in the chemical and biological literature, especially in conjunction with diffusion across thin films or membranes. One of the simpler conceptual models for the phenomenon is based on the postulate of ordinary molecular diffusion coupled with reversible homogeneous chemical reactions. The diffusional flux of a transferred chemical species is thereby "mediated" or altered by reversible combination with certain mobile "carrier" species which are indigenous to the film or membrane. Any such species, which is physically restricted to the carrier-mediated transport system, will be designated here as a "non-volatile" or "non-transferred" species.

In a recent general review of this subject [7], hereinafter referred to as Reference I, we have given a detailed discussion of the mathematical model for just such a carrier-mediated system.

The <u>steady-state</u> mathematical model involves a set of field equations, to describe the simultaneous diffusion and reaction in the interior of the system, which can be expressed in "vector" notation as

$$\underset{\sim}{D} \cdot \nabla^2 \underset{\sim}{C} = -\underset{\sim}{r}(\underset{\sim}{C}), \qquad (1)$$

where, for problems involving a total of S chemical species, $\underset{\sim}{C}$ and $\underset{\sim}{r}$ are S-dimensional "vectors" of species concentrations, and volumetric reaction rates respectively, and $\underset{\sim}{D}$ is an SxS matrix of diffusion coefficients (assumed to be constant here), with components C_s, r_s and D_s^m (s,m = 1,2,...S).

951

At the interface with the surroundings these equations are subject to an appropriate set of boundary conditions, representing the conditions of zero flux of the non-volatiles and some pre-scribed conditions on the boundary concentration or flux of the transferred species. As shown in Reference I, a system of this type is globally non-reactive in the steady-state which means that influx equals efflux for all transferred species. In the general problem, we may identify F volatile or transferred species and F' non-volatile species where S = F+F', and we may expect to have a number R < S of stoichiometrically independent reactions, corres-ponding to R' = S-R reaction invariants. As a consequence, there will be I' = R'-F = F'-R composition invariants which serve to characterize a particular carrier-mediated transport system (Refer-ence I).

Also in Reference I, a number of mathematical techniques for treating the above problem are summarized and evaluated. It appears that one of the most useful of these, for relatively rapid computa-tion, is an approximation technique based on an asymptotic, boundary-layer analysis for rapid or "near-equilibrium" reactions. This technique has its origins in an approximate "boundary-layer" method proposed by Kreuzer and Hoofd for treating a specific kinetic mo-del of myoglobin or hemoglobin facilitated diffusion of O_2 involving only one chemical reaction [1].

The method has subsequently been improved upon and applied to other single-reaction models of facilitated diffusion [2,8]. More-over, in Reference I it has been generalized and extended to handle problems involving multiple reactions and several transferred spe-cies but has not as yet been applied to specific systems.

Since the method has been shown to give useful and accurate approximations for flux in the case of a single reaction involving a single transferred species [2,8], we thought it worthwhile to pro-vide an application to an interesting physical system, in order to investigate competitive effects in the simultaneous transport of more than one species. In particular, as a model for the system O_2-CO-Hemoglobin (Hb), we shall treat the case of two (stoichiome-trically independent) reactions involving five species:

$$CO\uparrow + Hb = COHb \quad \text{and} \quad O_2\uparrow + Hb = O_2Hb, \tag{2}$$

where the arrows denote the volatile or transferred species. The reaction kinetics assumed here for the respective reactions are simply

$$\omega_1 \equiv -r_{CO} = k_1(C_1C_3 - \frac{1}{K_1}C_4), \quad [\text{moles/vol/time}],$$

$$\tag{3}$$

$$\omega_2 \equiv -r_{O2} = k_2(C_2C_3 - \frac{1}{K_2}C_5), \quad [\text{moles/vol/time}],$$

where the indices $s = 1,2,3,4,5$ refer, respectively, to the species CO, O_2, Hb, COHb and O_2Hb. Also, we shall assume a simple, "uncoupled" diffusion model ($D_s^m \equiv 0$ for $m \neq s$) with constant molecular diffusivities $D_s^m = D_s$ for $m = s = 1,2,\ldots,5$.

In the following section, we shall summarize the main ideas and the basic formulae involved, referring the interested reader to the appropriate references [2,7,8] for background and details of the underlying theory and derivations.

2. BASIC EQUATIONS

We are specifically concerned with the one dimensional form of (1), appropriate to diffusion through a homogeneous membrane or slab, of thickness L, say, lying in a spatial region $0 \leq x \leq L$. Also, we focus our attention on the case where the boundary concentrations on the volatiles have prescribed values, say $\bar{C}_v$ at $x = 0$, and $\underline{C}_v$ at $x = L$, for $v = 1,2,\ldots,F$. (Figure 1).

In the method referred to above, the membrane is conceptually treated as consisting of three rather distinct regions. The first is a central "equilibrium" <u>core</u> $0(\varepsilon L) < x < L - 0(\varepsilon L)$, wherein the "core" concentration fields of all species $\hat{C}(x)$, say, satisfy the conditions of reaction equilibrium,

$$\underline{r}(\hat{\underline{C}}) = 0, \tag{4}$$

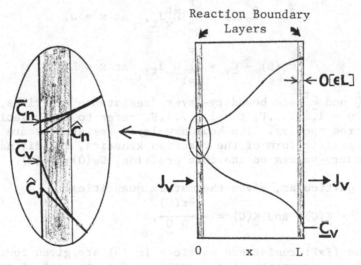

Reaction Boundary
Layers

Figure 1. A Schematic of Concentration Profiles Within the Film.

corresponding to $\omega_1 = 0$ and $\omega_2 = 0$ in (3). In addition, there are two reaction boundary-layer regions, $0 \leq x < O(\varepsilon L)$ and $0 \leq L-x < O(\varepsilon L)$, wherein the deviations from chemical equilibrium are important. Here, the dimensionless parameter, ε, assumed to be small in magnitude, represents a dimensionless singular-layer thickness of the form

$$\varepsilon = \sqrt{\frac{D^* C^*}{r^* L^2}} \tag{5}$$

where D^*, C^*, r^* are characteristic magnitudes of the diffusivity, concentration and reaction rate. In the example (3), of interest here, we might choose D^* to be any of the species diffusivities, with $r^* = k(C^*)^2$, where k is one or the other of the rate constants k_1, k_2 in (3) and C^* the total concentration of Hb (initially) present in the film. The theory we employ here is generally valid when the ε-parameters thus defined are (all) small, i.e. when all reactions are rapid compared to diffusion.

This situation is illustrated in Figure 1 where the concentration profiles for a typical volatile (v) and non-volatile (n) are shown schematically. Across the reaction layers there is, in general, a discontinuity in the species concentration fields, such that the equilibrium core concentration profiles $\hat{C}_v(0)$, $\hat{C}_v(L)$ differ from the prescribed boundary concentrations $\bar{C}_v = C_v(0)$ and $\underline{C}_v = C_v(L)$, by terms $O(\varepsilon L)$ or larger. This discontinuity may be regarded as a "driving force" for transfer across the reaction layer, and, as shown in Reference I, it is related to the transmembrane flux of volatiles by the set of linear equations

$$\bar{C}_v - \hat{C}_v(0) = \sum_{t=1}^{F} \bar{\Omega}_v^t J_t, \quad \text{at } x = 0,$$

$$\hat{C}_v(L) - \underline{C}_v = \sum_{t=1}^{F} \underline{\Omega}_v^t J_t, \quad \text{at } x = L, \tag{6}$$

where $\bar{\Omega}_v^t$ and $\underline{\Omega}_v^t$ are boundary-layer "resistance" matrices, and the indices $v = 1,2,\ldots,F$, $t = 1,2,\ldots,F$, refer to the volatile or transferred species. The boundary-layer resistances are determined by the specific form of the reaction kinetics, the diffusivities, and the boundary-values on the core profiles, $\hat{C}_v(0)$, $\hat{C}_v(L)$.

In particular, given the matrix quantities

$$\underset{\sim}{\Gamma}(C) = \underset{\sim}{D}^{-1} \cdot \underset{\sim}{K}(C) \quad \text{and} \quad \underset{\sim}{K}(C) = - \frac{\partial \underset{\sim}{r}(C)}{\partial \underset{\sim}{C}}, \tag{7}, (8)$$

then, the (FxF) resistance matrices in (6) are given formally by the appropriate components (i.e., restricted to the set of volatiles) of the following (SxS) matrices (with components $\bar{\Omega}_m^s$ and $\underline{\Omega}_m^s$),

$$\bar{\underset{\sim}{\Omega}} = [\Gamma\{\hat{\underset{\sim}{C}}(0)\}]^{-1/2} \cdot \underset{\sim}{D}^{-1} \quad \text{and} \quad \underset{\sim}{\Omega} = [\Gamma\{\hat{\underset{\sim}{C}}(L)\}]^{-1/2} \cdot \underset{\sim}{D}^{-1} \quad (9)$$

The matrix $\underset{\sim}{K}$ in (7) represents the linear-response coefficients or "pseudo-first-order" kinetic constants, for perturbations on a given concentration level $\underset{\sim}{C}$. Since $\underset{\sim}{K}$ is generally a singular matrix, the matrix-algebraic operations implied here in (9) are strictly formal. For the particulars of their exact interpretation, the reader is referred to Reference I. In the special case of interest here, two stoichiometrically independent reactions, the above relations can be greatly simplified since the reaction rates can be expressed as

$$\underset{\sim}{r} = \underset{\sim}{\nu}^1 \omega_1 + \underset{\sim}{\nu}^2 \omega_2 \quad (10)$$

where $\underset{\sim}{\nu}^1$ and $\underset{\sim}{\nu}^2$ are (vectors) of stoichiometric coefficients for the reactions; e.g., in the case of (3), the components ν_s^1, ν_s^2, $s = 1,2,\dots,5$, are given by

$$(\nu_s^1) = (-1, 0, -1, 1, 0),$$
$$(\nu_s^2) = (0, -1, -1, 0, 1) \quad (11)$$

Thus, the matrix $\underset{\sim}{\Gamma}$ of (7) can be expressed in the "dyadic" form

$$\underset{\sim}{\Gamma} = \underset{\sim}{\mu}^1 \underset{\sim}{\kappa}_1 + \underset{\sim}{\mu}^2 \underset{\sim}{\kappa}_2 \quad (\text{or } \Gamma_s^m = \mu_s^1 \kappa_1^m + \mu_s^2 \kappa_2^m). \quad (12)$$

where

$$\underset{\sim}{\kappa}_i = -\frac{\partial \omega_i}{\partial \underset{\sim}{C}} \quad (\text{or } \kappa_i^m = \frac{\partial \omega_i}{\partial C_m}) \quad (13)$$

and

$$\underset{\sim}{\mu}^i = \underset{\sim}{D}^{-1} \cdot \underset{\sim}{\nu}^i \quad (14)$$

for $i = 1,2$, where $\underset{\sim}{D}^{-1}$ is of course the matrix inverse for $\underset{\sim}{D}$. Owing to the simple form of (12), the resistance matrices in (9) can be expressed simply as

$$\bar{\underset{\sim}{\Omega}} = \underset{\sim}{\Omega}\{\hat{\underset{\sim}{C}}(0)\}, \quad \underset{\sim}{\Omega} = \underset{\sim}{\Omega}\{\hat{\underset{\sim}{C}}(L)\} \quad (15)$$

where

$$\underset{\sim}{\Omega}\{\underset{\sim}{C}\} = [a\underset{\sim}{\Gamma} - b\underset{\sim}{\Gamma}^2] \cdot \underset{\sim}{D}^{-1} \quad (16)$$

$$a = \frac{\lambda_1^{5/2} - \lambda_2^{5/2}}{(\lambda_1 \lambda_2)^{3/2}(\lambda_1 - \lambda_2)}, \quad (17)$$

$$b = \frac{\lambda_1^{3/2} - \lambda_2^{3/2}}{(\lambda_1 \lambda_2)^{3/2}(\lambda_1 - \lambda_2)}, \quad (18)$$

$\underset{\sim}{\Gamma}$ is given by (12), and λ_1, λ_2 are non-zero, presumably positive real eigenvalues or roots of the characteristic equation

$$\det(\underset{\sim}{\Gamma} - \lambda \underset{\sim}{1}) = 0. \quad (19)$$

In the relations (16) to (19), the quantities Γ, a, b,...,λ, are functions of the core concentrations $\hat{C}$, and the resistance matrices in (15) are, as indicated, to be evaluated at the extremities $x = 0,L$ of the core. In general, the two roots of (19) are given by

$$\lambda = \frac{\gamma_1^1 + \gamma_2^2 \pm \sqrt{(\gamma_1^1 - \gamma_2^2)^2 + 4\gamma_2^1\gamma_1^2}}{2} \tag{20}$$

where the γ_j^i are "scalar" products,

$$\gamma_j^i = \mu^i \cdot \kappa_j \equiv \sum_{s=1}^{S} \mu_s^i \kappa_j^s = \sum_{s=1}^{S} \mu_s^i \left(\frac{\partial \omega_i}{\partial C_s}\right)_{C_s = \hat{C}_s} \tag{21}$$

$i = 1,2$, $j = 1,2$. For the specific example (3), of interest here, these are found by (3), (11) and (13), to be

$$\gamma_1^1 = k_1\left(\frac{\hat{C}_3}{D_1} + \frac{\hat{C}_1}{D_3} + \frac{1}{D_4 K_1}\right) , \quad \text{and} \quad \gamma_1^2 = \frac{\kappa_1 \hat{C}_1}{D_3} \tag{22},(23)$$

with the remaining terms, γ_2^2 and γ_2^1 obtained from (22) and (23) by the permutation of indices:

$$1 \rightleftarrows 2, \ 4 \to 5, \ (3 \to 3)... \tag{24}$$

We recall that the D_s are the individual species diffusivities, $s=1,2,...$

In the general carrier-mediated diffusion problem, one has to terms $O(\varepsilon)$ the further relations for transmembrane flux of (F) volatile species

$$J_v = \sum_{s=1}^{S} \sum_{m=1}^{S} z_v^s D_s^m \left(\frac{\hat{C}_m(0) - \hat{C}_m(L)}{L}\right), \ v = 1,2,...,F, \tag{25}$$

in terms of driving forces across the core (Ref. I). The coefficients z_v^s ($s = 1,2,...,S$; $v = 1,2,...,F$) are "transport" numbers which, as shown in Reference I, are determined uniquely by the reaction stoichiometry. In the example (3), with volatiles $v = 1,2$, the relations (25) reduce to the simple form

$$J_1 = D_1\left(\frac{\hat{C}_1(0) - \hat{C}_1(L)}{L}\right) + D_4\left(\frac{\hat{C}_4(0) - \hat{C}_4(L)}{L}\right), \tag{26}$$

with J_2 obtained by the indicial permutation (24).

In the limit of infinitely rapid reactions ($\varepsilon \to 0$), Equation

(26) provides an expression for the equilibrium flux. In this case, the boundary values $\hat{C}_V(0)$, $\hat{C}_V(L)$ of the core profiles are merely equated to the imposed interfacial concentrations $\bar{C}_V$, $\underline{C}_V$, which corresponds to the neglect of terms on the right-hand side of (6). The non-volatile concentrations $\hat{C}_n(0)$, $\hat{C}_n(L)$ appearing in (26) must then be determined from the equilibrium relations (4) and a set of auxilliary relations, which arise as constraints on the amount present and the flux of certain reaction invariants. In general, there are a number $I' = R' - F = F' - R$ integral constraints and a set of R' flux conditions. For the example (3) at hand, $R' = 3$, $I' = 1$, and the relations in question can be taken as the constraints on the total amount of hemoglobin in all forms (Hb, COHb, O2Hb):

$$LC_{tot}\{Hb\} = \int_0^L (\hat{C}_3 + \hat{C}_4 + \hat{C}_5)dx = LC_0, \qquad (27)$$

and on the total flux of Hb, O2, and CO:

$$J_{tot}\{Hb\} = -[D_3\frac{d\hat{C}_3}{dx} + D_4\frac{d\hat{C}_4}{dx} + D_5\frac{d\hat{C}_5}{dx}] = 0 \qquad (28)$$

$$J_{tot}\{O_2\} = -[D_2\frac{d\hat{C}_2}{dx} + \frac{d\hat{C}_5}{dx}], \qquad (29)$$

a constant independent of x, and

$$J_{tot}\{CO\} = -[D_1\frac{d\hat{C}_1}{dx} + D_4\frac{d\hat{C}_4}{dx}], \qquad (30)$$

also a constant independent of x.

It can be shown (Reference I) that these relations together with the equilibrium relations from (3), namely,

$$\hat{C}_4 = K_1\hat{C}_1\hat{C}_3, \quad \text{and} \quad \hat{C}_5 = K_2\hat{C}_2\hat{C}_3, \qquad (31)$$

provide a sufficient number of equations to determine completely the equilibrium core profiles $\hat{C}_s(x)$, once the boundary conditions on volatiles $\hat{C}_V(0)$, $\hat{C}_V(L)$ are specified. Even in the case of complete equilibrium, where the latter can be equated to $\bar{C}_V$, $\underline{C}_V$, the solution is not a simple analytical task, essentially because of the integral constraint (27). However, the problem is greatly simplified if we assume that the diffusivities of the non-volatile or "carrier" species Hb, COHb, and O2Hb are all equal, i.e.,

$$D_3 \equiv D_4 \equiv D_5 = D_0, \text{ say}, \qquad (32)$$

For then, it follows from (27) and (28) that

$$\hat{C}_3 + \hat{C}_4 + \hat{C}_5 = C_0 \qquad (33)$$

everywhere in the film, $0 < x < L$. As a consequence, the equilibrium relations (31) give $\hat{C}_3$, $\hat{C}_4$, and $\hat{C}_5$ in terms of $\hat{C}_1$ and $\hat{C}_2$ as

$$\hat{C}_3 = C_o/[1 + K_1\hat{C}_1 + K_2\hat{C}_2] \tag{34}$$

with the corresponding forms for $\hat{C}_4$ and $\hat{C}_5$ following from (31).

Therefore, the expression for the flux of CO in (26) becomes

$$J_1 = \frac{D_1 \Delta \hat{C}_1}{L} + \frac{D_o K_1 C_o}{L} \Delta \left(\frac{\hat{C}_1}{1 + K_1\hat{C}_1 + K_2\hat{C}_2} \right) \tag{35}$$

where, for any quantity $C(x)$, $\Delta C \overset{\text{def}}{=} C(0) - C(L)$. A similar expression is obtained for the flux J_2 of O_2, by the permutation of indices (24). The usual type of expressions for the equilibrium flux are then obtained immediately, by taking $\hat{C}_v(0) = \bar{C}_v$ and $\hat{C}_v(L) = \underline{C}_v$ for $v = 1,2$.

Now, the main objective of this work is to account for departures from reaction equilibrium. Within the present method, this is accomplished by employing values for the quantities, $\hat{C}_v(0)$ and $\hat{C}_v(L)$, as obtained from Equations (6), instead of the imposed boundary concentrations $\bar{C}_v$, $\underline{C}_v$. This leads, then, to a set of _implicit_ relations, Equations (15) to (18), (20) and (22), for the resistance matrices appearing in (6), together with Equations of the type (35), for the fluxes J_1 and J_2 in (6). As emphasized in Reference I, this generally necessitates some type of iterative numerical solution to the equations at hand. This has been carried out in the present study and some of the results are discussed in the following section.

In closing here, we note that the formulae presented reduce in the limit of a single permeant or a single reaction to known results that have been applied in other contexts [2,8]. Generally these have been shown to agree well with "exact" numerical solutions of the equations.

3. RESULTS, DISCUSSION AND CONCLUSIONS

The set of equations cited in the preceding paragraph was solved iteratively on IBM-360. The iteration procedure was initiated under conditions such that the deviations from the equilibrium concentrations were small, that is,

$$\Omega \cong 0, \text{ in Equations (6).} \tag{36}$$

Subsequently, the parameters of the problem, carbon monoxide concentration $\bar{C}_1$, etc., were gradually altered, and the iteration process for each new set of parameters was initiated at core concentrations obtained from the previous set. The iteration procedure was assumed to have converged when the core concentrations (at the boundaries) obtained at the end of an iteration were within 0.01% of those at

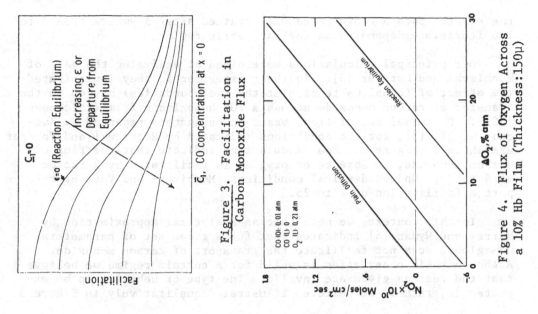

Figure 3. Facilitation in Carbon Monoxide Flux

Figure 4. Flux of Oxygen Across a 10% Hb Film (Thickness:150μ)

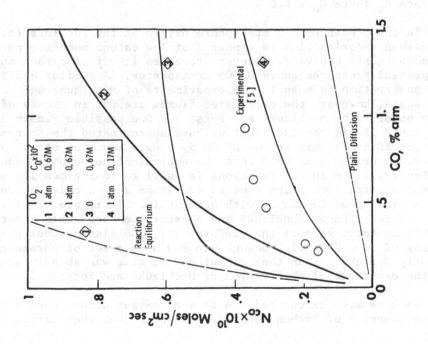

Figure 2. Flux of Carbon Monoxide Across a 10% Eb Film (Thickness:150μ)

the start.   Such a convergence was obtained after anywhere from 3 to 100 iterations depending on the parametric regime.

Our principal calculations were aimed at analyzing the data of Mochizuki and Forster [5].   In their experiments, they investigated the effect of (equal O_2 tension at the two boundaries) oxygen on the transfer of carbon monoxide across a 10% hemoglobin film (thickness:]50μ).   Our model calculations based on parameters reported by Kut-chai et al. [3], for the conditions indicated by Mochizuki and Forster are shown in Figure 2.   The calculations indicate that the flux of carbon monoxide, in absence of oxygen, is facilitated by a factor of 4 to 35.   Under identical conditions, Mochizuki and Forster report a facilitation of 2 to 25.

In this context, we note that an analytical approximation by Murray and Wyman [6] indicates that for a given set of parameters, hemoglobin does not facilitate the transport of carbon monoxide. Although their observation is valid for a certain regime we believe that the reasons given are invalid.   The type of behavior to be ex-pected is, rather, we believe, illustrated qualitatively in Figure 3.

The negligible facilitation observed in experiments by Witten-berg [9] was probably a result of high carbon monoxide partial pressure ($\bar{C}_1$) and not, as Murray and Wyman indicate, due to a condi-tion of rapid nearly-irreversible reaction, ε and $1/K_o \to 0$.   Indeed, for ε, $1/K_o \to 0$, one should observe a high degree of facilitation for moderate $\bar{C}_1$ (here $K_o = K_1 C_o$).

In the presence of 1 atmosphere oxygen at the two interfaces, the carbon monoxide flux is hindered at low carbon monoxide partial pressure.   This behavior, however, reverses itself at carbon monox-ide partial pressures above 0.005 atmospheres.   A similar qualita-tive observation is seen in the experiments of Mochizuki and Forster. Once again, however, the calculated fluxes are far in excess of those observed by Mochizuki and Forster.   One possible reason for this could lie in the fact that we have approximated the four-step reaction of hemoglobin and O_2 or CO by single reaction step.   Also, we have accordingly assumed that the equivalent hemoglobin concen-tration available in the reactions is equal to four times its actual molar concentration.   Also plotted in Figure 2 is a calculation based on 1:1 binding of CO and O_2 with hemoglobin.   It is seen that the experimental data of Mochizuki and Forster lie above this curve. In this context, we note that LaForce [4], who also assumes a 1:1 binding of CO and O_2 with hemoglobin and uses a set of parameters slightly different than that of Kutchai et al., was able to account for the experimental observation of Mochizuki and Forster.

We have also calculated the flux of oxygen across such a film in the presence of carbon monoxide.   We note that when carbon monoxide is maintained at 1% of an atmosphere and zero atmospheres, up-stream and downstream respectively, the flux of oxygen is greatly hindered, as is indicated in Figure 4. It appears that a partial pressure driving force of greater than 0.08 atmospheres is necessary in order to obtain a flux of oxygen in the direction of its gradient. The corresponding flux under conditions of chemical reaction equi-librium is also shown in Figure 4.

In concluding, we have found that present boundary-layer tech-nique provides a relatively rapid computation of carrier-mediated diffusion with competing reactions. We have illustrated the method with the example O_2-CO-Hb, for which some interesting effects are predicted. Using published kinetic and diffusion constants for the reactions, predictions qualitatively similar to the experimental measurements of Mochizuki and Forster [5] were obtained. In con-trast to the analysis of Murray and Wyman, we find that under appro-priate experimental conditions hemoglobin should facilitate the transport of carbon monoxide. Also, we find that at CO partial pressures above 1%, O_2 should further accelerate the transport of carbon monoxide. However, at low CO partial pressures, the presence of O_2 may inhibit the flux of CO. Further calculations, not includ-ed here, show that in the presence of O_2, the flux of CO is not as sensitive to downstream back pressure of CO, as is the case when O_2 is absent. Also, one can delineate conditions under which either CO or O_2 can diffuse against their concentration gradients due to interactions with a common carrier.

REFERENCES

1. Kreuzer, F. and J.L.C. Hoofd, Respir. Physiol., 8:208 (1970).
2. _____, Respir. Physiol., 15:104 (1972).
3. Kutchai, H., J.A. Jacquez and F.J. Mathers, Biophys. J., 10:38 (1970).
4. LaForce, R.C., Trans. Faraday Soc., 62:1458 (1966).
5. Mochizuki, M. and R.E. Forster, Science, 138:897 (1962).
6. Murray, J.D. and J. Wyman, J. Biol. Chem., 246:5093 (1971).
7. Schultz, J.S. and J.D. Goddard, (Reference I), Paper to appear (1973).
8. Smith, K.A., J.H. Meldon and C.K. Colton, AIChE J., 19:102 (1973).
9. Wittenberg, J.B., J. Biol. Chem. 241:104 (1966).

ACKNOWLEDGEMENTS

This work has been partially supported by P.H.S. Grant No. GM15152 and also by a P.H.S. Research Career Development Award to J.S. Schultz (No. 1 K04 GM 08271) from the Institute of General Medicine Sciences.

One of us (J.S.S.) would like to express gratitude to Professor F. Kreuzer (University of Nijmegen) for the interest generated in this problem.

OXYGEN MASS TRANSFER RATES IN INTACT RED BLOOD CELLS

B. H. Chang, M. Fleischman[1], and C. E. Miller[2]

Chemical Engineering[1] and Civil Engineering[2]

University of Louisville, Louisville, Kentucky

The overall objective of the preliminary work reported here is to characterize oxygen transport in red blood cells based on intact RBC data. The experimental method and analysis used is basically an engineering approach involving mass transfer rates and coefficients. However, the data can also be used in a transport phenomena approach to evaluate O_2 transport and reaction properties in RBC's. The long range information attainable from this study should be applicable to basic knowledge of O_2 transport in blood and in areas where RBC resistance to O_2 is a factor, e.g., oxygenators (1), in vivo mass spectrometer blood-gas analysis.

EXPERIMENTAL PROCEDURE AND DATA

Sheep cells were resuspended in buffered, isotonic saline solution to eliminate O_2 metabolism. The suspension was deoxygenated by bubbling N_2 gas which has been passed thru pyrogallol. O_2 was then bubbled thru the well stirred system, and the O_2 concentration in the saline continuously monitored with an oxygen electrode. Constant blood pCO_2 was maintained by all gases having the same fixed CO_2 percentage. The gases were also saturated with H_2O vapor prior to passing thru the blood.

Each experiment consisted of three different oxygenation runs; 1) saline only to provide the O_2 transfer coefficient from gas to saline 2) $NaNO_2$ treated RBC suspension to prevent O_2-Hb reaction and thereby yield information on diffusive cell resistance 3) suspended normal RBC's in which O_2-Hb reaction is occurring. Typical data is shown in Figure 1.

963

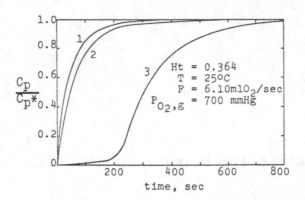

FIGURE 1. Typical Oxygenation Data for RBC in Saline Suspension

ANALYSIS OF DATA

One method of using the data derives from a balance on plasma oxygen

$$V_p(dC_p/dt) = \underbrace{K_p V_p(C_p^*-C_p(t))}-\underbrace{N_c(t)\,n\,V_p/(1-H)} \tag{1}$$
$$\underbrace{}_{\text{accumulation}}\ \underbrace{}_{\text{input from gas}}\ \underbrace{}_{\text{uptake by RBC's}}$$

For saline only, N_c is zero and the solution of equation 1 with $C_p(0)$ being experimentally set to zero is

$$\ln\{C_p^*/(C_p^*-C_p(t))\} = K_p t \tag{2}$$

Thus K_p can be determined by plotting the data according to equation 2, and a typical value is 0.0172 sec^{-1}.

In the presence of cells, equation 1 can be solved for the experimental O_2 uptake rate of both nitrited and reactive cells.

$$\underbrace{N_c(t)}_{\text{RBC } O_2 \text{ uptake}} = \frac{(1-H)}{n}\ \{K_p(C_p^*-C_p(t))-(dC_p/dt)_t\} \tag{3}$$

Integration of equation 3 yields the experimental, average O_2 accumulation of a cell. For inactive cells, this value can be used to yield the experimental average intracellular free O_2 concentration

$$C_c(t) = M_c(t)/V_c = \int_0^t N_c(t)dt/V_c \tag{4}$$

Typical experimental values thus far obtained are 1.24×10^{-9} moles O_2/sec for the maximum N_c and in the order of 3×10^{-8} moles/cell for M_c.

The value of M_c when equilibrium between the inactive cell suspension and gas phase is reached, is a direct measure of O_2 RBC solubility. Solubility can also be determined from reactive cell data if S_{Hb} is 100% at gas-suspension equilibrium.

$$M_c^* \text{ (dissolved)} = \underbrace{M_c^*(\text{exptl.})} - B \times Hb_T \qquad (5)$$
$$\text{reacting cells, } S = 1.0$$

A preliminary evaluation of the nitrited cell data indicates an RBC O_2 solubility of the same order as Van Slyke's estimate (2).

Basic RBC O_2 transport properties such as membrane permeability, intracellular diffusion coefficient and rate constants for the O_2-Hb reaction should be determinable by using the experimental data in an appropriate model of O_2 transport in the RBC. A possible model (3)is

$$\partial C_c/\partial t = D_c \nabla^2 C_c + R_{O2} \qquad (6)$$

$$\partial C_{HbO_2}/\partial t = -R_{O2} = f\ (Hb_T, S, C_c, k_f, k_r) \qquad (7)$$

with the appropriate condtions including those at the cell membrane involving h_m. For nonreacting cells, R_{O2} is zero and equation 6 can be solved for $N_c(t)$ and $C_c(t)$ as a function of D_c and h_m. The correct values of D_c and h_m should yield agreement between the predicted and experimental values of $C_c(t)$ and $N_c(t)$ for nitrited cells (3).

A similar approach can be used with the data from reacting cell suspensions and the solution to equations 6 and 7, to evaluate the rate constants for various kinetic models of the O_2-Hb reaction,e.g.

$$Hb + O_2 \underset{k_r}{\overset{k_f}{\rightleftharpoons}} HbO_2 \qquad (8)$$

Additional or alternatively the experimental cell O_2 uptake rates attributable to reaction can be determined and related to the kinetic model.

$$\qquad (9)$$
$$N_c^{(R)}(t) = \{\underbrace{N_c^{(3)}(t)} - \underbrace{N_c^{(2)}(t)}\} \text{ exptl.} = V_c R_{O2} = -V_c R_{HbO2}$$

uptake by reaction	total uptake by reacting cells	uptake by inactive cells

HOMOGENEOUS SYSTEM APPROACH

In the previous analysis, the RBC was treated as an O_2 sink, with and without reaction. Alternatively, the typical data previously shown can be empirically correlated as a homogenous system, and then deriving cell parameters. For the homogeneous assumption, equation 1 becomes

$$(dC_p/dt) = K_i' \ (C_p{}^* - C_{pi}(t)), \ i = 1,2,3 \qquad (10)$$

where $i = 1$ for saline only, $i = 2$ for nitrited cell suspensions. K_i' is thus an empirical overall mass transfer coefficient for the suspension of red cells based on plasma O_2 and can be evaluated by the same type of equation as equation 2. For saline only, K_i equals K_p as previously determined. The data as correlated by equations 2 and 10 are shown in figure 2. An interesting feature is that the value of K_3 after S_{Hb} is 1.0 at t_R, is close to the value of K_2 for inactive cells.

The experimental values of K_i' are a composite of K_p and a $K_{cell}(K_c)$, both for inactive and reactive cells. Considering saline as the continuous phase and RBC's as the dispersed phase, K_c can be estimated from (4)

$$K_i = K_p \ \{1 + \frac{2H \ (1-K_p/K_{ci})}{2(K_p/K_{ci}) + 1}\} \ /\{1 - \frac{H(1-K_p/K_{ci})}{2K_p/ \ K_{ci} +1}\} \quad (11)$$

Based on the data shown in figure 2, K_c for the nitrited cell is calculated to be $0.00752 \ sec^{-1}$. An interpretation has not been made of the meaning of K_c, but it is some measure of the cells resistance to diffusive O_2 transfer.

O_2 transfer in the cell can be considered to occur thru three resistances in series; plasma film, cell membrane, and cell interior. K_c is some measure of the overall conductivity of these three steps, and the plasma film portion k_p, should be calculable from (5)

$$Sh = k_p D/CDim = 2.0 + 0.60 \ Re^{0.5} Sc^{0.33} \qquad (12)$$

The experimental values of K_c and k_p can possibly be used in the model of intracellular O_2 transport, equation 6, to evaluate D_c and h_m.

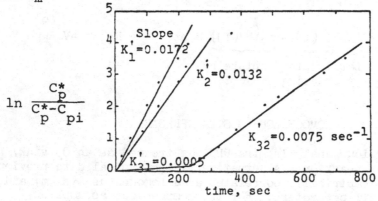

Figure 2.   Empirical Total O_2 Transfer Coefficients

For reactive or normal cells, K_{31}' and K_{cR} as determined from K_{31}' in figure 2 (O_2-Hb reaction occurring) and equation 11, are empirical measures of the overall conductivity of a reacting blood cell suspension, and as such might be useful for design purposes or rapid estimation of O_2 transfer rates. Interpretation of K_c is not presently clear, and it will depend upon the kinetic model chosen for the intracellular O_2-Hb reaction, e.g. equation 8. K_{cR} seems to be some measure of the forward reaction rate because deoxygenation experiments run in the same manner as the oxygenation experiments, yielded different results for reacting cells. Conversely, the experimental K_{cR} for deoxygenation of reactive RBC's should be a measure of the reverse reaction rate. This seems similar to some of Roughton's work (6).

Admittedly, the approach and results reported here are preliminary, and ultimately data will have to be taken on human cells at normal body conditions. Further experimental efforts will include continuous measurement of hemoglobin saturation to help analyze reacting cell data. To better evaluate the effect of the cell plasma film on RBC O_2 transfer, experiments will be run at different stirring speeds to vary the degree of turbulence. Further analysis will be necessary to interpret the empirical mass transfer coefficients attainable by this method. Experimental mass transfer coefficents will be compared to values which can be calculated from information in the literature (7).

NOMENCLATURE

a	- interfacial area per unit volume
B	- O_2 bound to fully saturated Hb, moles O_2/mole Hb
C_p, C_p^*	- saline dissolved O_2 concentration at t and in equilibrium with gas phase O_2, moles/unit vol.
$C_c, \overline{C}_c$	- free O_2 concentration in RBC, average concentration
C_{HbO_2}	- intracellular HbO_2 concentration
D_c	- intracellular O_2 diffusion coefficient
h_m	- RBC membrane O_2 permeability
H	- fractional hematocrit
Hb_T	- total hemoglobin concentration moles/unit volume of blood
K_p, K_c, K_{cR}	- overall coefficient for O_2 transfer from gas to saline solution, time^{-1}($K_p = K_L a$), general cell coefficient with O_2-Hb reaction
k_p	- mass transfer coefficient for plasma film of O_2
k_f, k_R	- forward and reverse rate constants for O_2-Hb reaction
M_c	- moles of O_2 accumulated by RBC
N_c	- O_2 uptake rate of RBC, moles/time
N	- red cell count, cells per unit volume of blood
RBC	- red blood cell

Re   - Reynolds number, dimensionless
RO_2, R_{HbO_2} - rate of O_2 depletion and HbO_2 formation by intracellular
                 O_2-Hb reaction, moles/time
S   - fraction or percent of total hemoglobin as oxygemoglobin
Sh   - Sherwood number, dimensionless
Sc   - Schmidt number, dimensionless
t   - time
V_p, V_c - total saline volume, volume of an RBC

REFERENCES

1. Spaeth, E. E. in Blood Oxygenation, D. Hershey, ed., Plenum
    Press (1970).

2. Sendroy & Van Slyke, J. Biol. Chem. 105, 597 (1934).

3. Fleischman & Hershey, CEP Symp. Ser. 99, Vol. 66, 77 (1970).

4. Van Vlack, Elements of Materials Science, 2nd Ed., Ch. 11,
    Addison-Wesley (1967).

5. Seagrave, Biomedical Applications of Heat and Mass Transfer,
    Iowa State Univ. Press (1971).

6. Roughton, F. J. W. in Progress in Biophysics and Biophysical
    Chemistry, Butler & Katz, Eds., Pergammon Press.

7. Middleman, Transport Phenomena in the Cardiovascular System,
    Wiley-Interscience (1972).

EFFECT OF CARBONIC ANHYDRASE ON THE FACILITATED DIFFUSION OF CO_2 THROUGH BICARBONATE SOLUTIONS

Shyam Suchdeo and Jerome S. Schultz

Dept. of Chemical Engineering, The University of Michigan

Ann Arbor, Michigan 48104

Over the last 10 years, several groups have given experimental evidence that carbon dioxide diffusion in aqueous solutions can be augmented by the presence of base, buffers, and catalysts for the hydration reaction of carbon dioxide with water (1,2,3). Although the mechanism for this augmentation has been presumed to be a carrier type of facilitation, no previous experimental study has covered a wide range of facilitation factors (augmented flux/Fick's Law flux) and at the same time presented a solution to the mathematical model which can be used to calculate the expected transport rates. This paper summarizes some of our efforts in this direction.

The experimental system employed here was similar to that used previously (4,5), that is, a liquid film is prepared by soaking a thin sheet of porous filter paper in the desired solution which is then interposed between two gas chambers. The gases metered through both chambers are analyzed with a gas chromatograph to determine the net transport rate of CO_2 across the liquid membrane.

In our analysis, the interfacial concentrations of dissolved CO_2 are assumed to be in equilibrium with the CO_2 partial pressures in the respective gas compartments. Also, we assume that only simple diffusion and chemical reaction processes are occurring in the film, and in addition that electrical field effects produced by concentration gradients within the film are negligible. Then, for each species (i) within the film, the unsteady-state reaction-diffusion material balances take the form:

$$\frac{\partial C_i}{\partial t} = D_i \frac{\partial^2 C_i}{\partial x^2} + r_i \tag{1}$$

where C = concentration, t = time, D = diffusivity, x = distance
($0 \leq x \leq L$) and r_i is the rate of production of species i by all
chemical reactions.

Table I gives the chemical reactions and species that are
involved in CO_2 facilitation system. A total of six reactions
involving 9 species can be identified. This implies that 9 equations of the type given in Equation (1) are necessary to define and
solve the problem.

However, other considerations show that further restrictions
on the steady-state model permit a significant simplification in
the mathematics. First, not all of the reactions are stoichiometrically independent, e.g. reaction 2 is equivalent to reaction
(1+3-6). Therefore, there must be a relation between the equilibrium constants for these reactions, i.e. $K_2 = K_1 K_3/K_6$. Also, by
a linear manipulation of the reactions given in Table I, we see
that in the steady state the net reaction of certain sets of
species is zero, as is shown in Table II. These steady state constraints result in a relation between the concentrations and
diffusion rates of these sets of species, shown in Table III.
Except for the first line, all the other equations result from the
fact that none of the species involved leave the membrane, i.e.

$$D_i \frac{dC_i}{dx} = 0 \quad x = 0, L \quad , \quad i=2,\ldots,9.$$

Additional approximations that are reasonable are (3):

[Na^+] = M_o at all x (M_o is the initial concentrations of

NaOH in the membrane); [H^+] $\ll$ [Na^+];

[OH^-] $\ll$ [HCO_3^-] + 2[$CO_3^=$]; $D_{HCO_3^-} = D_{CO_3^=}$;[HCO_3^-] + 2[$CO_3^=$]=[Na^+]

Further, reactions not involving molecular CO_2 (3 to 6 in Table I)
are very rapid and fortunately can be assumed to be at equilibrium.
The problem is then reduced to the solution of only one
differential equation, with the constraints of Table III, and
boundary conditions on partial pressure of CO_2.

The inclusion of carbonic anhydrase in the film introduces
an additional set of differential equations and constraints, which
makes the solution considerably more difficult. However, in the
experiments reported here the molar concentration of the enzyme
was less than 1/1000th the average concentration of dissolved CO_2,
and another several orders of magnitude less than the initial

TABLE I
Uncatalyzed Carbon Dioxide Reactions

1. $CO_2 + H_2O \underset{k_{-1}}{\overset{k_1}{\rightleftarrows}} H_2CO_3$ $\qquad$ $g_1 = k_1(CO_2) - k_{-1}(H_2CO_3)$

2. $CO_2 + OH^- \underset{k_{-2}}{\overset{k_2}{\rightleftarrows}} HCO_3^-$ $\qquad$ $g_2 = k_2(CO_2)(OH^-) - k_{-2}(HCO_3^-)$

3. $H_2CO_3 \rightleftarrows H^+ + HCO_3^-$ $\qquad$ $g_3 = k_3(H_2CO_3) - k_{-3}(H^+)(HCO_3^-)$

4. $HCO_3^- \rightleftarrows H^+ + CO_3^{--}$ $\qquad$ $g_4 = k_4(HCO_3^-) - k_{-4}(H^+)(CO_3^{--})$

5. $MOH \rightleftarrows M^+ + OH^-$ $\qquad$ $g_5 = k_5(MOH) - k_{-5}(M^+)(OH^-)$

6. $H_2O \rightleftarrows H^+ + OH^-$ $\qquad$ $g_6 = a_{H_2O} - (H^+)(OH^-)/K_6$

$$K_i = \frac{k_i}{k_{-i}} \qquad\qquad a_{H_2O}: \text{ activity of water}$$

TABLE II
Reaction Invariants (No Carbonic Anhydrase)

1. $r_{CO_2} + r_{H_2CO_3} + r_{HCO_3^-} + r_{CO_3^{--}} = (-g_1-g_2) + (g_1-g_3)$
$$+ (g_2+g_3-g_4) + (g_4) = 0$$

2. $r_{M+} + r_{MOH} = (g_5) + (-g_5) = 0$

3. $r_{H+} + r_{OH-} + r_{MOH} + 2r_{H_2O} + 2r_{H_2CO_3} + r_{HCO_3^-}$
$$= (g_3 + g_4 + g_6) + (-g_2 + g_5 + g_6) + (-g_5)$$
$$+ 2(-g_1-g_6) + 2(g_1-g_3) + (g_2+g_3-g_4) = 0$$

4. $r_{H+} + r_{M+} - r_{OH-} - r_{HCO_3^-} - 2r_{CO_3^{--}}$
$$= (g_3+g_4+g_6) + (g_5) - (-g_2+g_5+g_6)$$
$$- (g_2+g_3-g_4) - 2(g_4) = 0$$

TABLE III
Reaction Invariants and Integral Constraints

$$\sum_\sigma D_\sigma \ \sigma = a_i x + b_i$$

1. $\sigma = CO_2, H_2CO_3, HCO_3^-, CO_3^{--}$

2. $\sigma = M^+, MOH; \quad a_2 = 0; \ \frac{1}{L} \int_0^L (\Sigma\sigma)dx = \text{total}$

3. $\sigma = H^+, OH^-, MOH, 2H_2O, 2H_2CO_3, HCO_3^-; \ a_3 = 0; \frac{1}{L} \int_0^L (\Sigma\sigma)dx = $
$$\text{total hydrogen}$$

4. $\sigma = H^+, M^+, -OH^-, -HCO_3^-, -2CO_3^{--}; \ a_4 = 0;$
$$\frac{1}{L} \int_0^L (\Sigma\sigma)dx = 0, \text{ Total Net Charge Density}$$

bicarbonate concentration. Therefore, we have assumed that the
enzyme only provides an additional reaction mechanism and does not
contribute to the facilitation of diffusion by movement of enzyme
$-CO_2$ complexes. The kinetic mechanism and constants reported by
Kernohan (6) were used in this analysis.

In this situation, the diffusion-reaction equation at steady
state for carbon dioxide simplifies to (1 and 2 refer to CO_2 and
HCO_3^- respectively):

$$\frac{d^2C_1}{d\bar{x}^2} = \phi[\alpha_1^2\delta_1 + \alpha_2^2\,\delta_2\,] \text{ and } D_1C_1 + 1/2\,D_2C_2 = p\bar{x} + q \qquad (2)$$

where $\phi = [C_1 - \dfrac{K_4}{2K_1K_3}\dfrac{C_2^2}{M_o-C_2}]$

$$\delta_1 = 1 + \frac{k_2}{k_1}\frac{K_6}{2K_4}\frac{M_o-C_2}{C_2} \qquad \alpha_1^2 = k_1L^2/D_1$$

$$\delta_2 = \left[(1 + \frac{C_1}{K_s})(1 + \frac{C_2}{K_i})(1 + \frac{2K_4}{K_h}\frac{C_2}{M_o-C_2})\right]^{-1} \quad \alpha_2^2 = k_eE_oL^2/D_1K_s$$

with $\bar{x} = \frac{x}{L}$, C_1 given at $\bar{x} = 0$ and $\bar{x} = 1$ and $\dfrac{dC_2}{d\bar{x}} = 0$ at $\bar{x} = 0; 1$;
and K_s, K_i, K_h, k_e are constants for the enzymatic reaction.

The factor α_i^2 (Damköhler number) which appears in the differential equation is an indicator of whether diffusion or reaction
is a controlling factor. These equations have been solved here by
a number of techniques including quasilinearization (7), boundary
layer approximation (8,9), and asymptotic analytic methods (10,11).
The range of facilitation of CO_2 transport as a function of factor
α_1^2 is given in Figure 1 for one concentration of M_o. At low values
of the Damköhler number (that is, in the near-diffusion regime),
augmentation is directly proportional to α_1^2. Measurements in this
regime can be used to estimate the unfacilitated Fick's Law
permeability of CO_2 through reactive carbonate solutions (12). The
extrapolation of measured transport rates in this linear near-
diffusion range is shown in Figure 2. The extrapolated values for
the Fick's Law CO_2 flux were subtracted from the measured CO_2
fluxes, giving the facilitation in CO_2 transport over the entire
range of experimental observations.

A comparison between the measured CO_2 transport through carbonate solutes containing carbonic anhydrase and the calculated fluxes
using Equation (2) is shown in Figure 3. In the presence of
carbonic anhydrase, the second kinetic term in Equation (2) domin-
ates and therefore the characteristic Damköhler number is α_2^2.
According to graphs similar to Figure 1, the facilitation factor

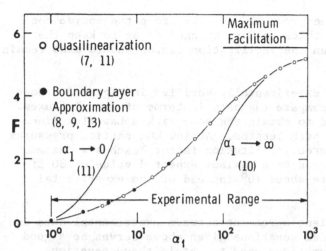

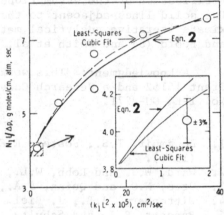

Figure 1. Calculated Facilitation Factor F versus α_1 for Diffusion Across IN NaHCO$_3$ Solution (p_0, CO$_2$ Pressure at x = 0, is 1 atm).

Figure 2. Near-Diffusion Model-CO$_2$ Permeability in IN NaHCO$_3$ Solution.

Figure 3. Carbon Dioxide Flux Across IN NaHCO$_3$ Film (membrane Area: 15.5 cm^2; p_L: CO$_2$ Pressure at x = L).

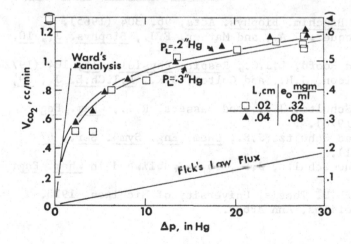

should depend only on the group α_2^2. If therefore the enzyme concentration and the film thickness are changed so as to keep the
group $E_o L^2$ constant, then the facilitation factor should also remain
constant.

This prediction is experimentally verified in Figure 3. Note
that in this figure to compare the data in terms of actual fluxes,
the scales were adjusted to obtain the same Fick's Law flux line
for different diffusion path lengths. At low CO_2 partial pressures
the experimentally measured facilitation factor reaches a maximum
value of ~20, which compares to a maximum expected value of ~80 if
the Damköhler number were about 10^8 instead of the experimental
range 10^4 to 3×10^6.

The apparent excellent agreement between experimental data and
predictions from the model equations given above serves as a good
verification of the assumptions used to obtain these equations.
The solid lines adjacent to the data were calculated (13) by approximate iterative analytical method similar to that given by Kreuzer
and Hoofd (8) and Smith et al. (9).

Acknowledgment: This work was partially supported by PHS
Grant 15152 and a Research Career Development Award to J.S. Schultz,
No. 1-K4-GM-8271.

References

1. Longmuir, I.S., Foster, R.E., and Woo, C-Y., Nature, 209, 393
   (1966).
2. Ward, W.J. and Robb, W.L., Science, 156, 1481 (1967).
3. Otto, N.C. and Quinn, J.A., Chem. Eng. Sci., 26, 949 (1971).
4. Wittenberg, J.B., J. Biol. Chem., 241, 104, (1966).
5. Bassett, R.J. and Schultz, J.S., Biochim. Biophys. Acta, 211,
   194 (1970).
6. Kernohan, J.C., Biochim. Biophys. Acta, 96, 304 (1965).
7. Kutchai, H., Jacquez, J.A., and Mather, F.J., Biophys. J., 10,
   38 (1970).
8. Kreuzer, F., and Hoofd, L.J.C., Respir. Physiol., 15, 104 (1972).
9. Smith, K.A., Meldon, J.H., and Colton, C.K., A.I.Ch.E. J., 19,
   102 (1973).
10. Goddard, J.D., Schultz, J.S., and Bassett, R.J., Chem. Eng.
    Sci., 25, 665 (1970).
11. Suchdeo, S.R. and Schultz, J.S., Chem. Eng. Symp. Ser., 67
    (114), 165 (1971).
12. Suchdeo, S.R. and Schultz, J.S., to be published in Chem. Eng.
    Sci.
13. Suchdeo, S.R., Ph.D. Thesis, University of Michigan, 1973,
    University Microfilms, Ann Arbor.

MODELING THE CORONARY CIRCULATION: IMPLICATIONS IN THE PATHO-

GENESIS OF SUDDEN DEATH

R.J. Gordon, W. Jape Taylor, and Harvey Tritel

Department of Chemical Engineering and Division of
Cardiology, Department of Medicine, University of
Florida, Gainesville, Florida 32601, U.S.A.

Numerous experimental and clinical studies of the coronary
circulation have been reported in recent years, but as yet these
studies have not been tied together in any sort of overall or
integrated picture. The objective of this paper is to present
some preliminary work on the formulation of such an integrated
picture, using the techniques of mathematical model building.
Such a development serves two purposes: first of all, the many
diverse experimental results may be succinctly and clearly summarized by their incorporation in the model; and secondly, the
model thus developed allows for the prediction of new results,
as well as yielding a better understanding of the pathogenesis
of such disorders as ischemic heart disease. In the following
discussion, we limit our considerations to the left ventricle,
the primary site of ischemic heart disease.

MODEL FORMULATION

Following Winbury,[1] it is assumed that coronary blood flow
to the left ventricle may be idealized as illustrated in Fig. 1.
Here P_A is the mean aortic pressure, P_{CA} the pressure at the junction of the subepicardial (EPI) and subendocardial (ENDO) branches
of the coronary artery, and P_V the pressure in the coronary sinus.
The pressure difference $P_A - P_{CA}$ along the extramural coronary
artery gives a direct measure of the severity of coronary artery
disease (atherosclerosis being limited to these extramural vessels).

The rate of blood flow to the subendocardium may be expressed
in the form

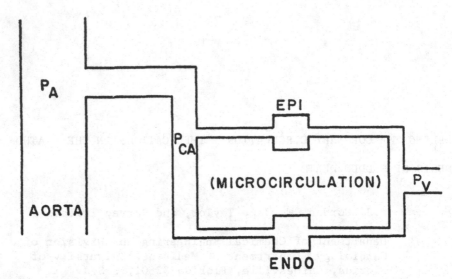

Fig. 1. Model of myocardial blood flow as suggested by Winbury.

$$Q_{ENDO} \left(\frac{ml.}{100\ gm.-min.}\right) = 60(1-F_{sys}) \frac{P_{CA} - P_V}{\eta x R_d} \qquad (1)$$

where F_{sys} is the fraction of the cardiac cycle which is systolic, $60(1-F_{sys})$ is the diastolic filling time per minute, η is the blood viscosity, and R_d the vascular resistance of the sub-endocardium.

Similarly, the rate of blood flow to the subepicardium is

$$Q_{EPI} \left(\frac{ml.}{100\ gm.-min.}\right) = 60 \frac{P_{CA} - P_V}{\eta x R_s} \qquad (2)$$

The blood viscosity η is given to a good approximation by the Vand Equation,[2]

$$\eta = \eta_p (1 + 0.025H + 7.35 \times 10^{-4} H^2) \qquad (3)$$

where η_p is the plasma viscosity and H the hematocrit.

Turning now to the vascular resistances R_s and R_d, these are assumed capable of decreasing--due to the stimulus of myocardial ischemia--down to 0.2 times their basal value. This figure is based on the reactive hyperemia studies of Gregg and coworkers with dogs.[3] The exact fraction chosen is unimportant since the following results and conclusions are rather insensitive to this quantity.

Another important effect which must be included in the model is the dependence of subendocardial resistance on the "transmural pressure" $P_{CA} - P_{LVED}$,[4-8] where P_{LVED} is the left ventricular end diastolic pressure. Available data suggests that for $P_{CA} - P_{LVED}$ less than about 40-50 mm of Hg, the subendocardium begins to show signs of ischemia, and clearly for $P_{CA} - P_{LVED} \cong 0$, subendocardial flow stops entirely.[7-9] This effect is included in the model by assuming that R_d is inversely proportional to $P_{CA} - P_{LVED}$, for transmural pressures less than 50 mm of Hg:

$$R_d \propto \frac{1}{P_{CA} - P_{LVED}} \qquad P_{CA} - P_{LVED} \leq 50$$

Such a function guarantees that $R_d \to \infty$ as $P_{CA} - P_{LVED} \to 0$. Clearly, more complicated functions could be introduced but the qualitative predictions are not greatly affected by the precise choice of R_d.

Finally, in order to evaluate P_{CA}, knowing the mean aortic pressure P_A, the pressure drop along the left coronary artery must be calculated. This is obtained from the expression

$$Q_{TOTAL} = Q_{ENDO} + Q_{EPI} = 60 \frac{P_A - P_{CA}}{\eta \times R_{CA} \times Athero} \qquad (4)$$

where R_{CA} is the undiseased vascular resistance of the left coronary artery, and Athero is a parameter representing the "degree of atherosclerosis." (Note that Athero x R_{CA} is the actual resistance of the coronary artery. Hence, increasing values of Athero (above 1.0) correspond to increasingly more severe coronary artery disease.)

Eqs. (1), (2), and (4) constitute a system of three equations and three unknowns. Consequently, Q_{ENDO}, Q_{EPI}, and P_{CA} may be calculated in a straightforward fashion.

MODEL PREDICTIONS AND DISCUSSION

Reduction in Coronary Perfusion Pressure

In the normal animal, and it is believed in man, the ratio of Q_{ENDO}/Q_{EPI} is approximately unity.[6] During gradual reduction in coronary artery perfusion pressure, Q_{ENDO}/Q_{EPI} at first remains unchanged. At sufficiently low pressures, however, this ratio begins to drop precipitously. This finding has been observed in numerous experiments.[4,8,10,11] Fig. 2 is a plot of relative coronary blood flow (base value is an arterial pressure of 100 mm Hg),

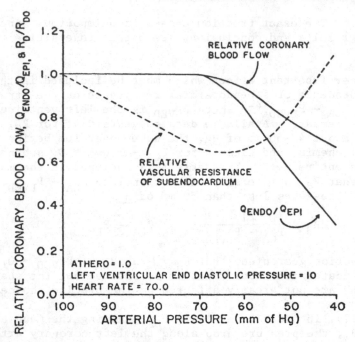

Fig. 2. Relative total left coronary blood flow, ratio of endocardial
   to epicardial flow, and relative endocardial vascular resistance
   versus arterial pressure.

Q_{ENDO}/Q_{EPI}, and the relative vascular resistance of the subendocardium vs. mean arterial pressure. As P_A is reduced, total flow and
Q_{ENDO}/Q_{EPI} at first remain unchanged. At $P_A \cong 65$ mm of Hg, however,
we start to see a decrease in Q_{ENDO}/Q_{EPI} with a lesser decline in
coronary blood flow. Further reductions in P_A lead to smaller and
smaller values of the ratio Q_{ENDO}/Q_{EPI}. These results are in excellent agreement with the existing data[6,8,10] (see especially
Griggs and Nakamura[8]).

Anemia

Bassenge et al.[12] have measured basal coronary blood flow as
well as total coronary flow during reactive hyperemia following
hematocrit reductions in unanesthetized dogs. Their results are
illustrated in Fig. 3 by the circles (reactive hyperemic response)
and triangles (coronary blood flow). The results of the model
are indicated by the solid lines. The agreement is excellent,
offering one more example of the predictive ability of the model.

Stress Testing

The model has been used to predict the response of the coronary circulation to both exercise testing and electrical pacing.
Coronary blood flow was found to increase linearly with heart rate
up to some value which decreased with increasing atherosclerosis,
and then to level off. The critical heart rate for the inducement

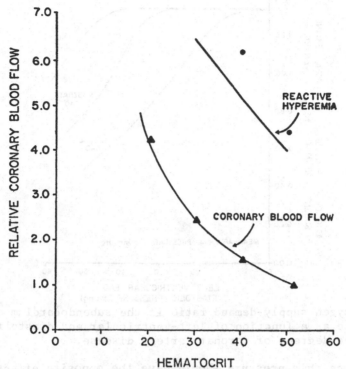

Fig. 3. Relative coronary blood flow and reactive hyperemic flow
versus hematocrit. Data of Bassenge.[12]

of angina was also found to vary inversely with the degree of atherosclerosis, and to be about 20 beats/min. lower for exercise testing
than for stress testing. All of these findings have been reported
clinically.

Coronary Artery Disease and Sudden Death

Fig. 4 illustrates the predictions of the model for the case
of coronary artery disease (CAD). Here, the oxygen supply-demand
ratio in the subendocardium of the left ventricle is plotted as a
function of left ventricular end diastolic pressure for three
levels of CAD. In each case the supply-demand ratio remains at a
value of 4/3 (corresponding to 75% O_2 extraction) up to some
critical left ventricular end diastolic pressure determined by the
severity of the disease. For example, for no CAD the "critical"
left ventricular end diastolic pressure required to induce subendocardial ischemia is quite large and far beyond the normal
physiologic range. On the other hand, for severe disease, fairly
small elevations in this pressure above normal can cause a significant imbalance between oxygen supply and demand. It would thus
be expected that in patients with severe coronary artery disease,
procedures which increase left ventricular end diastolic pressure
would tend to either precipitate angina or at least reduce the
anginal threshold in stress testing, and similarly procedures

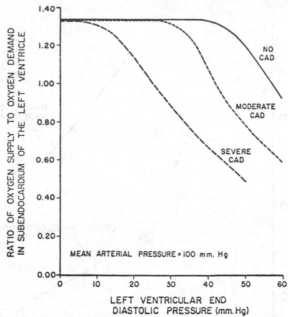

Fig. 4. Oxygen supply-demand ratio in the subendocardium of the left
  ventricle as a function of left-ventricular end diastolic pressure,
  for three degrees of coronary artery disease.

which reduce this pressure should have the opposite effect. A
series of recent papers by Parker and coworkers nicely confirm these
ideas.[13-15]
    Fig. 4 also suggests a possible mechanism for sudden cardiac
death from nonthrombotic causes. Let us suppose that the subendocardium of the left ventricle in a subject with severe CAD suddenly becomes ischemic, due to some external emotional or physical stress.
An ischemic myocardium requires a higher filling pressure to maintain constant cardiac output, and consequently the left ventricular
end diastolic pressure rises.[16] This in turn further reduces
blood flow to the subendocardium (as in Fig. 4), accentuating the
ischemia. Such a viscous circle can rapidly lead to a condition
of severe ischemia and imminent tissue necrosis, setting the
state for a fatal electrical disturbance. This type of picture
is very consistent with much of the recent pathological work of
Roberts,[17] who showed that in cases of sudden death or where
necrosis is limited to the subendocardium of the left ventricle,
coronary thrombi are infrequently found (< 10%); and also with
the recently reported frequent absence of myocardial infarction
in primary ventricular fibrillation.[18] Clearly, further experimental and clinical studies of this hypothesis are necessary, but
it appears certain that a critical (and frequently neglected)
factor in determining the extent of ischemia in the left ventricle
is the left ventricular end diastolic pressure.

References

1. WINBURY, M. M. Redistribution of left ventricular blood flow produced by nitroglycerine. Circ. Res. 29: (Suppl. I): 140-147, 1971.

2. MERRILL, E. W. Rheology of blood. Physiol. Rev. 49:863-888, 1969.

3. GREGG, D. E. Relationship between coronary flow and metabolic changes. Cardiol. 56:291-301, 1971/72.

4. BECKER, L. C., and B. PITT. Regional myocardial blood flow, ischemia, and antianginal drugs. Ann. Clin. Res. 3:353-361, 1971.

5. SALISBURY, P. F., C. E. CROSS, and P. A. RIEBEN. Acute ischemia of inner layers of ventricular wall. Am. Heart J. 66:650-656, 1963.

6. BUCKBERG, G. D., D. E. FIXLER, J. P. ARCHIE, and J.I.E. HOFFMAN. Experimental subendocardial ischemia in dogs with normal coronary arteries. Circ. Res. 30:67-81, 1972.

7. HIRSHORN, S., and G. A. KAISER. Effects of changes in left ventricular end diastolic pressure on the distribution of coronary blood flow and electrical activity of the heart. Curr. Top. Surg. Res. 2:463-475, 1970.

8. GRIGGS, D. M., JR., and Y. NAKAMURA. Effect of coronary constriction on myocardial distribution of iodoantipyrine-131 I. Am. J. Physiol. 215:1082-1088, 1968.

9. SUGIMOTO, T., K. SAGAWA, and A. C. GUYTON. Quantitative effect of low coronary pressure on left ventricular performance. Jap. Heart J. 9:46-56, 1968.

10. BECKER, L. C., N. J. FORTUIN, and B. PITT. Effect of ischemia and antianginal drugs on the distribution of radioactive microspheres in the canine left ventricle. Circ. Res. 28:263-269, 1971.

11. MOIR, T. W. Subendocardial distribution of coronary blood flow and the effect of antianginal drugs. Circ. Res. 30:621-627, 1972.

12. MESSMER, K., and H. SCHMID-SCHÖNBEIN, Eds. Hemodilution. S. Karger, Basel, 1972, pp. 174-183.

13. KHAJA, F., V. SANGHVI, A. L. MARK, and J. O. PARKER. Effect of volume expansion on the anginal threshold. Circulation 43:824, 1971.

14. PARKER, J. O., J. R. LEDWICH, R. O. WEST, and R. B. CASE. Reversible cardiac failure during angina pectoris. Circulation 39:745-757, 1969.

15.   PARKER, J. O., R. B. CASE, F. KHAJA, J. R. LEDWICH, and
       P. W. ARMSTRONG. The influence of changes in blood volume
       on angina pectoris. A study of the effect of phlebotomy.
       Circulation 41:593, 1970.

16.   GUYTON, A. C. Cardiac Output and Its Regulation.
       W. B. Saunders Co., Philadelphia, 1963, p. 358.

17.   ROBERTS, W. C. Coronary arteries in fatal acute myocardial
       infarction. Circulation 45:215, 1972.

18.   ALVAREZ, H., R. E. WILLIS, and L. A. COBB. Sudden cardiac
       death. Physiologic observations and therapeutic implica-
       tions (Abstract). Amer. J. Cardiol. 31:116, 1973.

DISCUSSION OF SESSION IV   (PART B)

Chairmen: Dr. Reinhard Wodick and Dr. Wolfgang Grunewald

DISCUSSION OF PAPER BY D. H. HUNT, D. F. BRULEY, H. I. BICHER AND
   M. H. KNISELY

Wissler: If we are to successfully model whole organs in which
there are gradients, or an entire mammal, we must have simplified
methods for describing microscopic phenomena. Computational limitations will determine the degree to which we can model such systems, and we cannot afford to place undue emphasis on one aspect
of the system.

Bruley: How much emphasis is placed on one aspect of the system
depends upon what you are trying to determine with your models.
If you are interested in examining phenomena at the cellular level,
you cannot make gross approximations about the system behavior with
lumped parameter models. This may require the developement of more
efficient numerical methods or other computer techniques to achieve
solutions.

Hunt: Whole organs may necessarily be characterized by a basic
anatomical unit within that organ. If that is the case, much useful information is garnered from defining gradients and point values
within the unit; hence, the computations require more complex models
containing partial differential equations.

DISCUSSION OF PAPER BY R. C. TAI AND H. CHANG

Goldstick: I have measured the oxygen diffusivity in the cornea
of the eye (rabbit) and the cremaster muscle (rat), and in both
of these structures, the oxygen diffusion behavior is that in a
homogeneous material. The diffusivity is about one-third of that
in water. The method is unsteady state. Could the fact that our
results indicating homogeneity, or at least, very slight heterogeneity, be related to the unsteady state nature of our measurements?

Chang: I am afraid I cannot answer this question to your satisfac-
tion. All we try to do here is to compute the permeability ratio
between extracellular material and cellular material, based upon
experimental data in the literature. The models are steady state,
and I do not know what a model based on unsteady state transport

983

would yield. The results, based on our models, give a heterogeneity factor (Ke/Kc) of approximately 5-10 if the value $\alpha \approx 0.2$ can be believed.

Longmuir: (1) If the intracellular diffusion coefficient is as low as you suggest, then cells must be almost completely anaerobic. Eggleston showed, thirty years ago, that even with a pO_2 inside cells equal to that in water, cells could be no more than 25 μ in diameter without becoming anoxic in the center.
(2) Do you know of the error in Hills derivation?

Chang: (1) Based on our models and the data which I quoted, the lump permeability of cellular material (including cell membranes) is about 1/10 to 1/5 that in the extracellular fluid. This number seems to be in the ball park.
(2) Yes, we have incorporated the correction made by Gold and Longmuir (1971) in our derivation.

Lübbers: (1) I would like to know about the reproducibility of your measurements and the calculated transport parameters?
(2) Did you control the form of red cells? How did this effect your parameters?
(3) What does the curve look like if you are using HLS-solutions instead of saline solutions?

Chang: (1) For the data shown, we did the same run twice and got good agreement. The other runs at different hematocrit, etc., have not yet been analyzed. The calculated transport properties, Dc, hm, have not yet been evaluated. A preliminary overall mass transfer coefficient for red cell without reaction, i.e., diffusive conductivity including plasma film, cell membrane and cell interior is 0.00792 sec.$^{-1}$.
(2) Not yet. We will run rigid and spherical cells in the future.
(3) Have not run as yet. I would guess that it should look much the same, since nitrated cells had little effect on plasma O_2 concentration time profiles.

DISCUSSION OF PAPER BY E. L. ROETMAN AND R. E. BARR

Groom, A. C.: In view of the serious fall in pO_2 behind the cornea, do you think that the solution would be rather worse for the case of man under water?

Barr: I am not familiar with the oxygen concentration and hence, the amount of exchange in water, say of a lake, between it and the cornea. I do know about contact lenses and in these instances, a very serious problem can arise.

Grunewald: Is the O_2-transport through the cornea enough to supply the whole lens, or does the lens also need O_2 which diffuses through the vitreous body?

Barr: I do not know. I was hoping you would be able to give some information about that.

Irwin: There is another avascular organ, the organ of Corti. Here oxygen appears to come from endolymph.

Do you think that the cornea may not obtain some oxygen from aqueous humor?

Barr: In certain circumstances, the cornea probably does obtain some oxygen from the aqueous humor. However, under circumstances of the eye exposed to air, it appears that the oxygen concentration gradient is from the cornea into the aqueous. Hence, it would seem that little, if any, oxygen is entering the cornea from the aqueous humor.

DISCUSSION OF PAPER BY A. C. GROOM AND M. J. PLYLEY

Ciuryla: (1) Have you made any measurements on exercise-conditioned animals?

(2) Also, frog capillaries are known to have diameters three times as large as the other species mentioned. Would you comment in the light of your findings?

Groom: (1) We have not made measurements on animals subject to exercise training nor on those adapted to high altitude. Capillary density has been reported to be higher under both these conditions, but this could well be due to a decrease in fiber size, rather than a change in the mean number of capillaries surrounding a fiber.

(2) Frog red blood cells are, of course, much larger than mammalian red cells and the capillaries are correspondingly larger. However, the muscle fibers are even larger, proportionately, than are red cells in the frog versus the mammal; thus the general similarity between capillary size to fiber size is not very different in amphibians versus mammals.

Rakusan: An additional explanation for higher values found in older literature might be the fact that the older authors did not prevent tissue shrinkage. Incidentally, I found the same results on comparing the capillary density in the heart muscle. Concerning your method: I would not rely on the control of reflection on the presence of red blood cells or unfilled capillaries. Direct straining of the capillary wall by PAS or alkalic phosphatase might be superior because it uncovers closed capillaries, also.

Groom: Yes, it is probably true that the pre-1950 values are affected by shrinkage, but this does not represent the whole picture. Our capillary density values agree well with these of Hammersen (1970) and both are significantly different than those pre-1950 values.

The use of Gomori tri-chrome staining allows us to evaluate our Microfil perfusion. Sections where red-cell filled capillaries were greater than 5% of counts were rejected (on assumption that there were an equal number of unfilled or plasma filled capillaries, this represents a poor perfusion). However, in sections where the percentage of red-cell filled capillaries was less than 5%, these vessels were included in our counts.

It is true that PAS and alkaline phosphatase reactions would uncover all closed capillaries and in this regard, is probably superior. However, our criteria for rejection, plus the simplicity of the technique with Gomori tri-chrome, make the above approach extremely worthwhile.

Longmuir: Could you not relate the variation in size of muscle fibers to oxygen consumption, and that there is thus, constancy of oxygen supply to the surface?

Groom: Yes, the metabolic rate of frog muscle fibers is, of course, very much less than that of mammalian fibers, and such a correlation as you suggest can, indeed, be made.

Lockard: If a capillary is shared by multiple muscle fibers, do you count it with each muscle fiber?

Groom: Yes, we do. We have measured separately the number of fibers sharing each capillary, but are reporting today, just the situation as it exists for each individual muscle fiber. In fact, the mean number of fibers sharing a vessel does vary in a rather systematic way with the mean number of vessels surrounding each fiber.

DISCUSSION OF PAPER BY P. D. WILSON

Guthermann: Doesn't this model require that the state variables interact linearly, a fact which, for oxygen consumption, is not true for large ranges of pO_2?

Wilson: The model does assume that the state dynamics may be described by a system of _linear_ differential equations and it is not likely that this is exactly true, especially in extreme regions of the state space. However, predictions are "updated" with each data acquisition, and all that is demanded is that the linear system be an acceptable approximation over the interval between data acquisitions. If this interval is small, it is not unreasonable to expect the linear approximation to be good. The empirical mean and variance of the prediction errors can be compared to those computed by the algorithm from the linear model. Important discrepancies may be attributed to errors in the model--possibly resulting from non-linearity in the state dynamics.

DISCUSSION OF PAPER BY B. F. CAMERON

Barnikol: How do you insure that, at the beginning of the registration, the blood is fully deoxygenated?

Cameron: In both techniques that I have used, the measurement is made on a suspension of red cells in phosphate buffer. The DCA (radiometer) is based on the method of Duvelleroy; the sample is placed in the chamber and degassed under 5% CO_2 in N_2 until the solution pO_2 is zero.
     For the Longmuir-Cameron technique, the blood is gassed with air or (for sickle cell anemia bloods) with 5% CO_2 in 30% O_2 before it is placed in the chamber containing buffer which has been equilibrated with the same gas.

Holland: I would like to know if your measurements were at 40°C, which is commonly given as rabbit body temperature, or did you do the measurements (as many of us have done) at 37°C.
     Secondly, I am surprised that the rabbits had so much effect from the carbon monoxide concentrations you used. The only value I have seen for M (relative affinity of CO and oxygen) of rabbit hemoglobin is low. Fox found a value of 40 at 20° which corrects to below 30 at body temperature if we apply to rabbits the temperature correction found for humans (Holland, Annals N. Y. Academy Science, Symposium on Biological Effects of Carbon Monoxide, 1970). Perhaps you have measured M in your rabbits?

Cameron: Our measurements were also at 37°. The rabbit for which data was presented here was somewhat below our normal P_{50} for rabbits under these conditions, which is about 30 mm Hg.

I do not have values for M. It would be very valuable to measure this quantity. These rabbits were not necessarily at equilibrium with 200 ppm CO; they had been respiring a gas containing this CO level in air for 30 minutes.

However, experiments in which normal rabbit blood (erythrocyte suspension) was equilibrated with 200 ppm CO - 5% CO_2 in air, the shift in P_{50} and n was essentially the same as in the CO-exposed rabbit. This would suggest that the value of M for rabbits is higher than that reported by Fox, but we need more experiments.

DISCUSSION OF PAPER BY R. L. CURL AND J. S. SCHULTZ

Moll: Did you assume the reaction rates to be finite or infinite?

Schultz: In this analysis, hemoglobin is assumed to be in equilibrium with the local dissolved free oxygen concentration. Therefore, the reaction rates are assumed to be very much faster than the diffusion rates and approaching infinite in magnitude in this respect.

DISCUSSION OF PAPER BY H. G. CLARK AND W. D. SMITH

Song: I am glad to find another outsider to please ourselves. How much CO_2 gradient do you think you can predict as a maximum value with your suggested energy?

Clark: We have not measured CO_2 concentration, but we believe that it should be coupled to the pH and oxygen tension. From energy considerations, we think that about 0.2 pH units is possible under conditions of vigorous exercise. At rest (i.e., cardiac output of 5000 ml/min.) there is probably not enough energy available to produce more than a few hundredths of a pH shift.

Longmuir: Were the pressures you used comparable to those employed by Wells to shift the dissociation curve?

Clark: The pO_2 was 20 Torr and pCO_2 was 40 Torr. There was an additional 10 cm of water pressure added to produce fluid flow in the capillary slit.

Barnikol: How great was the rate of hemolysis in your experiment?

Clark: Visual inspection of centrifuged blood showed no evidence
of hemolysis, but spectrographic analysis was not carried out.

DISCUSSION OF PAPER BY S. R. SUCHDEO, J. D. GODDARD AND J. S. SCHULTZ

Moll: (1) Analyzing facilitated diffusion of CO_2, we came to the
conclusion that the carbonate-bicarbonate mechanism does not work
at physiological pH, simply because the carbonate concentration,
i.e., the carbonate concentration gradient, is too small in this
pH range. Would you agree with that?
     (2) Shouldn't you take into account the electrical potentials?

Suchdeo: (1) The experiments conducted in this study were with
1 M Na_2CO_3 solutions and, therefore, outside the range of physiological solutions. The bicarbonate and carbonate concentration
profiles calculated here cannot be directly related to physiological conditions. We believe that the calculations for these interesting situations can be made by the methods presented here.
     In trying to estimate the effects of parameters, such as buffer
concentrations, it sould be remembered that equilibrium with respect
to the chemical reactions cannot be assumed at the boundaries. The
recognition of the nonequilibrium properties at the boundaries is
one of the main points of this presentation. Therefore, the actual
ratio of the concentrations of carbonate and bicarbonate ions may
not be given by the equilibrium relation and the presumed pH in the
system.
     (2) Under the conditions of the experiments reported here, we
estimated that electrical potentials generated by ionic diffusion
were not large enough to influence the calculated transport rates
of CO_2 through the membrane by more than 10%.

DISCUSSION OF PAPER BY B. H. CHANG, M. FLEISCHMAN AND C. E. MILLER

Mochizuki: How did you take into account the back pressure in the
treatment of oxygenation of red cells?

Fleischman: We have not yet accounted for the kinetics of the O_2-Hb
reaction, but when we do, the kinetics would follow either Adair
mechanism or, as a simplification:

$$Hb + O_2 \underset{kr}{\overset{kf}{\rightleftarrows}} HbO_2$$

$$RO_2 = k_f CO_2 \ C_{Hb} - kr \ C_{HbO_2}$$

Where $CHbO_2 = \underbrace{C_{Hb} \ total}_{\text{measured}} \times \underbrace{Satn.}_{\text{measured}}$

Session V

OXYGEN TRANSPORT PROBLEMS IN NEONATOLOGY

Subsessions: PLACENTAL OXYGEN TRANSFER

OXYGEN MONITORING

Subsession: PLACENTAL OXYGEN TRANSFER

Chairmen: Dr. Jose Strauss, Dr. Waldemar Moll and

Dr. Daniel D. Reneau

EXPERIMENTAL AND THEORETICAL INVESTIGATIONS OF OXYGEN TRANSPORT PROBLEMS IN FETAL SYSTEMS - INTRODUCTORY PAPER

Daniel D. Reneau

Department of Biomedical Engineering

Louisiana Tech University
Ruston, Louisiana

Adequate understanding of oxygen supply to the microenvironments of the neonatal system is dependent on the ability to make reliable measurements and the ability to accurately interpret the measurements. Experimental measurements in the placenta and fetal system range from overall measurements designed to produce clinical guidelines to detailed microscopic measurements designed to obtain a more basic understanding of fundamental processes. Both the planning and interpretation of measurements vary in scope from intuition based on observation to the beginnings of the mathematics of systems analysis.

Experimental studies related to the physiology of oxygen transfer rates in the placenta have yielded a variety of information that now makes possible a more complete understanding of placental hemodynamics, placental diffusing capacities, oxygen tension differences between maternal and fetal blood, and the acid-base status. Anatomical studies in the placenta have led to morphological determinations that give insight into spatial relationships affecting oxygen transfer rates and that indicate the magnitude of possible diffusion barriers. Biochemical studies on fetal and maternal blood have resulted in investigations on hemoglobin reaction rates, oxygenation velocity factors, and oxygen carrying capacities during health and disease. Physiological, anatomical and biochemical studies of the fetal system have given basic information on the distribution of blood flow to the different organs of the fetus, the respiratory status, and the metabolic demand. Oxygen deprivation studies have established threshhold values of oxygen dificiency leading to cardiovascular and brain pathological changes in the term primate fetus; and correlated studies have been conducted to determine the effect of hypothermia and hyperthermia during oxygen deficiency. The oxygen status in the arterial blood of the newborn is being continuously recorded with the use of recently developed implantable electrodes, scalp and skin electrodes, and umbilical catherization and cannula electrodes.

Theoretical mathematical studies have been conducted to obtain a more fundamental understanding of placental and fetal transport and for interpretation of experimentally measured results. The majority of the theoretical mathematical studies have been related to the steady-state transport of oxygen. A few studies have considered the simultaneous transport of oxygen and carbon dioxide and the associated pH effect. Dynamic analysis using unsteady-state mathematical models to simulate rates of response and regulatory behavior in neonatology is in its infancy. In addition mathematical studies of neonatal transport have been either designed for the study of a single organ such as the placenta or have lumped many fetal organs into a single unit.

Based on the present state of knowledge and on the rapid advances being made in microelectrode technology and spectrophotometric techniques, experimental investigations of the microenvironments of fetal brain, heart and other organs is now possible. Micromeasurements should be possible with respect to PO2, PCO2, pH, temperature, Na, K, and possibly glucose. Results and predictions obtained from the investigation of this multivariable fetalplacental system which interacts with the maternal system will require a systematic basis for analysis, interpretation and the efficient planning of future experiments.

Consequently, it is now not only possible but also feasible to develop a mathematical systems analysis of transport in the placenta-fetal system. We have attempted to anatomically divide and mathematically model the combined fetal-placental system in order to simulate the transport and interaction in each major organ of oxygen, carbon dioxide, pH, temperature, glucose and other metabolites. The approach is to allow dynamic as well as steady-state studies with superposition of control mechanisms when warranted. Systems analysis of the multivariable fetalplacenta system offers a new approach for comprehensive investigation and understanding.

This work is supported in part by USPHS Grant NIH-NS-08228 and The Frost Foundation

MICROPHOTOMETRY FOR DETERMINING THE REACTION RATE OF O_2 AND CO WITH RED BLOOD CELLS IN THE CHORIOALLANTOIC CAPILLARY

Masaji MOCHIZUKI, Hiroshi TAZAWA and Tsukasa ONO

Research Institute of Applied Electricity

Hokkaido University, Sapporo, Japan

The combination rate of O_2 and CO with the erythrocyte is one of the crucial determinants for assessing the gas exchange in the lung, the placenta and the chorioallantois. A stopped flow or rapid flow reaction apparatus has well been employed to measure the reaction rate of O_2 and CO with the red cell or hemoglobin solution. These flow methods have, however, some limitations in determining the reaction velocity as follows: 1) It is a difficult task to measure the reaction rate at a low Pco level comparable with that required for measurement of the pulmonary diffusing capacity for CO. 2) Pco in red cell suspension cannot be maintained constant throughout the reaction. The reduction of Pco becomes serious especially in a suspension with low Pco. 3) The apparatus requires a lot of blood. 4) In addition to a routine process, a microscopic examination is often required to ensure that no significant crenation or hemolysis occurs.

Since the problems of 1) to 3) are mainly caused by using the solution in which CO can be dissolved only slightly, these problems could be avoided by using a method in which CO gas directly reacts with the red cell. As for the problem of 4), it is most desirable to use a microscope for measureing the reaction rate. From these reasons and also from a requirement for checking the values measured previously by the flow reaction apparatus, we developed a new method of microphotometry. This technique has another advantage to make possible the determination of reaction rate in the single red cell even in the capillaries such as the chorioallantoic vascular plexus. Since the change in transmission measured in the red cell is much larger in the reaction with CO than in the oxygenation, in the present study we first attempted to apply this method to the CO combination reaction.

997

METHODS AND MATERIALS

1. The Outline of the Microphotometer Used

The most significant feature is to employ a microscope and to made CO gas directly contact with the red cells. The observation tube, the mixing chamber and reagent containers are therefore substituted by the microscope and its simple appendants. Fig. 1 shows a schematic diagram of the apparatus. The light emitted from a tungsten lamp (a) is condensed through lenses (f) and only the light whose wavelength is shorter than 460 mμ is reflected perpendicularly by a dichroic mirror (g). Since the optical lenses used do not pass the light with a wavelength shorter than 350 mμ, the bandwidth of injecting light ranges from 350 to 460 mμ. As Sirs stated (3), the dispossession of the extra light is necessary to eliminate the possible effect of incident light on the dissociation of CO from COHb. In addition, for testing the photodissociation effect we reduced the light intensity by using a rotating sector (d) which chopped the light beam from the source.

The reaction of CO with the red cell is made to occur in an air-tight cuvette (i) placed on a movable stage. A blood sample is put on the transparent part made with slide- and cover-glasses and CO gas is led into the cuvette through syringes (j_1) and (j_2). The colour change of the red cell caused by the reaction is detected by a couple of photomultipliers (q_1 and q_2) after dividing the transmitted light into two parts by a half mirror (o) and passing through interference filters (p_1 and p_2). Then, the output signal from each photomultiplier is amplified with a differential amplifier (r) which is balanced prior to the measurement. The balance before setting a blood sample was achieved by changing the applied voltage to the photomultipliers and after setting the sample the balance was made again by regulating the voltage through a potentiometer. The red cell is monitored through an eye piece (n) throughout the experiment.

Light source and rotating sector A conventional 300-watt projecter lamp is used as a light source. The lamp is placed in a lamp house which is kept at a constant temperature by circulating water (b). The light is emitted through a small window of 2 cm wide. Beside the window a rotating sector is mounted to reduce the light intensity, thus, to test the light effect on the reaction velocity. The sector is rotated at 21 revolutions per second in order to avoid the concomitance with a commercial frequency of 50 Hz. Because there are two windows in the sector plate, the light is injected at a rate of 42 Hz and at a duty ratio of approximately 2/15. The red cells are exposed to the light for 3 msec every 24 msec. The colour signal with 42 Hz carrier pulses is detected by a low pass filter.

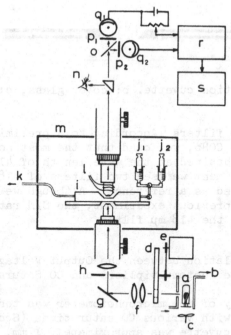

Fig. 1. A schematic diagram of the apparatus used.

Iris diaphragm The incident light is necessary to illuminate
once the whole area of microscopic vision for bringing the blood
cells into the best focus and at another time to illuminate the
single cell to measure the reaction rate in it. For satisfying
both the requirements the iris diaphragm (h) is mounted on the
light path. The center of the diaphragm is shown by a hole
drilled the center of a plastic plate which is fixed behind the
diaphragm. The red cell to be measured is brought into this
circle by moving the stage and then into focus. Then, the iris
diaphragm is narrowed to make the light beam illuminate only one
red cell.

Reaction cuvette The reaction cuvette must be air-tight and
have a space consisting of transparent glasses at the site where
red cells are placed. In addition, there must be an air channel
through which the reacting gas is introduced. Fig. 2 shows a
structure of the cuvette with two, cover-glass (A) and slide-glass
(B), parts. The small transparent chamber is provided when both
the parts (A) and (B) are tightly put together by a silicon
grease. The test gas is then flowed into the reaction space
through a vinyl tubing (e) from 5 ml syringes. Three syringes
are used for the CO-free gas, and for two gas mixtures with low
Pco ranging from about 1 to 7 mmHg and high Pco of about 70 mmHg.

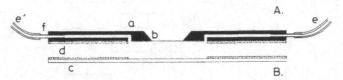

Fig. 2.   Reaction cuvette. b: cover glass, c: slide glass.

Interference filters   According to a preliminary study on
the absorption by COHb, we found that the most notable colour
change could be obtained at the wave length of 418 mμ. Thus, in
the present experiment we used two filters of 418 mμ and 402 mμ;
the latter was used as a reference. Although we used a filter
of 422 mμ in the previous experiment, the S/N ratio was a little
improved by using the 418 mμ filter.

2.   Relation between the Output Voltage of
the Photomultipliers and CO Saturation.

The linearity of the microphotometer was tested using hemoglobin solutions with various CO saturations (Sco's). Since the
thickness of the cuvette was approximately 1 mm, the stock hemoglobin solution which was prepared from freshly drawn blood was
diluted at ratios of 1:1000, 1:1500 and 1:2000 by a buffer solution so that the output voltage became similar to that of the
single red cell. The hemoglobin solution was divided into two
parts; one was equilibrated with CO-free gas and the other with
CO gas mixture of low Pco. These two hemoglobin solutions were
mixed at various ratios to prepare hemoglobin solutions with various Sco's which were determined separately with a conventional
spectrophotometer. Then, the solution was anaerobically introduced into the cuvette and the output voltage was recorded. Fig.
3 shows a calibration curve of the output voltage plotted against
the Sco.

3.   Preparation of the Materials.

Gas mixtures   Three kinds of gas mixture were used for each
experiment. First, gas mixtures with various O_2 fraction and
about 5.5 % CO_2 with N_2 balance were prepared. The Po_2 in these
gases were maintained at 105, 150, 220 and 355 mmHg. These gas
mixtures were used to equilibrate the red cell sample with Po_2
prior to the CO reaction as well as to prepare other gases with
low and high levels of Pco. Secondly, gas mixtures with low Pco
were made by adding a slight amount of CO into the CO-free gas
mixtures mentioned above, where the mixing ratio was varied to
provide the various Pco's. This mixture was used to measure the
reaction velocity of CO with the red cell, where the Pco ranged

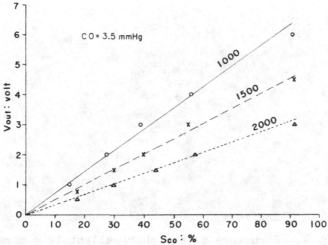

Fig. 3. Relation between output voltage and CO saturation.

from about 1 mmHg to 7 mmHg. The third gas mixture with high CO fraction of about 10 % was used to obtain 100 % Sco level. The gas analyses were performed with a Scholander gas analyzer for O_2 and CO_2 and an infrared CO analyzer for CO.

Blood sample The reaction was measured both in human red cells and in chicken embryo's red cells in the chorioallantoic capillary. In the former experiment, an extremely small amount of blood was put on the slide glass and then covered with a silicon membrane with 25 μ thickness to avoid drying. The excess blood was expelled by pressing the membrane with a gauze so that a thin layer of red cells could be obtained. After putting together the two parts of the reaction cuvette and placing it on the movable stage, the red cells were brought into focus and the iris diaphragm was narrowed so as to illuminate the single cell. Then, the photomultipliers were balanced by regulating the potentiometer and the three gas mixtures were flowed into the cuvette one after another.

The reaction rate was measured in each individual sample by varying both the Po_2 and Pco. The relation of the reaction rate to Pco was obtained at Po_2 of 105 mmHg for human red cells and 56 mmHg for chicken embryonic red cells, and that to Po_2, at about 4 mmHg Pco. Although the blood was usually stored in an ice box for about 1 week in order to attain one serial experiment, it was confirmed that the reaction rate showed no detectable change during these consecutive days.

Chorioallantoic membrane The chorioallantoic membrane grows just beneath the shell membrane and the red cells flow through the

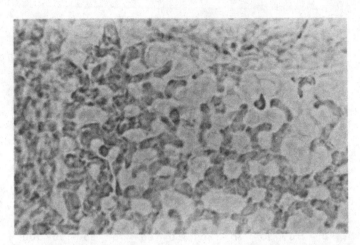

Fig. 4. Structure of the chorioallantoic membrane
observed through a rapid freezing method.

dense capillary net as shown in Fig. 4. Since the chorioallantoic
membrane can easily be sampled maintaining the red cells in it,
the reaction rate with CO in the capillary is measured in the conditions similar to in vivo. In order to estimate the diffusion
velocity through the capillary wall in comparison with the silicon
membrane, the measurements were performed in the following two
cases; 1) that the silicon membrane was put on the chorioallantois
and 2) that no material was put on the chorioallantoic membrane.

RESULTS

1. Change in Sco and Its Rate Factor.

After equilibrating the red cell with Po_2 in a CO-free gas,
the replacement reaction of O_2 by CO was begun by flowing the gas
mixture with low Po_2. The reaction of CO with the red cell needed
a rather long time at a low Pco level. After a lapse of about half
a minute the same gas was flowed again into the cuvette to check
if the reaction further proceeded or not. If no rapid change in
Sco was observed, the 10 % CO gas was introduced in succession to
obtain 100 % Sco level. A typical pattern showing the reaction is
shown in Fig. 5, which was recorded in the chorioallantoic membrane of 18 days old chicken embryo. Since prior to the measurement the membrane was placed in room air, the red cell was fully
saturated by O_2. First, the membrane was exposed to a CO-free gas
with 46 mmHg Po_2. Thus, the So_2 decreased to the level corresponding to 46 mmHg Po_2. At 15 sec, the CO gas mixture of 3 mmHg
Pco was rapidly led into the cuvette to cause the CO reaction.
The colour change due to the reaction lasted for a longer time

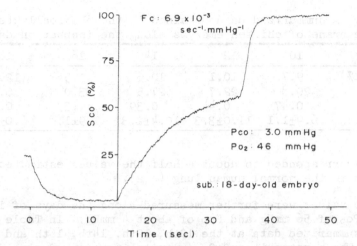

Fig. 5. Replacement reaction pattern of O_2Hb to $COHb$.

than 1 min. At 35 sec, the gas with high Pco was introduced to
increase rapidly the Sco to 100 % level. The change in Sco was
always increased in parallel with Pco. Therefore, for obtaining
the relationship of the reaction rate to Po_2 we calculated the
velocity factor of reaction according to Mochizuki as follows;

$$(1) \qquad Fc = \frac{O_2 \ capacity}{hematocrit} \cdot \frac{dS}{dt} \cdot \frac{1}{100 \cdot Pco}$$

The Fc value decreased, as expected, as Po_2 increased as shown in
Fig. 6, where the Fc values are obtained in the red cell contained
in the chorioallantoic capillary of 16 days old chicken embryo.
The values are expressed approximately by the following experimental equation;

$$(2) \qquad Fc = 2.25/(Po_2 + 250)$$

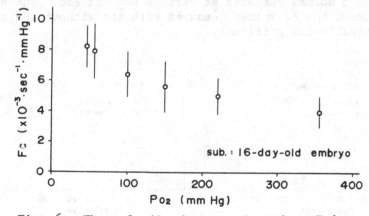

Fig. 6. The velocity factors at various Po's.

Table 1. A summarized data obtained in the chorioallantoic
          membrane of chicken embryos along the incubation days.

Age (days)	10	12	14	16	18
O_2-cap(vol%)	9.7	10.1	10.6	11.5	12.2
Ht (%)	20.8	22.7	27.5	33.0	36.6
K	0.47	0.44	0.39	0.35	0.33
Fc	6.9±1.1	11.0±3.3	10.4±2.3	7.9±1.8	6.0±1.5

The values corresponded to about a half the values estimated by
the authors in the normal human lung (1,2).

The Fc values were further measured along the days of incubation at Po_2 of 56 mmHg and Pco of about 4 mmHg. In Table 1 are
shown the summarized data at the 10th, 12th, 14th, 16th and 18th
days (4). The hematocrit and O_2 capacity increase during development as shown in the table. Since the former increase was however greater than the latter, the CO volume combining with 1 ml
red cells (K) decreased with the incubation days. Thus, the Fc
value in $sec^{-1} \cdot mmHg^{-1} \times 10^{-3}$ rather decreased at a later stage of
incubation, showing a maximum value at the 12th to 14th days.

2. Effect of Light Intensity on the Reaction Rate.

The effect of the incident light on the CO combination was
examined in human red cells, where the light intensity was reduced
to about 1/8 by using the rotating sector. In Fig. 7 are shown
two Sco-time curves obtained with and without rotating the sector,
where Po_2=105 mmHg and Pco=3.4 mmHg in both cases. As shown in
the Fc values depicted in the figure, no significant difference
was observed. Such a measurement was carried out in the red cells
sampled from 3 normal subjects at various Po_2 and Pco. The difference between the Fc values measured with and without the sector
was statistically insignificant.

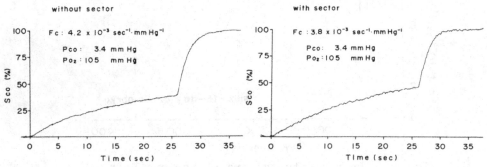

Fig. 7. Sco patterns measured with(right) or without sector(left).

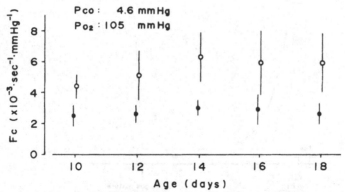

Fig. 8. Fc values measured with or without silicon membrane.
Open circles show the results measured without silicon membrane.

3. Effect of Silicon Membrane on the Reaction Rate.

In order to compare the diffusion velocity through the
capillary wall of the chorioallantoic membrane with that through
the silicon membrane we measured the reaction rate in two cases
that the chorioallantoic membrane was covered with the silicon
membrane and not covered. This difference in Fc value between
both cases is shown in Fig. 8, where the Fc values were measured
at the 10th, 12th, 14th, 16th and 18th days of incubation and at
$Po_2=105$ mmHg, and $Pco=4.6$ mmHg. The usage of the silicon membrane
retarded markedly the velocity to about a half the value measured
with no membrane. Between both the chorioallantoic and silicon
membranes there must be a layer of buffer solution with a certain
thickness. Therefore, we cannot estimate accurately the diffusion
velocity through the chorioallantoic membrane. However, it may
be provable that the diffusion velocity through the capillary wall
is of the same order of that of the silicon membrane.

DISCUSSION

A microphotometric method was developed to make possible the
observation on the oxygenation and CO combination reaction of the
red cell directly in the capillary. In the present study it was
revealed that the CO reaction could be obtained in the chorioallantoic capillary of chicken embryos and that the reaction rate
measured was fairly reproducible. Furthermore, it was confirmed
that Pco for the CO reaction could be reduced to the value less
than 1 mmHg and that the velocity factor Fc was independent of
the Pco at a range lower than 7 mmHg.

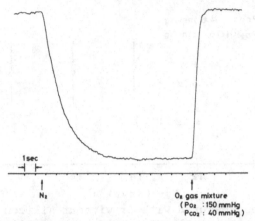

Fig. 9. A record of deoxygenation and oxygenation.

As shown in Fig. 5, the change in transmission was obviously detected in the process of deoxygenation at the same measuring wave length. Thus, it was suggested that this method can be applied to the study on the oxygenation and deoxygenation of the red cell. In Fig. 9 is shown an example of the record obtained in the red cell in the chorioallantoic capillary during successive processes of deoxygenation and oxygenation. The Po_2 was rapidly converted from about 150 mmHg to 0 mmHg at a constant Pco_2 of 40 mmHg, and then converted again to 150 mmHg. We found that a considerably long time was required for the deoxygenation than for the oxygenation.

In order to eliminate the effect of the incident light on the dissociation of CO from carboxyhemoglobin, Sirs used a narrow band filters (3). In the present apparatus a band width of the incident light was narrowed by using a dichroic mirror and condenser lenses, but it had still a width of 350 to 460 mµ. However, we could verify the effect on the reaction rate by changing the light intensity by use of the sector. Although we still have problems as mentioned above, this apparatus will become a very useful tool for the kinetic study of the red cell in situ.

REFERENCES

1) Fukui, K and Y. Kakiuchi: Jap. J. Physiol. 20, 332-347, 1970.
2) Fukui, K and M. Mochizuki: Mono. Series Res. Inst. Appl.
    Electr. No.20, 69-78, 1972.
3) Sirs, J. A. and F. J. W. Roughton: J. Appl. Physiol. 18, 158
    -165, 1963.
4) Tazawa, H.: Respir. Physiol. 13, 160-170, 352-360, 1971.
    17, 21-31, 1973.

MATHEMATICAL ANALYSIS OF COMBINED PLACENTAL-

FETAL OXYGEN TRANSPORT

Eric J. Guilbeau and Daniel D. Reneau

Department of Biomedical Engineering

Louisiana Tech University, Ruston, Louisiana 71270

During contractions of labor, uterine blood flow varies directly with the strength and duration of the contraction (3,5,6).
In addition to observations in sheep and dogs, radioangiographic
studies have shown that in both monkeys and humans, placental
interviilous blood flow is retarded in synchrony with the uterine
contractions of labor (2,13). Prolonged partial asphyxia resulting from strong oxytoxin induced contractions have been shown to
produce varying degrees of brain damage in term monkey fetuses
(11,12).

To our knowledge, no experimental measurements of the oxygen
concentration of fetal blood leaving the placenta and returning to
the fetus have been made on a continuous basis during labor. Fetal
arterial oxygen concentration is a better measure of placental efficiency and fetal well being than uterine or placental blood flow.
The difficulties of continuously monitoring fetal arterial oxygen
concentration during spontaneous labor explain the absence of this
type of measurement in the literature. In our previous analyses
(9,10) the fetal exchange unit was modeled as if all exchange occurred in a single concurrent system, and the fetal blood was assumed to return to the placenta with a constant arterial oxygen
concentration. Although these assumptions had no effect on the
conclusions reached in the previous studies, quantitative values
were affected as pointed out in the original publications. Following reductions in maternal placental blood flow, oxygen concentrations in the fetal blood leaving the placenta are reduced and
subsequently cause the oxygen concentration of the fetal blood
leaving the fetus to also decrease -- provided fetal oxygen consumption is maintained at its normal value. This effect can further exhibit itself by causing additional decreases in the oxygen

concentration of fetal blood leaving the placenta and returning to
the fetus. During transitory changes this can be a dangerous
pyramiding effect.

The purpose, of this analysis is to expand our initial study
so that the placental exchange unit includes the effects of concurrent and countercurrent flow and the effect of fetal oxygen consumption. Continuous solutions of this combined placental-fetus
model are analyzed in an attempt to understand more completely the
effects of cyclic variations in uterine blood flow on fetal blood
oxygenation.

MATHEMATICAL METHOD

In previous mathematical analyses of oxygen transport, distributed parameter analyses have been used to study the time response of the microscopic exchange unit (terminal villous) of the
human placenta (7,8,9,10,14,15). Based on a revised multivillous
flow arrangement fashioned after the arrangement of Bartel's (1),
it was hypothesized that within a single terminal villous of the
human placenta not only concurrent but also countercurrent as well
as cross current flow may exist. Due to the complexities involved
in solving sets of simultaneous, coupled, non-linear, partial differential equations under conditions of countercurrent flow, the
terminal villous was modeled and solved from a distributed point of
view as if all transport occurred in a single concurrent exchange
system. Fetal placental arterial PO_2 was maintained at a constant
normal value and conclusions were limited to trends predicted by
placing various changes on the parameters of the model.

In the present analysis, the fetus and placenta are treated
as an interconnected system. The placenta is represented by two
microscopic exchange units in which all oxygen exchange occurs.
These exchange units represent the concurrent and the countercurrent sections of a single terminal villous. Fetal blood enters
the concurrent section of the terminal villous and received oxygen
by simple diffusion from maternal blood flowing in an annular-like
space down the axial length of the terminal villous. The fetal
blood leaving the concurrent exchange section then enters the
countercurrent section where additional oxygen is received. The
fetal blood then leaves the placental exchange unit and returns by
way of the umbilical circulation to the fetus where oxygen is consumed by various fetal tissues. After having its oxygen supply
either partially or totally depleted by fetal tissue, fetal blood
then returns to the concurrent section of the placental exchange
unit to close the loop. Prior open loop studies now become more
realistic in a closed loop analysis. In a transient analysis of this
system the oxygen concentration of the fetal arterial blood can
vary as changes occur in parameters affecting placental oxygen
exchange.

In the mathematical analysis of this combined placental-
fetus system, the concurrent capillary system was "finite differenced" by dividing it into a series of compartments. The non-linear
partial differential equations which were used in the previous distributed parameter, mathematical analysis could then be replaced by
a system of non-linear ordinary differential equations (4). The
countercurrent exchange system was treated in the same manner, and
the resulting sets of non-linear, ordinary differential equations
for the nth compartment are given as follows for the concurrent and
countercurrent exchange sections, respectively:

<u>Concurrent Exchange Unit:</u>

Maternal Equation,

$$\frac{d\lambda'_{mn}}{dt} = \frac{1}{2} \frac{Q_m}{V_m} \left(\lambda'_{mn-1} - \lambda'_{mn+1} \right) - \frac{H'_n}{V_m N_m} \tag{1}$$

Fetal Equation,

$$\frac{d\lambda'_{fn}}{dt} = \frac{1}{2} \frac{Q_f}{V_f} \left(\lambda'_{fn-1} - \lambda'_{fn+1} \right) + \frac{H'_n}{V_f N_f} \tag{2}$$

where,

$$H'_n = KA_s \left(P'_{mn} - P'_{fn} \right) \tag{3}$$

<u>Countercurrent Exchange Unit:</u>

Maternal Equation,

$$\frac{d\lambda''_{mn}}{dt} = \frac{1}{2} \frac{Q_m}{V_m} \left(\lambda''_{mn-1} - \lambda''_{mn+1} \right) - \frac{H''_n}{V_m H_m} \tag{4}$$

Fetal Equation,

$$\frac{d\lambda''_{fn}}{dt} = \frac{1}{2} \frac{Q_f}{V_f} \left(\lambda''_{fn+1} - \lambda''_{fn-1} \right) + \frac{H''_n}{V_f N_f} \tag{5}$$

where,

$$H''_n = KA_s \left(P''_{mn} - P''_{fn} \right) \tag{6}$$

The relationship between fractional saturation and partial pressure was represented quantitatively by the Hill equation. The number of equations which must be used to accurately represent the exchange unit depends upon the number of compartments into which the system is divided. This number must be determined by trial and error for each individual system. Each compartment in the concurrent exchange system is represented by an ordinary differential equation for the maternal stream and an ordinary differential equation for the fetal stream. A third equation is required to represent the oxygen transferred from one stream to the other, and two additional expressions relate the fractional saturation of fetal and maternal bloods with their respective partial pressures. An analogous situation exists for each compartment of the countercurrent exchange unit.

The oxygen requirements of the fetus were represented by a single ordinary differential equation as follows:

$$\frac{d\lambda'_{fi}}{dt} = \frac{Q_{Fb}}{V}\left(\lambda''_{fo} - \lambda'_{fi}\right) - \frac{A}{N_f V} \tag{7}$$

The equations for the concurrent and countercurrent exchange sections of the placenta and the fetal equation were arranged in a continuous manner with all compartments interconnected. The resulting continuous, unsteady-state model was then solved to yield not only axial oxygen concentration profiles in the placental capillaries, but also, transient changes in fetal blood oxygen concentration.

RESPONSE OF THE COMBINED MODEL TO MATERNAL VELOCITY STEP CHANGES

Figure I shows the response of the combined placental-fetus model to step changes on maternal blood velocity. The fetal, end-capillary oxygen partial pressure is plotted versus time, expressed in terms of the number of times the fetal blood has passed through the system. Comparison of Figure I which represents the combined placental-fetus model with results of our previous distributed parameter analysis in which arterial oxygen concentration was held constant, revealed some interesting differences. Including the effects of fetal oxygen consumption causes a downward shift in the predicted time responses. For all step changes on maternal blood flow, the combined placental-fetus model predicted lower fetal, end-capillary oxygen concentrations than the original analysis. The time required to reach the new steady state value for the fetal end-capillary oxygen concentration is different for the combined placental-fetus model and the concurrent model. Figure I shows that for the case where fetal oxygen consumption is included the new steady state concentration has not been completely reached

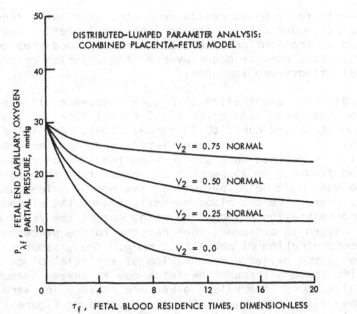

FIGURE 1 FETAL END CAPILLARY RESPONSE TO STEP CHANGES ON
MATERNAL VELOCITY

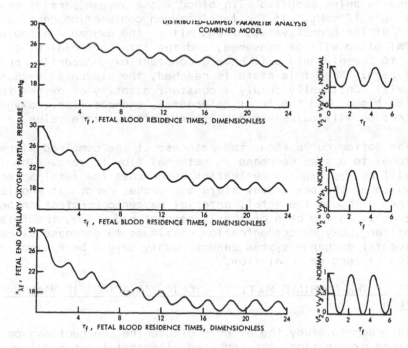

FIGURE 2 FETAL END CAPILLARY OXYGEN PARTIAL PRESSURE RESPONSE TO
SINUSOIDAL MATERNAL VELOCITY CHANGES OF VARIOUS AMPLITUDE

after twenty fetal blood residence times. Not only, then, does
the combined model predict greater changes in fetal, end-capillary,
oxygen concentration for identical maternal blood step decreases,
but also, these changes occur over a longer period of time when
the fetal effects are included.

Additional examination of Figure I shows a difference in the
shapes of the upper two curves (0.75 normal flow and 0.50 normal
flow) and the third curve (0.25 normal flow). The third curve
shows that the new steady state, fetal, end-capillary, oxygen concentration is reached much sooner when the step decrease on maternal blood flow was to 25 percent of normal flow than when the step
decrease was to 75 or 50 percent of the normal value (upper two
curves). Fetal, venous blood normally leaves the placenta with an
oxygen concentration of about 30 mmHg, enters the fetus where part
of this oxygen is consumed, then returns to the placenta with an
oxygen concentration of about 11.5 mmHg. Thus, about an 18.5 mmHg
decrease in the oxygen concentration of the fetal blood occurs
during its passage through the fetus due to oxygen consumption by
the fetal tissues (the values given are those which were used in
the mathematical analysis). The third curve of Figure I shows that
the fetal blood, end-capillary, oxygen concentration decreases
from the normal value of 30 mmHg to about 12 mmHg. The fetus in
this case is being supplied with blood whose oxygen partial pressure is only 12 mmHg. If the fetal oxygen consumption remains
constant at its normal value, then, all of the oxygen available in
the fetal blood will be consumed, and the fetal arterial blood returning to the placenta will have a constant oxygen partial pressure of zero. When this state is reached, the placental exchange
system will continually supply a constant quantity of oxygen to
the fetal blood, and the fetal, placental, end-capillary, oxygen
concentration will quickly reach its new steady state value.

The bottom curve shows the response of the combined placental
fetus model to a step decrease in maternal flow to zero. The fetal,
end-capillary, oxygen concentration approaches the fetal arterial
value which at the new steady state approaches zero. It should be
noted that although the fetal, arterial oxygen concentration becomes zero as in the case of the 25 percent reduction, the fetal,
end-capillary, oxygen concentration continues to decrease since
the placental exchange system cannot supply oxygen because of the
condition of zero maternal flow.

RESPONSE OF THE COMBINED MODEL TO CYCLIC VARIATIONS IN MATERNAL
BLOOD FLOW

In order to study the effects of uterine contractions on
fetal blood oxygenation, the combined placental-fetus model was
solved using sinusoidal velocity functions of various amplitudes.
Figure 2 shows the response of the combined placental-fetus model

to three sinusoidal velocity functions with the same 45 second period. The upper curve gives the fetal, end-capillary, oxygen concentration response to a maternal velocity fluctuation in which the minimum value is one-half of the normal value. The middle curve represents the response of the combined placental-fetus model to a maternal blood velocity fluctuation in which the maternal blood velocity cycles between the normal value and a value which is twenty-five percent of the normal value. The lower curve shows the response when the maternal blood flow varies between the normal value and zero. With an increase in the amplitude of the maternal blood flow variation, the fetal, end-capillary, oxygen concentration is reduced to lower and lower values. The upper two curves continue to decrease after twenty-four fetal blood residence times, i.e., the pseudo-steady state has not been attained. Comparison of these curves with the lower curve shows an interesting difference. The lower curve reaches the pseudo-steady state after about 17 fetal blood residence times. As in the step change studies, fetal oxygen consumption normally accounts for an 18.5 mmHg decrease in the fetal blood oxygen concentration. Hence, when the fetal venous oxygen partial pressure decreases below this value, the fetal arterial oxygen concentration falls to a constant value of zero. The pseudo-steady state is quickly reached as shown in the lower curve of Figure 2. These theoretical results demonstrate that cyclic variations in maternal blood flow can decrease fetal blood oxygenation significantly.

CONCLUSIONS

Based upon the results of the previous distributed parameter concurrent analysis (9) and the present analysis the following conclusions may be made:

1. The lumped-distributed parameter, transient, mathematical analysis is an adequate means of describing the placental-fetus system if radial gradients are not of importance and if the effects of axial diffusion may be neglected.

2. Transient mathematical analysis of placental oxygen transport can only be interpreted adequately when the fetus is included in the system and the placental-fetus system is solved as an interconnected continuous unit.

3. Cyclic variations in maternal blood velocity of sufficient amplitude can theoretically result in cyclic variations and reductions in fetal blood oxygen concentration.

4. Future mathematical analysis of placental oxygen transport should strive not only to simulate all of those factors which interact to affect fetal blood oxygenation at the level of the placenta but also should include those factors in the fetal circulation.

5. Continued experimental and theoretical analysis of the placental oxygen transfer problem should be conducted to gain a more quantitative representation of the effects of uterine contractions on placental blood flow and fetal blood oxygenation during labor so that the clinician may have available a more adequate basis for evaluation.

NOMENCLATURE

A   Metabolism Rate of Fetal Tissue, $cm^3_{O_2}$/second

A_{sn} Surface area of the nth placental exchange unit compartment, cm^2

K   Mass transfer coefficient, $cm^3_{O_2}$/cm^2-second-mmHg

N   Oxygen capacity of blood, $cm^3_{O_2}$/cm^3 blood

P   Oxygen partial pressure, mmHg

QFb   Fetal umbilical, volumetric flow rate, cm^3 blood/second

Q   Volumetric blood flow rate, cm^3 blood/second

t   Time, seconds

V   Volume, cm^3

V_2 Maternal blood linear velocity, cm/second

V'_2 Dimensionless maternal blood velocity, V_2/V_2 normal

GREEK SYMBOLS

λ Fractional saturation of blood

τ Number of residence times, dimensionless

SUBSCRIPTS

n   Refers to the nth compartment

f   Fetal value

m   Maternal Value

i   Inlet value

o   Outlet value

SUPERSCRIPTS

'   Refers to concurrent exchange system

"   Refers to countercurrent exchange system

REFERENCES

1.   Bartels, H., W. Moll, and J. Metcalf. Physiology of Gas
      Exchange in the Human Placenta. Am. J. Obstet. Gynec..
      84:1714-30, 1962.

2.   Borell, U., I. Fernström, L. Ohlson, and N. Wiqvist.
      Influence of Uterine Contractions on the Uteroplacental
      Blood Flow at Term. Am. J. Obstet. Gynec. 93:44-57, 1965.

3.   Brotanek, V., C. H. Hendricks, and T. Yoshida. Changes in
      Uterine Blood Flow During Uterine Contractions. Am. J. Obstet.
      Gynec. 103:1108-16, 1969.

4.   Franks, Roger G. E. Mathematical Modeling in Chemical
      Engineering New York, John Wiley & Sons, Inc., 1967.

5.   Greiss, F. C. Effect of Labor on Uterine Blood Flow.
      Am. J. Obstet. Gynec. 93:917-923, 1965.

6.   Greiss, F. C. A Clinical Concept of Uterine Blood Flow
      During Pregnancy. Obstet. Gynec. 30:595-604, 1967.

7.   Guilbeau, E. J., A Steady and Unsteady State Mathematical
      Simulation of Oxygen Exchange in the Human Placenta.
      Dissertation, Louisiana Tech University, Ruston, Louisiana
      1971.

8.   Guilbeau, E. J., D. D. Reneau, and M. H. Knisely, A Dis-
      tributed Parameter Mathematical Analysis of Oxygen Exchange
      From Maternal To Fetal Blood In The Human Placenta. In:
      Blood Oxygenation ed. by D. Hershey, New York, London,
      Plenum Press, 1970.

9.   Guilbeau, E. J., D. D. Reneau, and M. H. Knisely. The
      Effects of Placental Oxygen Consumption and the Contractions
      of Labor on Fetal Oxygen Supply - A Steady and Unsteady
      State Mathematical Simulation. In: Respiratory Gas Exchange
      in the Placenta ed. by L. Longo and H. Bartels, Department
      of H.E.W., NICHHD, (in press).

10.   Guilbeau, E. J., D. D. Reneau, and M. H. Knisely. A
      Detailed Quantitative Analysis of O_2 Transport in the Human
      Placenta During Steady and Unsteady State Conditions. In:
      Chemical Engineering and Medicine, Advances in Chemistry
      Series, American Chemical Society, (in press).

11.   Myers, R. E. Two Patterns of Perinatal Brain Damage and
      Their Conditions of Occurance. Am. J. Obstet. Gynec.
      112:246-576, 1972.

12.   Myers, R. E., R. Beard, and K. Adamsons, Brain Swelling in
       the New Born Rhesus Monkey Following Prolonged Partial
       Asphyxia. Neurology 19:1012-1018, 1969.

13.   Ramsey, E. M., G. W. Corner, and M. W. Donner. Serial and
       Cineradioangiographic Visualization of Maternal Circulation
       in the Primate Hemochorial Placenta. Am. J. Obstet. Gynec.
       86:213-225, 1963.

14.   Reneau, D. D., Jr., D. F. Bruley, and M. H. Knisely. A
       Mathematical Simulation of Oxygen Release, Diffusion and
       Consumption in the Tissues and Capillaries of the Human
       Brain. In: Chemical Engineering in Medicine and Biology
       ed. by D. Hershey, New York, London, Plenum Press, 1967.

15.   Reneau, D. D., D. F. Bruley, and M. H. Knisely. A Digital
       Simulation of Transient Oxygen Transport in Capillary-Tissue
       System (cerebral gray matter). A.I.Ch.E. J. 15:916-925,
       1969.

This work was supported in part by USPHS Grant NIH-NS-08228 and
The Frost Foundation.

PLACENTAL FUNCTION AND OXYGENATION IN THE FETUS

Waldemar Moll, M.D.

Department of Physiology

University of Regensburg, W.-Germany

The oxygenation in the newborn depends to some degree on the impact of labor. It has been shown repeatedly that fetal blood gases stay (as determined in the scalp blood) rather high and constant before labor. In the second stage of labor (the expulsive stage of labor) however, fetal O_2 tensions and saturations drop. According to Wulf and co-workers (1967) the O_2 tension in the scalp blood falls from 23 to 17 Torr. Fetal hypoxia and eventually fetal anoxia develop. A low oxygenation is finally found in the newborn baby.

The general feeling is that fetal hypoxia in the expulsive stage of labor is due to dysfunction of the placenta in this period, caused by uterine contractions. The current concept for the fetal hypoxia is this: uterine contractions throttle maternal placental flow. A transient fall of placental gas exchange results. Fetal p_{O2} drops. The drop provides an additional sink for O_2; placental O_2 transfer may reach its previous level. Eventually, fetal arterial concentrations may drop deep enough to cut short O_2 supply. Placental transfer falls.

Uterine contractions

Maternal placental flow

Placental O_2 transfer

Fetal arterial concentrations

O_2 supply to fetal tissue

In this paper flow reduction, placental function, and tissue O_2 supply in the expulsive stage of labor shall be described. The description first refers to normal conditions, secondly to pathological conditions when severe anoxia occurs. I will try to derive quantitatively the interrelationships between placental transfer, fetal arterial partial pressures, and O_2 supply to fetal tissues. For a quantitative understanding a model of the fetoplacental unit is discussed.

PLACENTAL BLOOD FLOW DURING LABOR

It has been proved qualitatively by Borell et al. (1964) that maternal placental blood flow in human is reduced during labor. The entry of blood into the placenta may be completely blocked by a uterine contraction. No quantitative data are available in human. In the Rhesus monkey, however, Martin (1973) established a quantitative relationship between intrauterine pressure and percent reduction of maternal flow. He found that the uterine conductance falls linearly with the intrauterine pressure reaching almost zero when the intrauterine pressure is 60 Torr.

As a first approximation we may assume this relationship to be true even in human. Doing so, we can derive placental flow from the well-known intrauterine pressures in human.

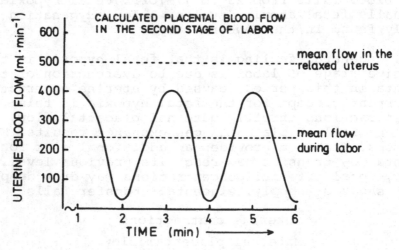

Fig. 1: Calculated maternal placental flow in human during the expulsive stage of labor.

The calculated maternal placental blood flow during
the expulsive stage of labor is shown in Fig. 1. According to this (admittedly rough) approximation placental
blood flow drops down during a vigorous contraction. It
may reach zero at the height of the contraction. The
mean flow during labor is calculated to be around
250 ml/min, i.e. around half the value in the relaxed
uterus (500 ml/min according to Metcalfe et al., 1953).

A FETOPLACENTAL MODEL

To get a quantitative understanding on the effect
of decreased uterine blood flow on placental function
and the oxygenation of the fetus and the newborn respectively we may study a simple model of the fetoplacental
unit we proposed two years ago (Fig. 2). This model makes
allowance for the arrangement of placental vessels and
the rate and distribution of cardiac output. In the model the fetal heart pumps mixed arterial blood across
the fetal tissue and the placenta. In the tissue, fetal
blood releases oxygen and returns via the fetal body
veins. In the placenta, the blood takes up oxygen as de-

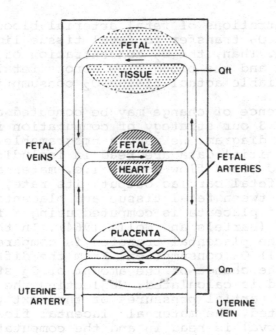

Fig. 2: A model of the fetoplacental unit

termined by placental blood flow and diffusion capacity
and returns via the umbilical veins. The blood of the
two different sorts of veins yields the mixed arterial
blood of the fetus. According to findings of Dawes et al.
(1954) in sheep we are not very wrong assuming complete
mixing of arterialized and venosized blood.

Let us consider now on the model what happens in
the baby during labor. For a given moment a definite arterial p_{O2} prevails in fetal and maternal blood. When a
contraction wave is on its way the actual maternal placental flow is lower than in the moment before. A reduced transfer in the placenta results. The placental
gain of O_2 now is lower than the O_2 consumption (O_2 loss)
of the tissue. Therefore, in the considered time interval, the O_2 stored in the fetal blood, i.e. O_2 concentration in the fetal blood will drop off. This drop off
is inversely related to the blood volume in the fetus.

At the next time interval new O_2 partial pressures
prevail. The new O_2 concentrations and the new flow rates of the next moment define the O_2 transfer of the second time interval and the next shift of fetal O_2 concentrations.

O_2 concentrations of fetal arterial blood may drop
thus deep that O_2 transfer to fetal tissue limits tissue
O_2 consumption. Than, the O_2 concentration of the mixed
arterial blood and the blood flow across fetal tissue
define the variable actual tissue O_2 consumption.

This sequence of change may be computed automatically. In Fig. 3 our strategy of computation is indicated by a block diagram. We used a programable desk cal-
culator. The initial data are read in including arterial
blood gases and blood flow rates, i.e. maternal placental flow, the fetal cardiac output, its rate, and its
distribution between fetal tissue and placenta. The O_2
transfer in the placenta is computed using a finite difference method (Bartels and Moll, 1964). In the non-
anoxic state the placental transfer is compared with the
normal (maximal) O_2 consumption. From the difference of
those values the change of the amount of O_2 stored in
the fetal blood is calculated. Neglecting the circulation
time of the O_2 partial pressures of the next time interval are estimated. The maternal placental flow of the
next time interval is read in and the computation process
is repeated yielding the situation for the next time interval.

In the anoxic state the actual O_2 consumption is derived from the arterial concentration and the blood flow across the fetal tissue. Using this value the new fetal p_{O2} is calculated.

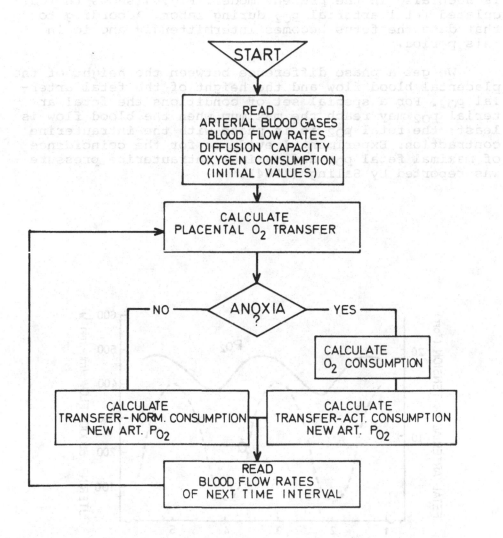

Fig. 3:   Block diagram for computing the oxygenation
          in the fetoplacental unit

CYCLIC CHANGES OF FETAL ARTERIAL pO2 DURING LABOR

As I described above uterine blood flow cycles upwards and downwards. What is the effect of this cycling on the fetal p_{O_2}? Reneau and Guilbeau (personal communications) postulated that a cyclic change of fetal p_{O_2} is to be expected in this condition. Indeed, such a cycling is seen also in the present model. Fig. 4 shows the calculated fetal arterial p_{O_2} during labor. According to that data the fetus becomes intermittendly anoxic in this period.

We get a phase difference between the height of the placental blood flow and the height of the fetal arterial p_{O_2}. For a special set of conditions the fetal arterial p_{O_2} may reach the maximum when the blood flow is least; the fetal p_{O_2} is in phase with the intrauterine contraction. Experimental evidence for the coincidence of maximal fetal p_{O_2} and maximal intrauterine pressure was reported by Saling (1964).

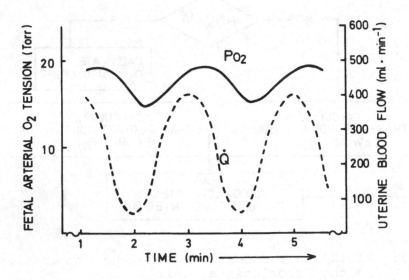

Fig. 4:   Calculated fetal arterial p_{O_2} during labor.

THE MEAN DROP OF THE FETAL ARTERIAL pO2 IN THE
EXPULSIVE STAGE OF LABOR

What is the mean drop of the p_{O2} in fetal arterial
blood when maternal placental flow is reduced to one
half of its previous value? Is it the value we actually
find in the fetus?

In Fig. 5 fetal arterial p_{O2} found in the model is
related to the mean placental flow. The O_2 uptake of the
fetoplacental unit is around 25 ml/min/kg according to
Metcalfe et al. (1953). Fetal cardiac output in human
seems to be 800 ml/min (Winsberg, 1972). The diffusion
capacity of the placenta is assumed to be 1.5 ml/min/Torr
as this value results in a reasonable fetal arterial p_{O2}.

When the uterus is relaxed placental flow is about
500 ml/min according to Metcalfe et al. (1953). When the
maternal placental flow drops to 250 ml/min (the value
we approximated above) fetal p_{O2} falls from 23 to 18 Torr.
Similar results were obtained by Longo et al. (1972) for
sheep. We may conclude that the fetal hypoxia in the expulsive stage of labor seems to be adequately explained
by the effect of uterine contractions of the maternal
placental flow. We need not to assume additional factors
as changes of fetal cardiac output or of the placental
diffusion capacity for explaining the fetal hypoxia during normal labor in human.

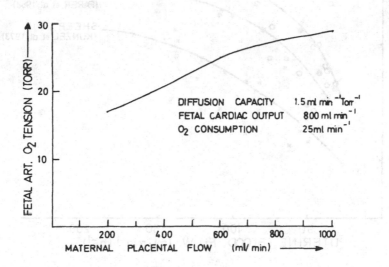

Fig. 5: Calculated mean fetal arterial p_{O2} related to
         maternal placental flow.

PLACENTAL O$_2$ TRANSFER TO THE FETUS DURING LABOR

Using the available data on fetal cardiac output in human (Winsberg, 1972) we have to assume that the placental O$_2$ transfer is decreased during the normal expulsive stage of labor. The situation is aggravated when the intensity of uterine contractions is abnormally high or when the uterine perfusion is primarily decreased as in the EPH gestosis.

Fetal anoxia is indicated by a decrease of O$_2$ consumption. Such a decrease of O$_2$ consumption is observed in the fetus more frequently than in the adult. In animals (see Fig. 6) the relation between uterine blood flow and uterine O$_2$ uptake is established fairly well. It can be shown that uterine O$_2$ uptake is reduced when the uterine blood flow falls below 80 ml/min/kg. As the picture is rather uniform in the 3 species (Rhesus monkey, sheep, Guinea pig) we may assume that it is also true in human. According to the data, a reduction of fetal O$_2$ uptake is to be expected when maternal placental flow falls to 250 ml/min, i.e. around 50 ml/min/kg.

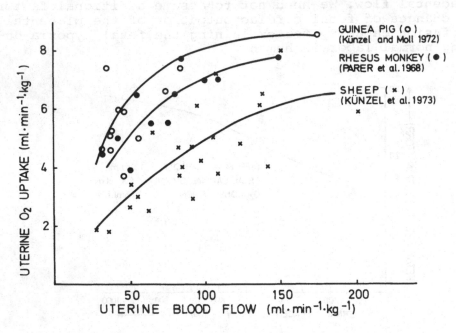

Fig. 6:   Placental O$_2$ transfer related to uterine blood flow.

THE EFFECT OF FETAL CIRCULATORY REACTIONS ON PLACENTAL
O_2 TRANSFER AND FETAL ARTERIAL pO_2

It has been found by various authors that rate and
distribution of fetal cardiac output change during severe fetal anoxia in animals.

According to studies on the model a fall of cardiac
output further reduced the O_2 transport to fetal tissue,
regardless whether a redistribution of fetal cardiac
output occurs or not. The effect of this fall is small
as long as fetal placental flow still exceeds the reduced maternal placental flow.

Puzzling phenomena may occur as far as the fetal
arterial pO_2 is concerned. When cardiac output falls
without any redistribution taking place the pO_2 of the
fetal mixed arterial blood may rise even if the baby
gets increasingly worse (see Fig. 7). This may be explained as follows: a fall of fetal placental flow results in a higher oxygenation of fetal blood. A fall
of the blood flow across fetal tissue does not alter
the venous concentration if this concentration has already reached a value near zero. The O_2 concentration
of the mixed arterial blood will therefore rise.

The pO_2 of the mixed arterial blood is an ambiguous
sign. A high pO_2 may indicate an excellent O_2 supply as
well as a desolate situation.

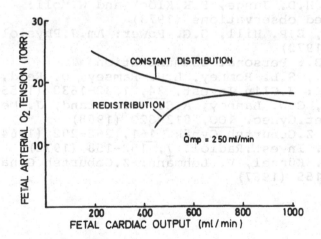

Fig. 7:   Fetal arterial pO_2 during reduced maternal
          placental flow related to fetal cardiac output.

Behrman and co-workers (1970) reported that cardiac output is redistributed when it falls during hypoxia. Studying Rhesus monkeys during anoxia they found that placental blood flow was cut in half while the blood flow across the heart and the brain remained constant.

This redistribution results in a different behavior of the fetal arterial p_{O2}. Now fetal p_{O2} drops rapidly together with anoxia. Again this shows that intricate relations are to be expected between fetal p_{O2} and the adequacy of fetal oxygen supply. What we find in fetal arterial blood seems to depend on fetal circulatory reaction.

REFERENCES

Bartels, H. and W. Moll: Pflügers Arch.ges.Phys. 280, 165-177 (1964)

Behrman, R.E., M.H. Lees, E.N. Peterson, C.W. De Lannoy, A.E. Seeds: Am.J.Obstet.Gynec. 103, 956-969 (1970)

Borell, U., I. Fernström, L. Ohlson, N. Wiquist: Am.J.Obstet.Gynec. 89, 881-890 (1964)

Dawes, G.S., J.C. Mott, J.O. Widdicombe: J.Physiol. 126, 563 (1954)

Guilbeau, E.J. and D.D. Reyneau: Personal communications

Künzel, W. and W. Moll: Z.Geburtsh.Perinat. 176, 1o8-117 (1972)

Künzel, W., H.D. Junge, F.K.Klöck and W.Moll: Unpublished observations (1973)

Longo, L.D., E.P. Hill, G.G. Power: Am.J.Physiol. 222, 730-739 (1972)

Martin, Ch.B.: Personal communication

Metcalfe, J., S.L. Romney, L.H. Ramsey, D. Reid, S. Burwell: J.Clin.Invest. 34, 1632-1638 (1955)

Parer, J.T., C.W. Lannoy, A.S. Hoversland, J. Metcalfe: Am.J.Obstet.Gynec. 100, 813-820 (1968)

Saling, E.: Z.Geburtsh.Gynäk. 161, 262-292 (1964)

Winsberg, F. Invest.Radiol. 7, 152-158 (1972)

Wulf, H., W. Künzel, V. Lehmann: Z.Geburtsh.Gynäk. 167, 113-155 (1967)

MORPHOLOGICAL DETERMINANTS IN O_2-TRANSFER ACROSS THE HUMAN AND

RHESUS HEMOCHORIAL PLACENTA

Uwe E. Freese, M.D.

Dept. of Ob.-Gyn., Pritzker Sch. of Medicine, University

of Chicago

The transfer of gases across the membranes of the hemochorial placenta is accomplished by simple diffusion and therefore, ought to
be a simple process. In fact, however, the determination of rate,
time, sites, efficiency and mode of transfer are extremely difficult
to assess because of the large number of variables involved in this
process and because of technical obstacles.

A closer look at the problem reveals that there are two extremely active hemodynamic systems separated by a complex, active
biological membrane, the syncytiocapillary membrane. These two
systems are divided up into an unknown number of exchange units
which vary in size, configuration and number from one moment to the
next, belonging to a closed, morphologically more stable system on
the fetal and to an open, less stable one on the maternal side.

The blood in these two systems moves with different velocities
and the amounts passing through the placenta on the fetal and in the
intervillous space on the maternal side differ markedly overall,
from time to time and from one exchange unit to the next.

While diffusion characteristics of a membrane, such as the
syncytiocapillary, is an important factor to obtain maximum transfer, convection, like in most biological transport, is likely to
determine limitations.

The purpose of this paper is to examine the two hemodynamic
compartments, their topographical relationship to each other and
the areas where most of the exchange is likely to occur.

THE FETAL COTYLEDON

More recently, the fetal cotyledon of the human and primate placenta has been clearly defined as the functional fetal unit of the placenta.(3,4,5,7,8,13,14) In the human, about half of the approximately 200-240 cotyledons are specifically structured as a result of their topographical relationship to the spiral artery (6).

First and most importantly, the central portion of the cotyledon is so loosely structured that, in vivo, it is almost completely devoid of villi. Measurements show great variations in the dimensions of these central spaces but values of 0.5x1.7cm±3-5mm for each figure are obtained in the human. These central spaces are the site of entry of maternal blood(11). (Fig.1)

Secondly, the periphery of these cotyledons is composed of a few layers formed by anchoring villi, which are sparsely populated with terminal villi, therefore forming channels measuring about 200μ in width, extending around the entire periphery of the cotyledon (8).

Thirdly, in between these two compartments, the central and the peripheral spaces, is the bulk of the terminal or resorption villi densely packed and separated by channels measuring 20-100μ and rarely exceeding these values.

According to Moll and Freese (11), blood pressures in the central and peripheral spaces are homogeneous, yet, of course, lower in the latter. These findings then fulfill the prerequisites for a homogeneous perfusion of all placental levels from the decidua to the sub-chorionic region and from the center to the periphery of each cotyledon.

THE MATERNAL SPIRAL ARTERY

Spiral arteries at the placental site differ considerably from those in adjacent areas, in that they are much larger (Fig.2). While averaging 0.5-1mm in diameter, at 28 weeks gestation, they enlarge to 1.5-2mm in diameter at term. Their terminal portions are not dilated according to our findings (8). Significant is the topographical relationship between some of these arteries and certain fetal cotyledons. Their number varies from case to case, but about 100 spiral arteries are located beneath the center space of an equal number of fetal cotyledons. This relationship identifies cotyledon and spiral artery as inseparable components forming a truly functional unit (6,8). (Fig.3)

As can be seen from (Fig.4) disturbances such as the formation
of thrombi in the center space will eliminate the entire unit
through cessation of the blood flow into these cotyledons.

THE IVS

The intervillous space is a vast network of channels which are
capillary-like in size and width.(Fig.5) Larger spaces are found
only in the center of cotyledons in their periphery and in the sub-
chorial region. Maternal blood occupies this compartment and varies
in amount according to the size of the placenta and intactness of
the intervillous space. Its point of entry is the central space of
those cotyledons which are supplied by a spiral artery. The pressure under which it is released is unknown, nevertheless, the maternal pressure must be considered to be the vis a tergo for the
blood flowing through the intervillous space.

As pointed out above, central and peripheral spaces of the cotyledon represent areas of homogeneous pressure. This seems to be a
sine qua non, considering the density, the total size and the capillary size and nature of that portion of the intervillous space
which is interposed between center and periphery of the cotyledon.
It is this portion where the exchange takes place for the most part.
The capillary-like channels measure usually between 10-100μ in diameter, are lined by trophoblast of the villi and toward term by increasing amounts of fibrin. The considerable variations in channel
sizes, their different distance away from the orifice of the spiral
artery and the differing velocity of maternal blood in these capillary-like channels, provide formidable problems for those trying to
determine rate and site of gas transfer (2,9,10). It seems to me,
based upon anatomical and cineangiographical studies, that the multivillous stream system discussed by Bartels, Moll and Metcalf (1)
is still the closest to real system. It appears to be of little re-
levance whether or not we are dealing with a pure concurrent or
countercurrent system as long as the flow of maternal blood is unin-
terrupted at the point of entry, i.e., into the center space. Rather than being crowded randomly through the intervillous space, the
maternal blood perfuses the cotyledon concentrically from the center
space. Differences in velocity must occur because of narrowing or
blockage of channels by fibrin or clumped corpuscular elements of
the maternal blood. The total surface of the capillary sized intervillous spaces, between the central and the peripheral spaces, however, appears to be adequate to prevent backing up of maternal blood.

Unquestionably, the maternal red cells establish contact with
surfaces of several villi and accordingly a blood sample, moving
from one cleft to the next, will constantly change its gas content.
(Fig.6) Preliminary data obtained by Reynolds, Freese and Caldeyro-
Barcia, et al. (12), in the Rhesus monkey, indicate decreasing val-

ues for PO_2 toward the periphery of the cotyledon and toward the chorionic plate and the decidua. The same can be said for blood pressure and pH.

While the direction of movement of maternal blood in the narrow network of the intervillous space, in relation to the direction of fetal blood streams, appears to be of little importance, the basic entry, perfusion and exit pattern of maternal blood is paramount, for it governs the amount, direction and initial velocity of the entering maternal blood.

From a hemodynamic view point, the role of those cotyledons, not supplied by a spiral artery, is of real interest. Supplied with maternal blood from adjacent units (cotyledon and spiral artery), the velocity must be slow and the blood pressure low. No information on these cotyledons is available.

VILLOUS CAPILLARIES

Emerging from the precapillary network of vein and arteries, they extend toward term well into the periphery underlying the trophoblast. The syncytium is often extremely thin with nuclei retracted into flanking positions on each end of the syncytiocapillary membranes. This membrane is known to average about 5.5μ or less in thickness. The biological activity of this membrane, which possesses two basement membranes, one for the trophoblast and one for the fetal capillary endothelium, is largely determined by its thickness. The villous capillaries follow an irregular course approximating the syncytical layer in several places. This anatomical arrangement seems to render any discussion about flow direction obsolete. (Fig. 7)

There is no evidence that the fetus in response to hypoxic conditions exerts any control over compensatory mechanisms of these capillaries. All indications are that villous and villous-vascular changes are in response to local phenomena.

SUMMARY

The hemochorial placenta of the human and Rhesus monkey has, to a large extent, predetermined sites and initial pathways for the entering maternal blood. Except for the central and peripheral cotyledonary spaces and the subchorionic region, the intervillous space is composed of capillary-size spaces created by densely packed terminal or resorption villi. (Fig.8)

The maternal blood is discharged from the spiral arteries into

the center space, perfuses the cotyledon in a concentric manner and
returns into maternal veins through peripheral cotyledonary spaces
(6). Perfusion homogineity is facilitated by two homogeneous pressure
areas, the central space and the peripheral spaces of the cotyledon.
Exchange areas are not clearly definable morphologically, but they
seem to be confined principally to the syncytiocapillary membranes.
The flow system within the intervillous space is most compatible
with a multivillous scheme. As is true for most biological transport, convection represents the limiting factor and the gas transport from the mother to the fetus and vise versa is not exempted
from this basic concept.

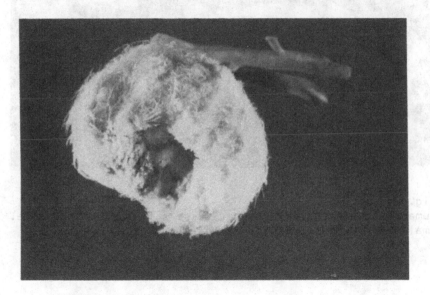

Fig. 1
Human fetal cotyledon. Central and peripheral spaces surround
dense core consisting of terminal villi and capillary size
intervillous spaces.

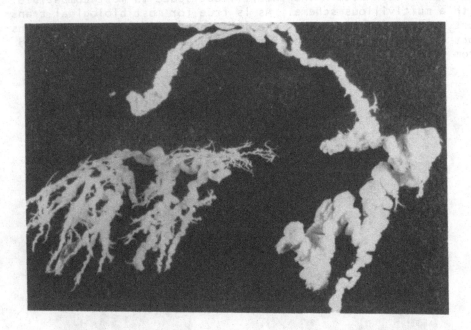

Fig. 2
Human spiral arteries. Larger vessels from placental site.
Smaller from opposite uterine wall.

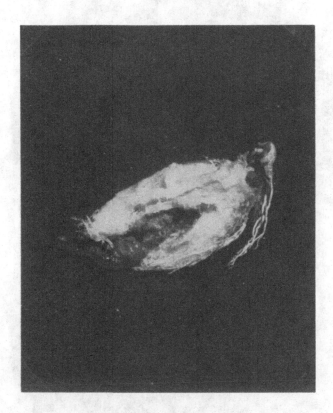

Fig. 3
Human term spiral artery projecting into center space of fetal
cotyledon.

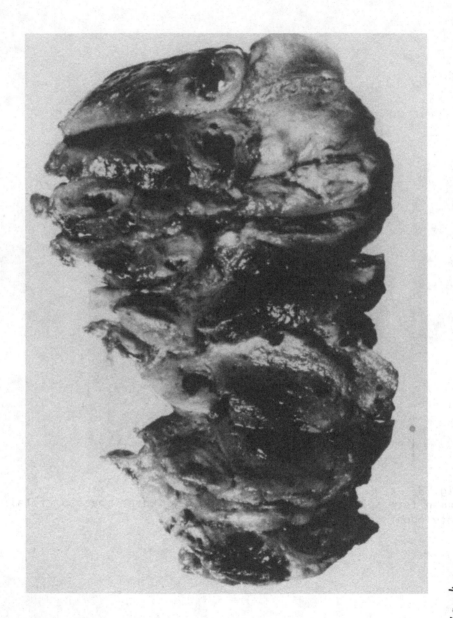

Fig. 4
Human term placenta showing numerous empty and thrombosed central spaces of cotyledons.

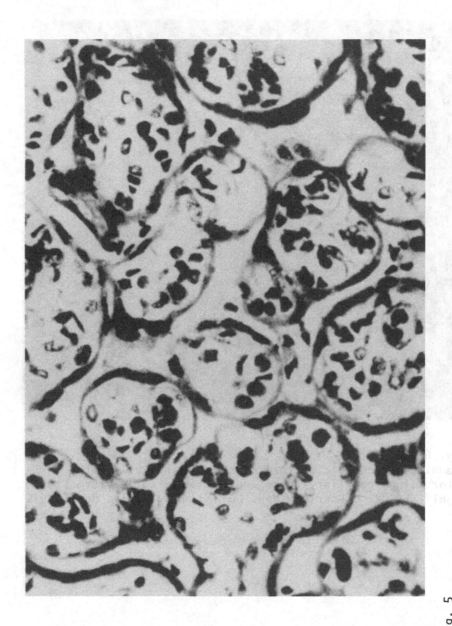

Fig. 5
Crossection of term human placenta. The narrow intervillous spaces and syncytiocapillary membranes are identified.

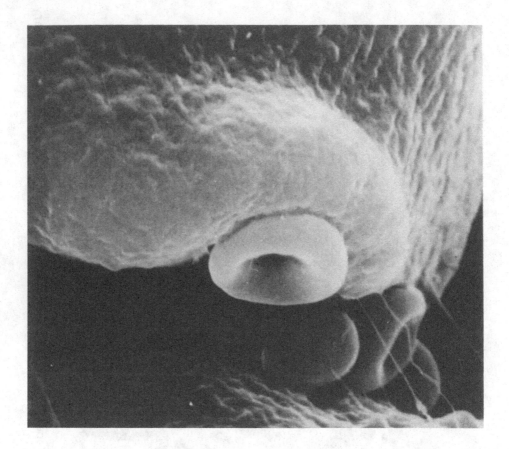

Fig. 6
Scanning microscopic picture of two terminal vill with
intervillous capillary space and four maternal erythrocytes.
Magnif. 5000x.   (Courtesy Hans Ludwig, M.D., Univ. Munich)

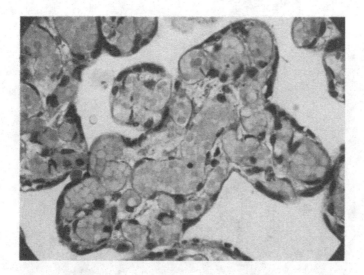

Fig. 7
Terminal villi from human term placenta identifying fortuous
course of fetal capillaries and their proximity to syncytiocapillary membrane. H.E. Stain. Magnif. 400x.

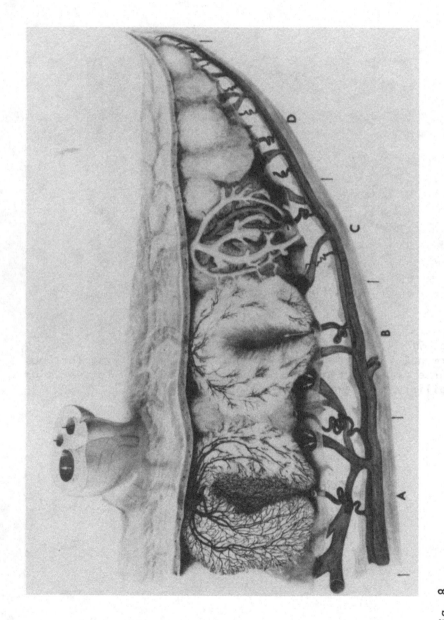

Fig. 8
Scheme of human term placenta and its topographical relationship to maternal spiral arteries and veins.

REFERENCES

1. Bartels,H., Moll,W. & Metcalf,J., Physiology of gas exchange in the human placenta. Am. J. Ob.& Gyn. 84:1714-1730, 1962.

2. Bartels,H., Moll,W. Passage of inert substances and oxygen in the human placenta. Pflugers Arch.280:165, 1964.

3. Bøe,F. Studies on the human placenta. ACTA Obstet. Gynec. Scand. 47:420-435, 1968.

4. Crawford,J.U. Vascular anatomy of the human placenta. Am. J. Ob.& Gyn. 84:1543, 1962.

5. Freese, U.E. The fetal-maternal circulation in the placenta. Am. J. Ob.& Gyn. 94:354, 1966.

6. Freese, U.E. The uteroplacental vascular relationship in the human. Am. J. Ob.& Gyn. 101:8, 1968.

7. Freese, U.E., Ranniger, K. & Kaplan, H. The fetal-maternal circulation of the placenta. Am. J. Ob.& Gyn. 94:361, 1966.

8. Freese, U.E. & Mociolek, B.J. Plastoid injection studies of the utero-placental vascular relationship in the human. Ob.Gyn. 33:160, 1969.

9. Hill, E.P., Power, G.G.& Longo, L.D. A mathematical model of carbon dioxide transfer in the placenta and its interaction with oxygen. Am. J. Physiol. 224: 283-299, 1973.

10. Longo, L.D., Hill, E.P. & Power, G.G. Theoretical analysis of factors affecting placental O_2-transfer. Am. J. Physiol. 222: 730-739, 1972.

11. Moll,W. & Freese, U.E. Hamodynamik des intervillosen Raumes der Primatenplazenta. Perinatale Medizin, Band III, 1972.

12. Reynolds, S.R.M., Freese, U.E., Bieniarz, J., Caldeyro-Barcia, R., Mendez-Bauer, C. & Escarcena,L. Multiple simultaneous intervillous space pressures recorded in several regions of the hemochorial placenta in relation to functional anatomy of the fetal cotyledon. Am. J. Ob.& Gyn. 102: 1128-1134, 1968.

13. Wigglesworth, J.S. Vascular anatomy of the human placenta and its significance for placental pathology. J. Obstet. Gynaec. Brit. Cwlth. 76:979, 1969.

14. Wilkin, P. Morphogenese. In: Le Placenta Human, (ed. Snoeck) 23-70. Masson, Paris.

UMBILICAL AND UTERINE VENOUS PO$_2$ IN DIFFERENT SPECIES DURING LATE GESTATION AND PARTURITION

M. Silver and R. S. Comline

Physiological Laboratory, Cambridge, England

1. VEIN-TO-VEIN PO$_2$ DIFFERENCES ACROSS THE PLACENTAE OF RUMINANTS, HORSE AND PIG

A large PO$_2$ difference (17-21mmHg) between uterine and umbilical venous blood is found in the cotyledonary placenta of ruminants, whereas in the equine placenta the corresponding vein-to-vein difference is only 2-4mmHg (Silver, Steven & Comline, 1973). The pig is another species with a diffuse placenta like that of the mare but without the specialised microcotyledons of the latter; simple vascular networks are found on either side of the pig placenta (Tsutsumi, 1962). Preliminary experiments have now been carried out in this species to determine the transplacental blood gas gradients and the changes which occur when maternal arterial levels are altered. In acute experiments under sodium pentobarbitone anaesthesia, vinyl catheters were inserted into uterine and umbilical vessels without disturbance of the fetus. The umbilical vessels were catheterised through small branches which were only a short distance from the cord. Because each fetus was small even near term, repeated observations were made on different fetuses rather than on a single individual. In each fetus examined the PO$_2$ gradient between maternal and fetal blood was as great or greater than that seen in ruminants; vein-to-vein PO$_2$ differences were 20-24mmHg. When the sow was ventilated with different O$_2$ mixtures the changes in both umbilical and uterine venous blood were small and the vein-to-vein gradient remained virtually unchanged.

In Fig 1, the results obtained in the present experiments

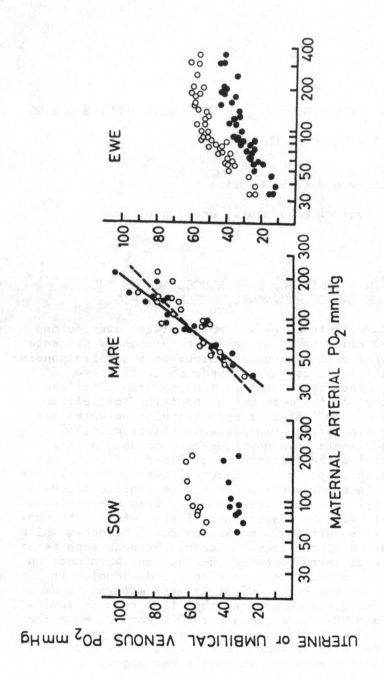

Fig. 1. The relation between the PO$_2$ in maternal arterial blood (log scale) and that in the uterine vein (o) and umbilical vein (●). A comparison between results for the pig (6 fetuses; 2 sows) and data previously published (Comline & Silver, 1970b) for 7 ewes and mares. All experiments were carried out on acute, anaesthetised preparations.

may be compared with previous findings in the sheep and mare under similar conditions (Comline & Silver, 1970b). The small changes in umbilical and uterine venous PO_2 in both pig and sheep fetus, over a wide range of maternal arterial levels, are in sharp contrast to the situation in the fetal foal where umbilical vein PO_2 can exceed that in the uterine vein at high maternal arterial PO_2, while the reverse occurs during hypoxia.

These observations suggest that the specialised microcotyledonary exchange units, which appear to be a unique feature of the equine placenta (Tsutsumi, 1962; Steven, 1968), are essential for its efficient gas exchange. However, before further comparisons between the horse and the pig can be made, information about the rate and direction of maternal and fetal placental blood flow in the latter is required.

2. CHANGES IN UMBILICAL PO_2 DURING PARTURITION IN THE COW

The technique for the insertion and maintenance of chronic intravascular catheters in the pregnant ewe and fetus (Meschia, Cotter, Breathnach & Barron, 1965) has been extended and modified so that catheters may be placed in various parts of the fetal and maternal circulations in sheep, cows and horses for sequential sampling during late gestation and parturition (Comline & Silver, 1970a; Silver et al., 1973). In the sheep observations during parturition were made on peripheral maternal and fetal blood; the latter was withdrawn from hind limb catheters threaded into the vena cava and lower aorta (Comline & Silver, 1972). Fetal blood gas and metabolite levels remained remarkably stable in these preparations during late gestation and throughout parturition. Only during the last 0-15 min of labour was there a fall in fetal blood PO_2, and in some animals no change occurred until after delivery. Nevertheless, increased fetal plasma lactate concentrations during the last hour or so of labour suggested the occurrence of some anaerobic activity in the fetal tissues. It was argued that transient changes in fetal blood PO_2 might well have escaped detection by the method and site of sampling. It was essential therefore to investigate changes in umbilical venous as well as arterial blood during the various stages of labour, and where possible to correlate myometrial activity with umbilical and uterine blood gas levels. Some of these problems have now been investigated in the cow, in which catheterisation of the main umbilical vessels through cotyledonary branches is easier than in the sheep and large numbers of blood samples can be taken without detriment to the fetus or mother.

Catheters were inserted into uterine and umbilical vessels

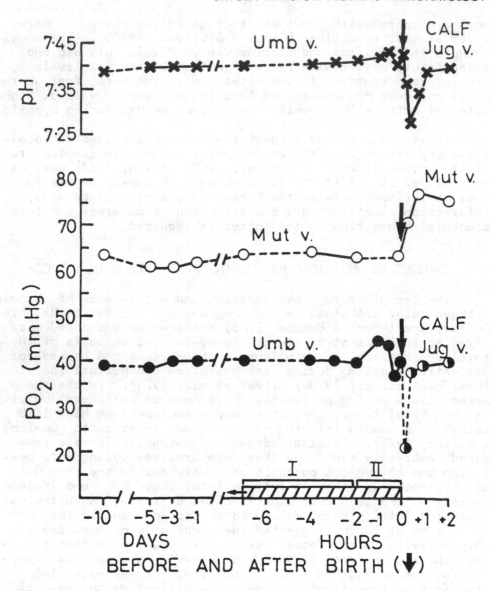

Fig. 2.   Sequential changes in uterine (o) and umbilical (•)
venous PO_2 in late gestation and during parturition. Changes
in fetal and neonatal blood pH (x) are shown above. The lst
stage of labour (I) was followed during the last 7 hr; 2nd
stage (II) lasted about 2 hr. Sampling from the uterine vein
was continued after birth while jugular blood samples (o) were
taken from the calf after delivery.

in 9 cows under general anaesthesia (chloral + N_2O/O_2) at 250-260 days gestation (term 280 days). A saline-filled balloon was placed in the allantoic cavity to record uterine activity during parturition. After recovery from operation, daily blood gas, pH and intra-uterine pressure measurements were made. At parturition, which occurred 7-21 days later, more frequent measurements were made although not all catheters remained patent throughout labour.

The onset of myometrial activity was generally detected 12-18hr before delivery. The contractions increased in magnitude during this period and abdominal straining began 1-2hr before birth. Maximal pressure changes were about 30mmHg (myometrial) and 50-80mmHg (myometrial + abdominal). In all animals but one the PO_2 levels in umbilical and uterine blood remained virtually unchanged both during late gestation and throughout parturition until a few minutes before delivery. Fig 2 shows the results from one of these cows; the vein-to-vein PO_2 gradient across the placenta remained stable during labour. Slight changes in umbilical venous PO_2 occurred near birth, but the major alterations in both PO_2 and pH occurred immediately after delivery. Table 1 summarises the data on umbilical PO_2 changes from all experiments and shows that only those PO_2 decreases which occurred 0-5min before birth were statistically significant.

Table 1 <u>CHANGES IN UMBILICAL PO_2 DURING PARTURITION IN THE COW</u>

VESSEL	n	PO_2 (mmHg) BEFORE LABOUR (MEAN ± S.E.)	MEAN PO_2 CHANGES (±SE) DURING LABOUR		
			1st STAGE	2nd STAGE	
			5 - 10 Hr	1 - 2 Hr	1 - 5 Min BEFORE BIRTH
UMB.V.	6	36.1 (± 2.0)	+ 0.4	- 1.7 (± 1.9)	- 4.3 * (± 1.5)
UMB.A.	4	25.4 (± 2.3)	- 0.5	- 2.4 (± 1.2)	- 6.2** (± 0.8)

* $p < 0.05$
** $p < 0.01$

In most animals the length and resistance of the catheters precluded rapid sampling. However, in three fetuses umbilical venous blood samples were timed to coincide with different phases of uterine activity. Little or no change in umbilical venous PO_2 was found when samples were taken during 20-30
mmHg myometrial contractions and occasionally an increase in
PO_2 was found as the uterus relaxed and the pressure fell.
During the much larger pressure changes found in late 2nd
stage a fall in umbilical venous PO_2 of 8-20% was generally
detectable during the period of maximal abdominal and myometrial pressure.

The present findings in the cow largely confirm those previously reported for the sheep and show that in this species
also, placental exchange can continue unchanged during the
first and second stages of labour in spite of the variations in
uterine blood flow which are generally associated with myometrial contractions (Greiss, 1965). Preliminary observations on
the pregnant mare suggest that here too umbilical venous PO_2
remains high until birth. Whether the small PO_2 fluctuations
observed in calf umbilical blood during labour are sufficient
to excite fetal chemoreceptor mechanisms (Dawes, Duncan, Lewis,
Merlet, Owen-Thomas & Reeves, 1969) is not yet clear. Some
restriction of the peripheral fetal circulation during labour,
perhaps induced by chemoreceptor stimulation, could explain the
occurrence of high fetal plasma lactate concentrations at this
time, but direct evidence on this point is still lacking.

This work was supported by the Milk Marketing Board and
the Horserace Betting Levy Board. The authors would like to
thank the many members of the laboratory technical staff for
their help with experiments. They are particularly indebted
to Mr. R. Proudfoot, Mr. R. L. Tindall and Mr. P. Hughes.

References

Dawes, G. S., Duncan, S. L. B., Lewis, B. V., Merlet, C. L.,
    Owen-Thomas, J. B. & Reeves, J. T. (1969) J.Physiol.,Lond.
    201, 105.
Comline, R. S. & Silver, M. (1970a) J.Physiol.,Lond. 209, 567.
Comline, R. S. & Silver, M. (1970b) J.Physiol.,Lond. 209, 587.
Comline, R. S. & Silver, M. (1972) J.Physiol.,Lond. 222, 233.
Greiss, F. (1965) Am.J.Obstet.Gynec. 93, 917.
Meschia, G., Cotter, J. R., Breathnach, C. S. & Barron, D. H.
    (1965) Q.Jl exp.Physiol. 50, 185.
Steven, D. H. (1968) J.Physiol.,Lond. 196, 24P.
Tsutsumi, Y. (1962) J.Fac.Agric.Hokkaido(imp)Univ. 52, 372.

THRESHOLD VALUES OF OXYGEN DEFICIENCY LEADING TO CARDIOVASCULAR

AND BRAIN PATHOLOGICAL CHANGES IN TERM MONKEY FETUSES

Ronald E. Myers

Laboratory of Perinatal Physiology, NINDS

National Institutes of Health

The present paper discusses the effects of varying degrees
of oxygen deprivation on fetal cardiovascular performance and on
nervous system integrity. Apart from the central role played by
oxygen deficiency in producing fetal brain damage and death, the
vital signs changes associated with lack of oxygen serve as one
of the principal methods for the moment-by-moment surveillance
of the fetal condition in utero during labor and delivery. Until
recently, the relationship between changes in fetal heart rate
and fetal asphyxia has remained largely speculative. However,
recent studies in our Laboratory have defined a more precise relationship between the severity of fetal oxygen deprivation and
changes in fetal vital signs (1).

Blood samples taken from the abdominal aorta of term monkey
fetuses (under barbiturate anesthesia) typically show a pO_2 of
26-30 mm Hg, a hemoglobin saturation with oxygen of 60-70%, and
an oxygen content of 10-12 volumes percent. Blood samples withdrawn from either carotid artery, on the other hand, show higher
oxygen values because of the preferential shunting of the more
highly oxygenated blood which arrives over the inferior vena cava
to the left side of the heart and ascending aorta as compared to
the passage of the less oxygenated blood from the superior vena
cava to the right side of the heart and through the ductus
arteriosus to the descending aorta. Thus, unlike in the adult,
significant variations exist in the oxygen values of blood according to its site of sampling both on the arterial and the venous
sides of the circulation.

The effects of progressive oxygen deprivation on the vital
signs and central nervous system integrity of anesthetized term

monkey fetuses appear in figure 1. The oxygen values in the
abdominal aortic blood may decline through a wide range while no
changes are observed in the vital signs (mild asphyxia). The
heart rate and blood pressure remain stable and alterations in
CNS function fail to appear. However, as the pO_2 values approach
15-16 mm Hg, the hemoglobin saturation with oxygen lies in the
range 25-30% and the oxygen content drops to 3-4 volumes percent,
slowing of the fetal heart and decreases in arterial blood pressure appear for the first time. As the blood oxygen values decline still further, the fetal vital signs also diminish in such
a fashion that the levels of blood oxygenation seem to directly
define the values of the fetal heart rate and blood pressure.
Figure 2, for example, depicts the relatively fixed relation between the arterial blood oxygen contents (when below 4.5 volumes

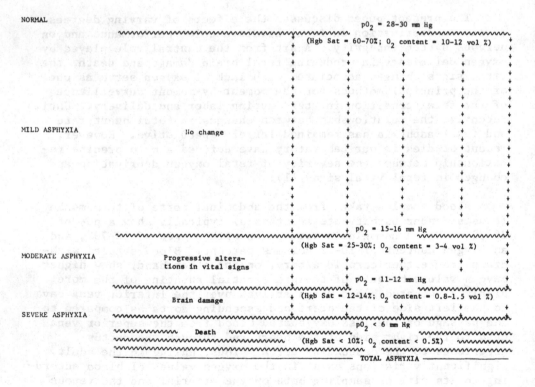

Figure 1

percent) and the heart rate of fetuses whose mothers are subjected to carbon monoxide (2). Despite the significant changes
in fetal vital signs, the fetal oxygen values may be maintained
at moderate asphyctic levels for prolonged periods (beyond 1-2
hours) without producing brain injury or other abnormalities in
the fetus. However, when the pO_2 of the fetal arterial blood
approaches 11-12 mm Hg, the hemoglobin saturation with oxygen
ranges from 10-14% and the oxygen content is 0.8-1.5 volumes
percent, and such a degree of asphyxia is maintained beyond 10-
15 minutes, evidences for brain damage regularly occur. Furthermore, as the asphyxia progresses beyond these levels, more severe
brain damage appears and fetal death occurs in increasing propor-
tions of cases.

Thus, the primate fetus can sustain significant oxygen de-
privations before changes in vital signs appear. However, vital
signs changes regularly occur prior to the appearance of that
asphyxial severity which produces brain injury. When that
severity of asphyxia occurs where vital signs changes do appear,
a relation exists between the severity of the asphyxia beyond
this amount and the magnitude of the changes in the fetal vital
signs. Such a proportionality exists from the time that the

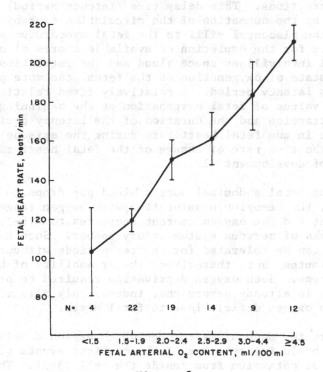

Figure 2

fetal arterial blood pO_2s drop to 15-16 mm Hg until the asphyxia becomes total when fetal rates of 60-70 beats/minute persist even though the blood pressure has declined to negligible levels. Direct heart examination under these circumstances (in the absence of a mechanical work output) either shows no visible signs of contraction or slight twitches of the atrial myocardium may appear. Nonetheless, electrocardiographic patterns continue to be generated for long periods and these ECG patterns repeat at the basal anoxic rate of the heart.

The uterine contractions of labor impose episodic perturbations of the state of the fetus. Each contraction leads to impairments in maternal blood perfusion of the placental intervillous spaces. Such uterine contractions, therefore, even in the normal circumstance, lead to brief episodes of some oxygen deprivation. When the fetus is already asphyxiated and exhibits some changes in vital signs, such further episodic perturbations of its state of oxygenation are associated with corresponding episodic exaggerations of its bradycardia and hypotension. These bursts of exaggeration of the bradycardia roughly correspond in their durations with the durations of the corresponding uterine contractions. However, there exists a minimum 20 seconds delay in their onset as timed from the initiation of the eliciting uterine contractions. This delay time (latency period) is accounted for by the summation of the circulation time of the fetal blood from the placental villi to the fetal myocardium and the time required for the depletion of available stores of oxygen in the maternal intervillous space blood and the fetal tissues. The better the state of oxygenation of the fetus, the more prolonged will be this latency period. A relatively fixed relation exists between the values of fetal oxygenation at the beginning of a uterine contraction and the duration of the latency period, the total change in the fetal heart rate during the episode of bradycardia and the time rate of change of the fetal heart rate during its period of development (1).

When the fetal abdominal aortic blood pO_2 drops to levels of 10-12 mm Hg, the hemoglobin saturation with oxygen ranges around 12-14 percent and the oxygen content decreases to 0.8-1.5 volumes percent, risks of nervous system injury appear. Such depressed oxygenation can be tolerated for shorter periods without injury (up to 10 minutes) but, thereafter, the probability of brain damage develops. Such oxygen deprivation required to produce brain injury is already severe and, indeed, only slightly greater increases in oxygen deficit lead to fetal death.

What are the mechanisms through which oxygen deprivation produces fetal brain injury? One of the earliest events of asphyxia is the loss of potassium from inside the cell (3-5). The severity

of fetal partial asphyxia which produces brain injury also causes
increases in water and chloride and decreases in sodium and potassium of grey matter (6). These changes are associated with dramatic shifts of fluid from the extra- to the intra-cellular
spaces of the brain (7). Gross examination of the brain reveals
convolutional flattening and herniation of the cerebellar tonsils
and vermis (8). Swelling of grey matter tissue leads to impairments in capillary blood circulation in restricted or more widespread cortical areas (9). Such perfusion impairments and
ultimately stasis of flow causes tissue necrosis and ultimately
conversion of zones of necrosis into areas of ulegyria, white
matter sclerosis, status marmoratus of the basal ganglia, por-
encephaly, or cystic brain degeneration (10-13). Thus, the
sequence of events which produces this family of brain injury
patterns starts out as a disturbance in electrolyte and water
distribution.

Why does oxygen deficiency cause changes in fetal vital signs
and to electrolyte and fluid shifts in the brain? Such oxygen deprivation seems likely to lead to alterations in the redox potentials of the tissue and to the development of states of tissue
hypoergy (Gr. hypo = under; ergon = work) or of diminished energy
availability to tissues. That the slowing of the heart and the
lowering of the blood pressure cannot be attributed to alterations
in parasympathetic tonus may be seen from the fact that atropine
administration fails to alter in any way the magnitudes or other
characteristics of the episodes of fetal bradycardia which follow
uterine contractions. Still more convincing indication that the
declines in vital function with asphyxia result from deficits in
energy availability is that, as the asphyxia deepens, the cardiovascular and CNS functions decline without interruption until they
cease altogether.

One may suppose that the diminished availability of oxygen
ultimately leads to a diminished availability of high energy
phosphate (creatine and other pyro-phosphates) in the individual
tissue cells. However, paradoxically, Lowry and co-workers have
shown that CNS function may fail dramatically at a time when the
supplies of high energy phosphate in the brain tissue have not
greatly altered (14). Thus, the degradation of nervous tissue
function and the appearance of significant electrolyte and fluid
shifts appear caused by circumstances other than tissue high-
energy phosphate depletions. A second and related paradox is
that epinephrine injections lead to explosive increases in heart
action under circumstances where the mechanical work output and
other parameters of heart function have degraded to zero levels
because of oxygen lack. Such epinephrine-induced bursts of
mechanical work output by the heart may appear and achieve full
expression in the continued absence of oxygen. Epinephrine thus

is capable of releasing or providing the circumstances under which
still-preserved energy stores (in the form of pyro-phosphates?)
may be utilized despite the continued absence of oxygen. Furthermore, these remarkable epinephrine effects are largely prevented
by prior injections of prostaglandin F-2 α. These facts together
suggest that both cardiac and cerebral functions may totally cease
despite the presence of continued considerable local stores of
energy. Finally, this stored energy (which may support a considerable mechanical work output by the heart over a sustained
period) can be released in the continued absence of oxygen by such
agents as epinephrine. These findings lead to the speculation
that these latter chemical messengers or mediators may exert their
dramatic tissue effects by releasing energy to the cell outside of
the usual oxygen-dependent electron transport systems acting
through changes in activity of adenyl-cyclase and cyclic AMP
levels.

REFERENCES

1. Myers, R. E., Mueller-Heubach, E., and Adamsons, K.: Pre-
   dictability of the state of fetal oxygenation from quantitative analysis of components of late deceleration. Amer. J.
   Obstet. Gynec. 115: 1083-1094, 1973.

2. Ginsberg, M. D., and Myers, R. E.: Fetal brain damage
   following maternal carbon monoxide intoxication: An experimental study. Acta Obstet. Gynec. Scand. (in press).

3. Bito, L. Z., and Myers, R. E.: On the physiological response
   of the cerebral cortex to acute stress. J. Physiol. (London)
   221: 349-370, 1972.

4. Vyskocil, F., Kritz, N., and Bures, J.: Potassium-selective
   microelectrodes used for measuring the extracellular brain
   potassium during spreading depression and anoxic depolariza-
   tion in rats. Brain Res. 39: 255-259, 1972.

5. Van Harreveld, A., and Schadé, J. P.: Chloride movements in
   cerebral cortex after circulatory arrest and during spreading depression. J. Cell Comp. Physiol. 54: Suppl. 1, 65-77,
   1959.

6. Selzer, M. E., Myers, R. E., and Holstein, S. B.: Prolonged
   partial asphyxia: Effects on fetal brain water and electrolytes. Neurology 22: 732-737, 1972.

7. Bondareff, W., Myers, R. E., and Brann, A. W.: Brain extracellular space in monkey fetuses subjected to prolonged
   partial asphyxia. Exp. Neurol. 28: 167-178, 1970.

8. Myers, R. E., Beard, R., and Adamsons, K.: Brain swelling in
    the newborn rhesus monkey following prolonged partial as-
    phyxia. Neurology (Minneap.) 19: 1012-1018, 1969.

9. Reivich, M., Brann, A. W., Shapiro, H., and Myers, R. E.:
    Regional cerebral blood flow during prolonged partial
    asphyxia. In Proc. IVth Salzburg Conference on Cerebral
    Flow, Salzburg, Austria, 1970.

10. Myers, R. E.: Two patterns of perinatal brain damage and
    their conditions of occurrence. Amer. J. Obstet. Gynec. 112:
    246-276, 1972.

11. Myers, R. E.: Atrophic cortical sclerosis associated with
    status marmoratus in a perinatally damaged monkey.
    Neurology 19: 1177-1188, 1969.

12. Myers, R. E., Valerio, M. G., Martin, D. P., and Nelson,
    K. B.: Perinatal brain damage: Porencephaly in a
    cynomolgous monkey. Biol. Neonate (in press).

13. Myers, R. E.: Cystic brain alteration after incomplete
    placental abruption in monkey. Arch. Neurol. 21: 133-141,
    1969.

14. Duffy, T. E., Nelson, S. R., and Lowry, O. H.: Cerebral
    carbohydrate metabolism changes in acute hypoxia and
    recovery. J. Neurochem. 19: 959-977, 1972.

PLACENTAL OXYGEN GRADIENTS DUE TO DIFFUSION AND CHEMICAL REACTION

Robert A. B. Holland

School of Physiology and Pharmacology
University of New South Wales
Kensington, N.S.W. 2033, Australia

This paper is a report of preliminary computations showing that gradients of oxygen tension are likely to exist in the human placenta due to slowness of oxygen equilibration between maternal and fetal blood. Calculations of likely gradients have been recently made by Hill, Power, and Longo (3) using a placental model in which blood flow was of the concurrent pattern, that is to say maternal and fetal capillaries ran together for a distance and the flow was in the same direction in each. These workers found that there was very little tension gradient if the capillary transit time in each circulation was taken as 1.7 seconds. The calculations whose results are presented in this paper differ from those of Hill et al (3) in two important respects; firstly they consider not only the concurrent model of the placenta but also a pool model and the multivillous stream system model, and secondly they use recently determined results for oxygen reaction rates with partly saturated human adult and fetal red cells.

In making the calculations, certain values had to be taken for the diffusing capacity of the placental membrane and for the maternal and fetal placental capillary blood volumes. These were taken from the work of Longo, Power, and Forster (6) who measured CO exchange between maternal and fetal blood in sheep during hyperbaric oxygenation. They found that the diffusing capacity of the placental membrane for CO (DmCO) was approximately 2.6 ml min^{-1} torr^{-1} and that the maternal and fetal capillary blood volumes (VcM and VcF) were both about 10 ml. These values were for a sheep fetus of 3 kg, about the weight of a full term human fetus. Their figures were consistent with previous estimations by the same authors (5) of the CO diffusing capacity of the placenta

1055

(DpCO) in sheep and dogs under more physiological conditions.
Power (8) found that in placental tissue oxygen was 1.14 times as
soluble as CO and so DmCO may be multiplied by 1.14 x (28/32) =
1.05 to obtain DmO_2. In these calculations DmO_2 was taken as 3 ml
min^{-1} torr $^{-1}$ which is a compromise between the figures obtained
using Power's ratio of solubilities and using the ratio of 1.30
for their solubilities in water. It was also assumed that the
volume rates of maternal and fetal flow were equal.

Two likely values were taken for the maternal and fetal
capillary transit time. The first was 1 second which was calculated
by Longo, Power, and Forster (6) from known values for blood flow
and from their values for VcM and VcF; the second was 1.7 seconds
which Hill, Power, and Longo (3) considered to be a better estimate.

The oxygen capacity of the maternal blood was taken as 18 Vols
per cent; that of the fetal blood was taken as 21.8 Vols per cent.
Standard curves for the equilibrium between oxygen and whole blood
(dissociation curves) were used (1) the P50 or tension for half
saturation being 26.6 torr in maternal blood and 22.1 torr in fetal
blood. No attempt was made to allow for the Bohr effect. This was
partly for simplicity in the already complex models; and partly
because preliminary work in this laboratory suggests that the early
stages of the reaction between CO_2 and fetal red cell suspensions
is relatively slow because of the lower concentration of carbonic
anhydrase in fetal red cells. We thus do not know what the
magnitude of the Bohr effect will be at any given time after the
maternal and fetal capillaries have come into contact. The oxygen
tensions in the uterine and umbilical arteries were taken as 100
torr and 17 torr respectively, the figures given by Bartels (2).
Flow rates were assumed constant and uniform and no account was
taken of the placental oxygen consumption.

The values used for reaction rate of oxygen with partly
saturated adult and fetal red cells were those recently found in
this laboratory (4). Measurements were made using a stopped-flow
apparatus with optical recording of the course of the reaction.
The values for θ (ml oxygen taken up by 1 ml red cells per min, per
torr oxygen tension gradient into the red cells) for adult cells
were about 60% of the values previously reported by Staub, Bishop,
and Forster (9) over the range 0 to 75% saturation. The latter
authors used the continuous flow apparatus and followed the
progress of the reaction with an oxygen electrode. The discrepancy
between the two sets of results is probably due to the presence of
a stagnant layer on the oxygen electrode causing overestimation of
θ in their experiments. The results of Holland et al (4) for
fetal θ were a little lower than for adult θ, the value being 1.4
ml ml^{-1} min^{-1} $torr^{-1}$ at zero saturation and changing little up to
75% saturation. The stopped-flow apparatus was not able to measure

θ at high saturations so it was assumed that θ could be predicted
from the concentration of deoxyhemoglobin which fell as oxygen
saturation increased. The prediction of θ was consistent with the
finding that θ was little affected by saturation in the range
0-75% (4,9) but fell sharply in the high range of saturation (9).

The calculations were of the forward integration type similar
to the integrations performed by Staub, Bishop, and Forster (9) in
the lung. Where Dp is the diffusing capacity of the placenta as a
whole, we make use of the relationship (6)

$$\frac{1}{Dp} = \frac{1}{(\theta M)(VcM)} + \frac{1}{Dm} + \frac{1}{(\theta F)(VcF)} \tag{1}$$

From the initial saturations in maternal and fetal blood, the
initial values of θ may be calculated and hence the initial value
of Dp. The initial oxygen tension gradient between maternal and
fetal red cells may also be broken up into three components;
maternal red cell to maternal plasma, maternal plasma across
placental membrane to fetal plasma, and fetal plasma into fetal
red cells. The magnitude of the components of the total gradient
at any time are proportional to the magnitudes of the terms on the
right hand side of *equation (1)*. Using these relationships, the
time, Δt, for a small quantity of oxygen, ΔO_2, to be transferred
across the placenta was calculated. This transfer of oxygen
changed the maternal and fetal saturations and hence the values of
θ; so for the next transfer of oxygen, there was not only a
smaller total gradient, but also a new distribution of the gradient
which could also be calculated from *equation (1)*. Using the new
gradient and the new value for Dp, the time for the next transfer
of ΔO_2 was calculated. The process was repeated until equilibrium
was reached or the sum of the transfer times exceeded the capillary
transit time. In all cases the final value found for oxygen tension
in fetal blood leaving the placenta was close to 29 torr, a value
for which there is some evidence.

Results of Calculations

Concurrent Flow Placenta. In this type of placenta, there
was a 5 torr gradient of oxygen tension remaining after 1 second.
Full equilibration would have given a lower maternal and higher
fetal oxygen tension with a resultant rise in fetal oxygen
saturation of approximately 5%. After 1.75 seconds, maternal and
fetal blood were close to full equilibration. These are
essentially the findings of Hill et al (3) who used the higher
values (9) for the reaction rates of oxygen with red cells. The
similarity of the findings is explained by the fact that most of
the resistance to oxygen transfer is in the placental membrane
rather than in the capillaries.

     Pool Placenta. In this type of placenta the fetal
capillaries are considered as passing through a pool of maternal
blood with a uniform oxygen tension. The maternal oxygen tension
must be taken as equal to the maternal uterine venous tension, and
the value for this was taken to be 33 torr (2). Another
implication of the maternal oxygen tension being uniform is that
VcM must be regarded as infinite, and the fetal blood is treated
as if passing through a lung with alveolar oxygen tension = 33
torr. In this situation all the oxygen transfer occurs at low
tension gradient, and it was found that equilibration was slow.
After 1 sec fetal tension was 24.5 torr and saturation was 57%;
after 1.75 sec, fetal tension was 28.5 and saturation was 66% and
equilibration was still incomplete at times greater than 2 seconds.

     Multivillous Stream System. This type of placental flow has
been postulated by Metcalf, Bartels, and Moll (7) as the best
representation of the human placenta. In this model the maternal
flow runs in one direction and fetal capillaries successively dip
into the maternal stream, run concurrently with it and then leave
the stream. A number of such fetal capillaries join to give mixed
fetal blood leaving the placental gas exchange region. As pointed
out by Metcalf et al, such an arrangement can give oxygen tension
in the umbilical vein higher than in the uterine vein. Integrating
in such a system with five fetal capillaries successively entering
the maternal stream, it was found that for a 1 second transit time,
there were oxygen tension gradients at the end of all fetal
capillaries and that oxygen saturation in the mixed fetal blood
leaving the placenta was 8.5% less than it would have been if full
equilibration had occurred. For a transit time of 1.75 sec, the
gradients were less and oxygen tension in the fetal blood leaving
the placenta was higher than in the maternal blood. However full
equilibration would still have improved fetal saturation by 3.5%.

                              Conclusions

     In the two placental types which are most likely to represent
the human placenta: the pool, and the multivillous stream, there
appears to be a limitation on oxygen transfer due to slowness of
diffusion and chemical reaction. These calculations do not exclude
shunts as a cause of one known gradient but do establish diffusion
limitation as a likely part of the whole picture. The diffusion
gradients are not great but a small tension gradient is more
important in the placenta than in the lung because placental
transfer occurs on the steep part of the oxygen-hemoglobin
equilibrium curve. Accurate estimates of the gradient are not
possible at this time because of the inherent difficulty of making
direct measurements of diffusion and blood flow in the human
placenta in vivo.

REFERENCES

(1) Altman, P.L. and Dittmer, D.S. Respiration and Circulation.
    Federation of American Societies for Experimental Biology,
    Bethesda Md., 1971.

(2) Bartels, H. Prenatal Respiration. North-Holland Publishing
    Co. Amsterdam and London, 1970. Page 50.

(3) Hill, E.P., Power, G.G., and Longo, L.D. A mathematical
    model of placental O_2 transfer with consideration of
    hemoglobin reaction rates. Am. J. Physiol., 222: 721-729,
    1972.

(4) Holland, R.A.B., Van Hezewijk, W., and Zubzanda, J. The
    velocity of oxygen uptake by partly saturated adult and
    fetal red cells. Proc. Aust. Physiol. Pharmacol. Soc.,
    3 (2), 181, 1972.

(5) Longo, L.D., Power, G.G., and Forster, R.E. Respiratory
    function of the placenta as determined with carbon monoxide
    in sheep and goats. J. Clin. Invest., 46: 812-828, 1967.

(6) Longo, L.D., Power, G.G., and Forster, R.E. Placental
    diffusing capacity for carbon monoxide at varying partial
    pressures of oxygen. J. Appl. Physiol., 26: 360-370, 1969.

(7) Metcalf, J., Bartels, H., and Moll, W. Gas exchange in the
    pregnant uterus. Physiol. Rev., 47: 782-838, 1967.

(8) Power, G.G. Solubility of oxygen and carbon monoxide in
    blood and pulmonary and placental tissue. J. Appl.
    Physiol., 24: 468-474, 1968.

(9) Staub, N.C., Bishop, J.M., and Forster, R.E. Importance of
    diffusion and chemical reaction rates in O_2 uptake in the
    lung. J. Appl. Physiol., 17: 21-27, 1962.

EXPERIMENTAL MODELS FOR IN VIVO AND IN VITRO INVESTIGATIONS ON PLACENTAL HEMODYNAMICS AND OXYGEN SUPPLY TO THE FETUS

Maurice Panigel

University Paris VI, Dept. Biology of Reproduction

Tour 32, 4 Place Jussieu, Paris 5, France

Gas exchange across the placenta is for obvious reasons of the utmost interest to all obstetricians and pediatricians, placental transfer of oxygen being essential to keep the fetus alive and to promote its development. Theoretic models have been constructed lately to represent respiratory gas transfer in the complex intra-uterine environment using mainly physiological parameters obtained by using "chronic preparations," thus avoiding the adverse effects of surgical manipulation or of anesthesia which produce important hemodynamic changes in the extremely sensitive umbilical or uterine vasculatures.

Placental hemodynamics vitally affect oxygen transfer from mother to fetus. Since the ultimate goal of all fetal or placental research is to assure the best developmental environment for the human fetus, it is of utmost importance to look for a suitable living experimental model that would approximate as much as pos-sible the human natural intrauterine environment. Preferential at-tention should be given to the study of circulatory dynamics in the placenta. The placental circulatory pattern is now known to be closely similar in pregnant women and in gestating monkeys (macaques and baboons). From anatomic, radiologic, radioisotope scanning and electron microscopic studies, it is now well established that these catarrhine monkeys are the best animal models to choose for the experimental study of placental hemodynamics and respiratory gases transfer (1).

Pilot studies described here deal with the search of suitable experimental models permitting extrapolation of the results to the human species to be of immediate practical use to the clinician interested by the physiopathologic problems related to prenatal respiration. Fetal hypoxia may result from impairment of maternal placental circulation due to administration of vasoconstrictive agents as well as to interruption of maternal blood supply to the placenta.

FETAL PLACENTAL DISSOCIATION IN THE RHESUS MONKEY

The surgical removal of the fetus or the experimental ligation of the fetal blood vessels leading to particular parts of the placenta does not terminate pregnancy or lead to placental shedding. The experimental interruption of the fetal placental circulation permits the comparison of uterine blood flow and oxygen consumption in rhesus monkeys during normal and "afetal" pregnancies with retained placentas. A net increase (of up to 64 mm Hg) is observed in the PO_2 of the uterine venous blood following fetectomy in the monkey (2). When the fetal blood circulation is stopped in only one disc of the bidiscoid rhesus monkey placenta through interplacental ligation, samples of the effluent venous blood can be taken from uterine veins leading directly from the area of the intact primary disc and separately from veins leading directly from the abnormal secondary disc (3). The venous blood draining the intact primary disc then manifests an abnormal depression of its PO_2 while that draining the secondary disc lacking a fetal circulation shows a major elevation of its PO_2 (up to 80 mm Hg). It is thus possible to dissociate oxygen consumption in utero of placental tissues and the placental oxygen transfer and consumption during normal pregnancy at comparable stages of fetal development.

EXPERIMENTAL PLACENTAL DETACHMENT IN THE RHESUS MONKEY

A need exists to develop an animal model in which a part of the placenta can be detached from its uterine insertion and the consequences of interrupting the maternal blood supply studied. Only monkeys showing an intervillous circulation of the maternal blood entering the placenta from spiral arteries orifices can serve as such experimental models. The use of the rhesus monkey offers particular advantages since this species possesses a bidiscoid placenta and digital premature detachment of one disc only can be performed while the other placenta remains still inserted. Moreover, due to the more restricted trophoblast penetration in the decidual tissues, experimental detachment can easily be effected by inserting a finger between the placenta and the uterine tissues. The effect of such a partial detachment of the placenta upon the fetus oxygenation has already been studied (4) as well as the effect of this surgical procedure on the placental ultrastructure (5). Secondary disc detachment is associated with a loss of maternal fetal respiratory gas exchange proportional in its extent to the part of the total placental tissue separated.

EFFECTS OF VASOCONSTRICTIVE DRUGS ON THE PLACENTAL CIRCULATION
OF THE RHESUS MONKEY

Another way of changing experimentally the uteroplacental blood flow and to reduce the maternal blood supply to the placenta, is to administer vasoactive drugs (epinephrine, 5-hydroxytryptamine, vasopressine) intra-arterially or intravenously to the fetus or to the mother and to follow by radioangiography techniques (6) the obliteration of intervillous space inflow (spurts) believed to be due to a direct vascular effect on the spiral uteroplacental arteries wall and not to the effect of these drugs on uterine muscle activity. As radioangiographic visualization does not lend itself to numerical expression (except changes in the number of spurts), we are now developing a new radiopharmacological technique correlating angiographic observations to quantitative results obtained through placental scanning with radioactive microspheres labelled with 99 m Tc and 113 m In. (7).

IN VITRO PERFUSION OF BOTH MATERNAL AND FETAL CIRCULATION
IN THE ISOLATED HUMAN PLACENTAL LOBULE

The study in vivo of placental materno-fetal exchange in the human species is, for obvious reasons, very difficult. We have developed a method to perfuse in vitro isolated lobules of term postpartum human placentas (8). The perfusion of maternal intervillous and fetal capillaries in the chorionic villi, has been applied to the pharmacological study of fetal placental vasomotricity (9), the study of the transfer of antipyrine, sodium and leucine (10), of catecholamines (11), and can be used to measure oxygen consumption by placental tissues as compared to its transfer in vitro across the human placental membrane (12). Rapid transfer of norepinephrine from the maternal to the fetal side has been demonstrated, although there was no evidence of stereospecificity of the norepinephrine transfer mechanism. Antipyrine diffuses rapidly across the placental membrane. Sodium diffuses much more slowly and leucine is known to be aided by an active transport mechanism. Recent studies (13) of the permeability in the sheep placenta to inert gases like argon, has proved that oxygen transfer may, contrary to argon, be at least partially carrier mediated by a microsomal and mitochondrial cytochrome (P 450) which can be inactivated by certain drugs and emulsifying agents. This finding has to be confirmed on human material in vitro or in vivo in nonhuman primates to determine if the same pharmacologic agents also influence, in these experimental conditions, the transplacental oxygen flux. The simple fact that the specific transport by a carrier of oxygen in the placenta is possible, shows that however elaborate and careful has been the building of mathematical models (on the assumption that oxygen transfer is a simple diffusion phenomenon), these models soon may have to be modified.

REFERENCES

1. Panigel, M. The fetoplacental complex. In Risks in the Prac-
    tice of Modern Obstetrics. S. Aladjem (Ed.). Chapter 8,
    Mosby, Saint Louis, 1972. pp. 168-191.

2. Parer, J. T., C. W. Lannoy and R. E. Behrman. Uterine blood
    flow and oxygen consumption in rhesus monkeys with retained
    placentas. Amer. J. Obstet. Gynec. 100:806-812, 1968.

3. Panigel, M. and R. E. Myers. Histological and ultrastruc-
    tural changes in rhesus monkey placenta following interrup-
    tion of fetal placental circulation by fetectomy or inter-
    placental umbilical vessel ligation. Acta anat. 81:486-506,
    1972.

4. Myers, R. E. Conditions leading to perinatal brain damage in
    non-human primate. In Fetal Evaluation During Pregnancy and
    Labor. Academic Press, New York, 1971. pp. 175-195.

5. Myers, R. E. and M. Panigel. Experimental placental detach-
    ment in rhesus monkey: Changes in villous ultrastructure.
    Med. Primatology (In press).

6. Misenhimer, H. R., S. I. Margulies, M. Panigel, E. M. Ramsey
    and M. Donner. Effect of vasoconstrictive drugs on the pla-
    cental circulation of the rhesus monkey, a preliminary report.
    Investigative Radiology 7:496-499, 1972.

7. James, E., E. M. Ramsey, M. K. Seigel, M. Panigel, R. Misen-
    himer and M. Donner. Work in progress.

8. Panigel, M. Placental perfusion. In Fetal Homeostasis, vol. 4.
    R. Wynn (Ed.). Appleton Century Crofts, New York, 1968.
    pp. 15-25.

9. Panigel, M. Placental perfusion experiments. Amer. J.
    Obstet. Gynec. 84:1664-1683, 1962.

10. Schneider, H., M. Panigel and J. Dancis. Transfer across the
     perfused human placenta of antipyrine, sodium and leucine.
     Amer. J. Obstet. Gynec. 114:822-828, 1972.

11. Morgan, D., M. Sandler and M. Panigel. Placental transfer
     of catecholamines in vitro and in vivo. Amer. J. Obstet.
     Gynec. 112:1068-1075, 1972.

12. Guiet, A. Les échanges materno-foetaux et la consommation de l'oxygène au niveau de lobules placentaires humains maintenus en survie in vitro. Mémoire de D. E. A., Université de Paris, 1969.

13. Gurtner, G. H. and B. Burns. Possible facilitated transport of oxygen across the placenta. Nature 240:473-475, 1972.

MATERNAL ALKALOSIS AND FETAL OXYGENATION

John H. G. Rankin, Ph. D.

University of Wisconsin Medical School

Departments of Physiology & Gynecology-Obstetrics

There are two conventional approaches available for
the investigation of placental transfer. The first is to
derive a mathematical model of the placenta. On the basis
of this model, the response of the placenta to a change can
be predicted. The second method is to perform experiments
on living animals, and observe the response of the animal to
a change. Sometimes the two methods are compatible and
sometimes they are not. Where there is disagreement, it is
often due to the complex nature of the animal response.
The response that was predicted from the mathematical model may
exist, the the ability of the investigator to observe this
response will be hindered by compensatory reflexes which mask
the predicted response.

We therefore believe that there is room for an intermediate
step. This intermediate step is a physical model. If a
physical model of a placenta can be constructed, then a change
can be induced in one parameter and a clear response can be
seen which will be free of the masking responses of the
animal.

In a previous publication, we have predicted, on the basis
of a mathematical model, that maternal alkalosis will induce a
diffusional limitation to placental oxygen transfer (8). When
we examine the literature on this subject, we find that maternal
alkalosis is frequently associated with fetal acidosis and
hypoxia (1). On the face of it, it would therefore appear that
the experiments upon the living animal and the mathematical
model are compatible. This compatibility is false. When one

examines the literature more closely, it becomes apparent that
the fetal acidosis and hypoxia that are associated with
maternal alkalosis are ascribed to vasoconstriction of the
umbilical (7) or uterine (3) vasculature. The response of the
animal to maternal alkalosis appears to be complex and it would
appear to be a useful situation in which to examine the response
of a physical model.

After many unsuccessful attempts to construct a model
placenta in which the diffusing capacity for oxygen was
sufficiently great as to permit a significant degree of oxygen
exchange during one pass, it was found that satisfactory
results could be obtained with a Dow Hollow Fiber Beaker Gas
Permeator (B/HFG-1, Cordis Laboratories). The permeator
consists of approximately 1000 silicone rubber copolymer
capillaries mounted in a beaker. The beaker has two ports
which enable gas or liquid to be passed through the body of
the beaker. The capillaries are bonded together at each end
so that fluid or gas can be pumped into one end and collected
from the other end. The wall thickness of the capillaries is
40 microns, the surface area is 500 sq. cm. and the interior
diameter is 100 microns.

The prediction that the maternal alkalosis would produce
a diffusional limitation to placental oxygen transfer was
based on the effects of alkalosis on a variable called d. This
variable is defined in the following manner:
$$d = Dp_{O_2}/QS$$
where Dp_{O_2} is the placental diffusing capacity for oxygen
(ml/(mm. x mm.Hg)), Q is the uterine placental blood flow
(ml/min) and S if the effective solubility of oxygen in
maternal blood (ml/(ml x mm Hg)).

In order to have the permeator operating under conditions
which are analagous to that of a near-term sheep placenta, it
is necessary that d for the permeator be similar to that of
the near-term sheep placenta. As the same blood would be used
in the exchanger as is used in the sheep placenta, then S will
be the same for both. It is therefore necessary to adjust the
ratio of diffusing capacity to blood flow.

Before this ratio can be adjusted, it is necessary to
measure the diffusing capacity for oxygen of the permeator
The measurement of diffusing capacity for oxygen can only
properly be performed in liquid in which there is a linear
relationship between P_{O_2} and oxygen content and under conditions
of disequilibrium. That is to say, when the P_{O_2} continues
to change along the entire length of the capillaries. If
equilibration occurs, then the diffusing capacity of the entire

capillary bundle is not measured. The obvious course is to attempt to satisfy these two conditions while pumping water through the capillary bundle. Unfortunately, the solubility of oxygen in water is so low that equilibrium rapidly occurs between the gas on the outside of the capillaries and the water on the inside. The second condition can, therefore, not be satisfied unless very large flow rates are used. Such flows may be destructive to the capillary bundle and may produce conditions which are not comparable to those seen during perfusion with blood. The problem was solved by using a fluorochemical fluid known as FC-80 (3m Company). The solubility of oxygen in FC-80 is reported by the 3M Company to be 48.8 volumes per volume per atmosphere. We observed that the fluid did not equilibrate with the gas in the permeat r at reasonably low flow rates.

When gas is passed through the beaker and fluid is passed through the capillaries, the capillary bundle acts as if it were a pool flow exchanger in which the capillaries dip into a pool of constant composition. Under these conditions, it is difficult to determine the ratio of flows and the concentration of oxygen in the gas. For this reason, it was decided to operate at very high gas flows. Under this condition, the exchanger was operating close to the limit at which the ratio of the flow in the capillary to the flow in the beaker was tending to zero. At this limit,it can be shown that for most types of exchangers, eg. concurrent, countercurrent, pool flow, the effectiveness is equal to $(1 - e^{-d})$ (2). Consequently, if the effectiveness of exchange is measured under these conditions, then the equation can be solved for the value of d. The value of the diffusing capacity can thus be experimentally determined because d is equal to $DO_2/S\, Q_m$, where S, the solubility of oxygen in the fluid is known and Q_m can be measured.

Several determinations of placental diffusing capacity were made at a number of flow rates and a constant value of 0.012 ml/(min x mm Hg) was obtained. For the near-term sheep with a placental diffusing capacity of 2.1 ml/(min x mm Hg) (3) and a uterine placental blood flow of 800 ml/min (6), the ratio of DP_{O_2}/Q_m is equal to about .003. A blood flow of approximately 4 ml/min in combination with the diffusing capacity of .012 would produce a similar DP_{O_2}/Q_m ratio in the placental model. Experiments were done at several blood flows in this general range.

The format of the experiment was to pump blood through the capillary bundle at a known flow rate. The blood had previously been equilibrated with air and the hemoglobin was therefore saturated with oxygen. The pH was alkaline because much of

the CO_2 had been removed. In-going and out-going blood samples
were then taken for each of 2 conditions.The first condition was
that the gas flowing through the permeator was nitrogen and
the second condition was that the gas flowing through the
beaker was nitrogen plus carbon dioxide. In the first case,
the blood would leave the exchanger, having given up some of
its oxygen but retaining its alkaline pH. In the second case,
the blood would leave the exchanger, having given up some oxygen,
but during the exchanger transit, carbon dioxide would have
diffused into the blood and the blood would become more acid.
It was therefore possible to compare the effect of pH on oxygen
transfer with no change in exchanger architecture, blood flow
or input concentration. All perfusions were performed at 37° C.
Percent saturation was calculated from the PO_2 using the relationships provided by Hellegers and Schruefer (4).

Data on several experiments are shown in Table 1. You can
see that during acidosis, when CO_2 is in the gas mixture, the
blood left the exchanger at a much lower pH than that at which
it entered. In every case, it can be seen that the percent
saturation of the leaving blood was lower in the case of
acidosis than with the equivalent alkalotic condition.

TABLE I

EFFECT OF pH ON OXYGEN TRANSFER IN A MODEL PLACENTA

FLOW	pH		PO_2 (mm Hg)		% SAT	
ml/min	ACID	ALK	ACID	ALK	ACID	ALK
17	7.205	7.723	33	27	51	74
4	7.060	7.778	32	23	53	66
2.4	6.965	7.895	29	18	27	59
25	7.191	7.889	34	23	53	74
17	7.258	7.929	40	19	55	65
4	6.930	7.940	35	21	36	72
2.4	6.852	7.963	30	13	23	41

A detailed analysis of one of these experiments is shown
in Figure 1. In this figure, there are diagrammed the two
experimental conditions. Where nitrogen gas and carbon dioxide
flowed through the beaker, the blood left the exchanger with a
percent saturation of 27% and gave up 73% of the bound oxygen.
When nitrogen gas perfused the beaker, the blood remained alkaline
throughout the transit and it can be seen that the blood left at
a saturation of 59%. In this case, only 41% of the bound oxygen
was exchanged. In these two conditions, the only difference is
the pH of the blood flowing through the exchanger and 41% of the
oxygen exchanged in alkalosis and 73% exchanged in acidosis. A
comparison of the amount of oxygen exchanged during alkalosis

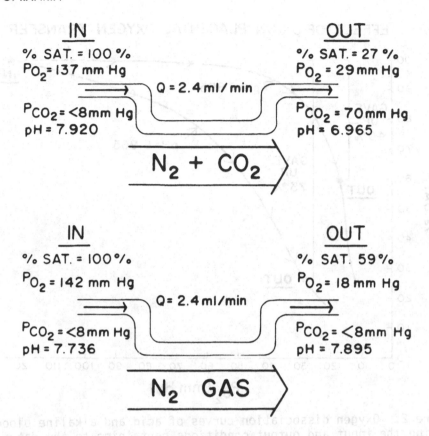

IN

% SAT. = 100 %
P_{O_2}= 137 mm Hg

Q = 2.4 ml / min

P_{CO_2}= <8mm Hg
pH = 7.920

OUT

% SAT. = 27 %
P_{O_2} = 29 mm Hg

P_{CO_2} = 70mm Hg
pH = 6.965

$N_2 + CO_2$

IN

% SAT. = 100%
P_{O_2} = 142 mm Hg

Q = 2.4 ml/min

P_{CO_2} = <8mm Hg
pH = 7.736

OUT

% SAT. 59%
P_{O_2}= 18 mm Hg

P_{CO_2} = <8mm Hg
pH = 7.895

N_2 GAS

Figure 1. The results of one pair of experiments in which the capillary blood flow was 2.4 ml/min. Perfusing blood was equilibrated with air in a disc tonometer.

with that during acidosis demonstrates that the alkaline condition resulted in a 56% decrease in the amount of oxygen given up by the blood during the placental transit.

This experiment demonstrates that changes in the pH of maternal blood can cause very large changes in the amount of oxygen crossing the placenta. A graphical analysis of what happened in the exchanger is given in Figure 2. Here we have a plot of percent saturation versus P_{O_2} and the initial and final points of each blood. In the case of the acid blood, the percent saturation went from the point marked "in" which is the same for both cases, to the point marked "out" on the line marked pH = 6.965. This is not the path that would be taken as the blood

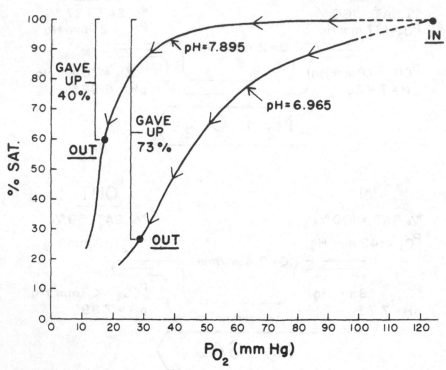

Figure 2. Oxygen dissociation curves of acid and alkaline blood showing the input and output conditions pertaining to the data given in Figure 1.

flows through the exchanger because the pH of the entering blood was 7.895 and during the transit in an undefined manner, the pH changed to 6.965. Be that as it may, the blood entered at a saturation of 100% and left at a saturation of 27%. This blood should be compared with the blood   when nitrogen alone was blown through the beaker. In this exchanger, the blood became slightly more alkaline as it passed through the exchanger due to the removal of some additonal carbon dioxide. The blood entered at a saturation of 100% and left at a saturation of 59% and a PO_2 of 18 mm. Hg. There are two factors which are causing the high saturation of this blood. One is the diffusional limitation to oxygen transfer produced by the alkaline blood. This is seen in the very steep section of the oxygen dissociation curve, along which exchange has occurred. The other factor is the decrease in the driving force for oxygen exchange. At a saturation of 59%, the acid blood would have a far higher PO_2. The left shifted curve has a decreased PO_2, such that when the blood is leaving the exchanger, there is a driving force of only 18 mm Hg. tending to promote the exchange.

In conclusion, we have shown that theoretical considerations have predicted that the placental exchange of oxygen should be compromised during maternal alkalosis. There is evidence in the literature that this is indeed the case and confirmatory evidence has been provided by experiments upon a model placenta which show that changes in the pH of maternal blood are capable of producing extremely large changes in the amount of oxygen which crosses the model placenta under a given set of conditions. The deleterious results of maternal alkalosis that have been observed in the literature may therefore be due to the effect of alkalosis on the oxygen dissociation curve and the observation does not necessitate the existence of vasomotor reflexes sensitive to acid-base disturbances.

REFERENCES

1. Dawes, G. S. Fetal and Neonatal Physiology. Year Book Medical Publishers p.112 - 1968.
2. Effectiveness is a term used in heat exchanger theory and in this context is defined as $(I-O)/I$ where I and O are the PO_2 in the inflowing and outflowing fluids respectively.
3. Fuller, E. O., P.M. Galletti, H.Y. Chou and E. C. Peirce II. Effect of PCO_2 on the vascular resistance of the pregnant sheep uterus. Physiologist V15 No. 3 p. 140 - 1972.
4. Hellegers A.E. and John J. P. Schruefer. Nomograms and empirical equations relating oxygen tension, percentage saturation and pH in maternal and fetal blood. Am. J. Obstet. & Gynec. 81:377-1961.
5. Longo, L. D., G. G. Power and R. E. Forster, II. Respiratory function of the placenta as determined with carbon monoxide in sheep and dogs. Circ. Res. V46:812-1967.
6. Metcalfe, J. H. Bartels and Waldemar Moll. Gas exchange in the pregnant uterus. Physiological Review 47:782 - 1967.
7. Motoyama, E.K. T. Fuchigami, C. J. Zigas and H. Cohen. Effect of pH and PCO_2 changes on the vascular resistance of the fetal placenta. Physiologist V15 No. 3 p.222 - 1972.
8. Rankin, J.H.G. The effect of permeability on placental oxygen transfer. Proceedings of the Barcroft Centenary Symposium, Cambridge, England, 1972. In press.

HEMOGLOBIN CONCENTRATION OVERESTIMATES OXYGEN

CARRYING CAPACITY DURING FAVIC CRISES

P. M. Taylor, L. G. Morphis and K. Mandalenaki

Depts. of Ped., Univ. of Pittsburgh Sch. of Med., and

St. Sophia Hosp., Athens Univ. Med. Sch., Athens, Greece

Acute hemolytic anemia in patients with red blood cells deficient in enzymes such as G-6-PD is characterized by excessive oxidation of hemoglobin (Hgb) to a variety of soluble heme pigments and to precipitates, known as Heinz bodies, attached to the inner coat of the red cell membrane. These pigments and precipitates do not carry oxygen. The cyanmethemoglobin method, widely used to determine blood Hgb concentration, detects not only Hgb but methemoglobin and other non-oxygen carrying heme pigments as well. In a previous study (P. M. Taylor, et al., Proc. Soc. Exper. Biol. Med., in press) we tested the hypothesis that non-oxygen carrying pigments might constitute a significant fraction of "hemoglobin," as measured by the cyanmethemoglobin method, during the course of acute hemolytic anemia induced in dogs by phenylhydrazine, a potent oxidizing agent.

We injected 8 dogs with phenylhydrazine and then compared changes in oxygen carrying capacity ($Cmax_{O_2}$) and Hgb concentration of blood drawn daily. Oxygen carrying capacity was measured directly by the micromethod of Roughton and Scholander (J. Biol. Chem. 148:541, 1943) and Hgb by the cyanmethemoglobin method. $Cmax_{O_2}$ expected from Hgb was calculated by multiplying Hgb by control oxygen binding capacity. Fig. 1 shows the time-course of the changes in Hgb and $Cmax_{O_2}$ following injection of phenylhydrazine on days 0 and 2 in a typical study. As anemia developed, $Cmax_{O_2}$ ($\triangle$) fell relatively more than Hgb (O) and therefore $Cmax_{O_2}$ as percent of that expected from Hgb ($\square$) also fell progressively. When functional anemia (decreased $Cmax_{O_2}$) and apparent anemia (decreased Hgb) were most marked, on day 8, $Cmax_{O_2}$ was but 54% of that expected from Hgb. For the series of 8 dogs the lowest

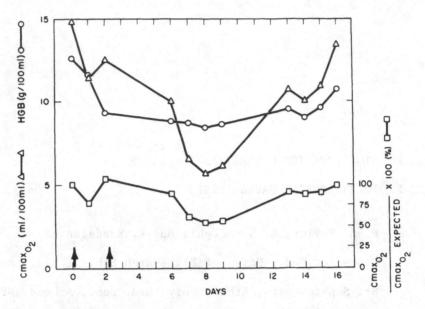

Fig. 1   Changes in $Cmax_{O_2}$ and Hgb following injection
of phenylhydrazine on days 0 and 2 (arrows).

measured $Cmax_{O_2}$ averaged 69% of that expected from Hgb. Our
hypothesis was supported by the fact that a mean of 31% of the pigments detected as Hgb did not carry oxygen when phenylhydrazine-
induced acute hemolytic anemia was most severe.

Heinz bodies were always in abundant supply in peripheral red
cells when anemia developed. We tested the possibility that Heinz
bodies might contribute to the non-oxygen carrying pigments detected as Hgb in these anemic dogs. Aqueous solutions of hemoglobin were prepared by the method of Mann and Romney (Amer. J.
Obstet. Gynec. 101:520, 1968). These buffered, dialyzed solutions
contained many Heinz body-laden red cell ghosts. In the left-hand
and middle columns of Fig. 2 the measured $Cmax_{O_2}$, as percent of
$Cmax_{O_2}$ expected from Hgb, of uncentrifuged aqueous solutions are
compared before injection of phenylhydrazine and at the height of
the anemia induced by phenylhydrazine. Arrows indicate mean values.
When anemia was most severe $Cmax_{O_2}$ was well below that expected
from Hgb in each study. Following ultracentrifugation, the $Cmax_{O_2}$
of the ghost-free supernatant (shown in the right-hand column of
Fig. 2) always increased to much more nearly that expected from
Hgb. Since ultracentrifugation of the solution resulted in both
reduction of non-oxygen carrying apparent Hgb and in loss of red
cell ghosts, many of which supported Heinz bodies, we concluded
that Heinz bodies contributed most of the non-oxygen carrying pigment detected as Hgb. Methemoglobin concentration was measured

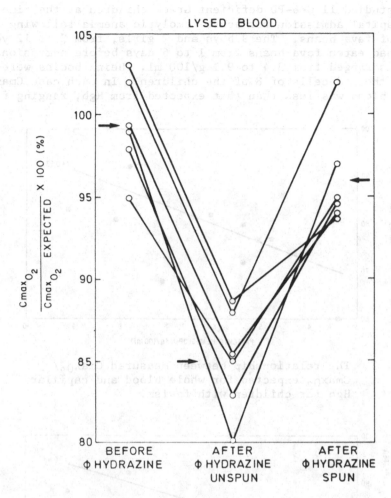

Fig. 2   The effect on $Cmax_{O_2}/Cmax_{O_2}$ expected of cen-
          trifuging aqueous Hgb solutions prepared from
          the blood of anemic dogs.

in 2 dogs, and in both (perhaps fortuitously) accounted for the
remainder of the non-oxygen carrying pigment in the spun solutions.

The acute experimental hemolytic anemia we studied in dogs
is similar both in time-course and severity to that experienced
by individuals with G-6-PD deficient red cells when exposed to an
oxidant such as Primaquin (J. Amer. Med. Ass. 149:1568, 1952) and
in certain G-6-PD deficient individuals following ingestion of
fava beans. We were thus anxious to learn whether blood hemoglobin
concentration overestimates blood oxygen carrying capacity in
G-6-PD deficient patients in hemolytic crises as it does in dogs
following injection of Phenylhydrazine.

We studied 11 G-6-PD deficient Greek children at the time of their hospital admission for acute hemolytic anemia following ingestion of fava beans. The 9 boys and 2 girls, from 2 to 12 years of age, had eaten fava beans from 1 to 6 days before admission. Hgb levels ranged from 3.4 to 9.1 g/100 ml. Heinz bodies were noted in the red cells of 8 of the children. In each case $Cmax_{O_2}$ of whole blood was less than that expected from Hgb, ranging from

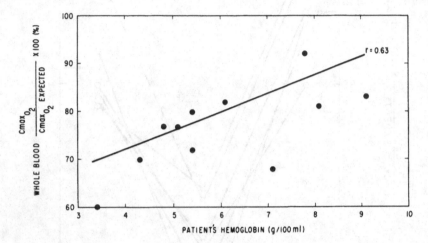

Fig. 3   The relationship between measured $Cmax_{O_2}$/$Cmax_{O_2}$ expected for whole blood and capillary Hgb for children with favism.

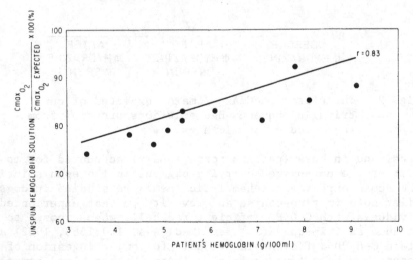

Fig. 4   The relationship between measured $Cmax_{O_2}$/$Cmax_{O_2}$ expected for unspun Hgb solution and capillary Hgb for children with favism.

61 to 92% (mean, 76%) of that expected. In absolute terms non-
oxygen carrying pigments detected as Hgb ranged from 0.5 to 2.0
g/100 ml (mean 1.1 g/100 ml) of apparent Hgb; the level of non-
oxygen carrying pigment was not systematically related to total
apparent Hgb. On the other hand, $Cmax_{O_2}$, as percent of that
expected from Hgb, was directly and significantly related to the
patient's Hgb (Fig. 3). Thus for children with favism the more
profound the anemia, the greater the fraction of apparent Hgb that
does not carry oxygen.

We have also studied one G-6-PD deficient boy whose acute
hemolytic crises followed sulfa drug therapy for infection. His
whole blood $Cmax_{O_2}$ was 61% of that expected from Hgb, and observations on aqueous solutions of his Hgb were similar to those reported below for children in favic crises.

As for whole blood there was a direct relationship between
degree of anemia and percent of non-oxygen carrying apparent Hgb
in unspun aqueous solution. (Fig. 4)

Fig. 5 presents $Cmax_{O_2}$, as percent of that expected from Hgb,
of unspun (●) and spun (↓) solutions; data from individual cases
are joined by vertical lines and plotted against the patient's Hgb.
The $Cmax_{O_2}$ of supernatant Hgb solution was always more nearly that
expected from Hgb than the $Cmax_{O_2}$ of the uncentrifuged solution,
increasing from a mean of 80% of that expected from Hgb in unspun

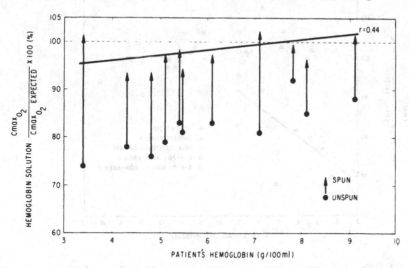

Fig. 5   The relationships between measured $Cmax_{O_2}$/
          $Cmax_{O_2}$ expected for unspun (●) and spun (↓)
          Hgb solutions and capillary Hgb for children
          with favism.

solutions to 96% of that expected from Hgb in the supernatant solution. Centrifugable material thus accounts for the bulk of the
non-oxygen carrying pigments detected as Hgb. Decrease in non-
oxygen carrying apparent Hgb following spinning of the aqueous
solution was as great for the 3 children whose red cells did not
contain Heinz bodies as it was for the 8 children whose red cells
did contain Heinz bodies. In other words a significant fraction
of apparent Hgb may be non-oxygen carrying during favic crises
even when Heinz bodies are not seen on peripheral blood smears.

It is also apparent from Fig. 5 that there is a direct
relationship between $Cmax_{O_2}$, as percent of expected from Hgb, of
the spun solution and the patient's Hgb. Thus, in general, the
lower the child's Hgb, the less likely was the $Cmax_{O_2}$ of the supernatant solution to attain the $Cmax_{O_2}$ expected from the Hgb concentration of that solution. This suggests that the more severe the
anemia, the greater the accumulation of soluble, non-oxygen carrying pigments such as methemoglobin. Unfortunately, methemoglobin
levels were not measured in this study.

The question of course arose as to whether the presence of
significant quantities of non-oxygen carrying pigments detected as
Hgb by the cyanmethemoglobin method might be a unique feature of

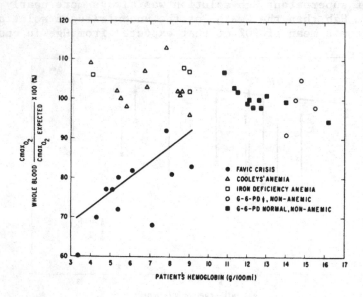

Fig. 6 $Cmax_{O_2}$ was that expected from Hgb for children
         with Cooley's anemia (Δ) and iron deficiency
         anemia ($\square$), and for non-anemic G-6-PD deficient
         children (O), in contrast to the lower-than-
         expected $Cmax_{O_2}$ for children with favism ($\bullet$).

acute hemolytic anemias associated with Heinz bodies in peripheral
red cells. A tentative answer to this question may be reached
from the data presented in Fig. 6, in which $Cmax_{O_2}$ of whole blood,
as percent of that expected from Hgb, is plotted against patient's
Hgb for children with Cooley's anemia (Δ) and iron deficiency
anemia ($\square$) and compared to the data previously presented for
children with favism ($\bullet$). The Hgb levels of the children with
Cooley's anemia and iron deficiency anemia were similar to the Hgb
levels of the children with favism. The $Cmax_{O_2}$ of the whole blood
of children with Cooley's and iron deficiency anemias was about
that expected from Hgb, in marked contrast to the lower-than-
expected $Cmax_{O_2}$ of children with favism. Thus non-oxygen carrying
pigments detected as Hgb, pronounced in favism, are not detected
in one variety of hemolytic anemia not characterized by Heinz
bodies (Cooley's anemia), and in one chronic, non-hemolytic anemia,
that of iron deficiency in childhood. Fig. 6 also presents data
on 4 non-anemic children with G-6-PD deficient red cells ($\bigcirc$)
whose $Cmax_{O_2}$ values were essentially within the range of the
control group of non-anemic children with normal G-6-PD levels ($\blacksquare$).

In conclusion, we have shown that non-oxygen carrying heme
pigments account for a portion of apparent hemoglobin in one experimental and one naturally occuring acute hemolytic anemia.
Blood hemoglobin concentration, as measured by the widely-used
cyanmethemoglobin method, thus underestimates the severity of
these anemias. Heinz bodies were always present in the canine
anemia and usually present in the human anemia. Most of the non-
oxygen carrying pigments detected as Hgb were removed from aqueous
Hgb solutions by ultracentrifugation.

DISCUSSION OF SESSION V   SUBSESSION: PLACENTAL OXYGEN TRANSFER

Chairmen: Dr. Jose Strauss, Dr. Waldemar Moll and
          Dr. Daniel D. Reneau

DISCUSSION OF PAPER BY D. D. RENEAU

Nemoto: I would like to make a comment on Dr. Panigel's statement
that the exchange of RBC's and antibodies across the placental
barrier need not be considered in modeling of maternal-fetal oxygen
and metabolite transfer.

    It is generally regarded that the placental barrier is similar
in permeability characteristics to the blood-brain barrier. That
is, it is quite permeable to charged molecules. The observation
that RBC can cross the barrier (which should be more clearly demonstrated), provides a clue to take a closer look at the modes of
solute transfer across the placenta.

DISCUSSION OF PAPER BY E. J. GUILBEAU AND D. D. RENEAU AND
    DISCUSSION OF PAPER BY W. MOLL

Brown: We have observed clinically that the asyphixiated newborn
infant often has a low cardiac output, as inferred by decreased
heart rate with oxygen tensions within the ranges of 50-100 Torr.
There is very little A-V difference, which implies poor tissue
extraction of O_2 and probably is a manifestation of shock which
began in vitro.

Silver, Marian: In sheep, cows and horses, fetal umbilical pO_2
stays the same during the gestation and during the various stages
of labor until the last few minutes before birth, despite quite
large changes in intra-uterine pressure. There seems no doubt
that changes in total uterine blood flow occur during uterine
contractions, but we have to be careful in interpreting total
changes in terms of what is happening in the placenta. There
may be much larger changes in myometrial and endometrial flow
relative to changes in placental flow, and so the overall effect
on placental exchange may be much smaller than has been calculated.
It is also possible that placental flow may be decreased more
severely in the human placenta during uterine contraction than
in the animal where the structure is very different.

Guilbeau: I agree with you. Fetal hypoxia seems to occur right
before the latest stage of labor. Your second point is important

too. The data I referred to were data on uterine blood flow.
However, most of the blood reaching the pregnant uterus seems
to cross the placenta.

DISCUSSION OF PAPER BY M. SILVER AND R. S. COMLINE

Strauss: This paper emphasizes the importance of identifying the
species and type of anesthetic used when applicable. We must agree
on the value of chronically implanted catheters and on the need to
develop suitable tissue electrodes, which can be implanted both for
acute and chronic experiments.

Silver: Obviously, the use of tissue or blood electrodes in the
chronic fetal preparation is desirable and might show small pO_2
fluctuations in the fetus during first and early second stage
labor, which are undetectable by our present methods. However,
as yet, there are no electrodes available which are sufficiently
stable to be implanted several weeks before birth.

DISCUSSION OF PAPER BY R. A. B. HOLLAND

Moll: How important is the capillary blood volume for placental O_2
transfer? Did you make calculations using different figures for
capillary blood volume?

Holland: The work of Longo, Power and Forster suggests that most
of the resistance is in the placental membrane and, therefore, the
reaction rates and pulmonary capillary blood volume are of less
importance. In the pool system, the maternal capillary blood volume
was taken as infinite but this required a low maternal pO_2 and this
gave a slowing of oxygen exchange.
     What I hope I have shown is that we need good measurements of
blood flow patterns and other features of the placental circulation
in order to know how great any oxygen gradient is.

Rankin: All of us who have worked with mathematical models know
that the results depend on the magnitudes of the selected parameters.
In reality, we have a range of values for each of these parameters,
not one specific value. Could you inject a measure of uncertainty
into your initial conditions and obtain a range of values as a result,
rather than one particular value?

Holland: I agree with that criticism. As yet, I have not varied all
the possible factors in my calculations but my message is that, using

good data from the literature, we find that there is likely to be an oxygen diffusion gradient due to slowness of diffusion and chemical reaction.

Longmuir: Gurtner and Burns have shown that a significant fraction of placental oxygen flux is apparently carried by cytochrome P-450, which is situated at 10-33A° intervals on the endoplasmic reticulum. What do you feel is the relation of this observation to your work?

Holland: I find the work of Gurtner and Burns hard to reconcile. I understand the P_{50} of cytochrome P-450 is about 1 Torr, in which case it is hard to see it as the vehicle for facilitated diffusion of the oxygen tensions found in the placenta.

DISCUSSION OF PAPER BY J. H. G. RANKIN

Moore: In your assessment of fetal oxygenation, what do you feel is the in vivo contribution of a right hemoglobin-oxygen dissociation curve shift (decrease in Hb-O_2 affinity) in the maternal blood? I have observed increased levels of 2,3 DPG in the blood of 3rd trimester pregnant women as has Mikael Rørth. In other words, do you feel maternal right curve shifts make or can make a significant contribution to fetal oxygenation.

Rankin: I definitely think that right shifted maternal curves would improve placental oxygen transfer. It would be incorrect, however, to encourage the induction of maternal acidosis in a compromised pregnancy until a strong series of animal experiments have proved that there are no unforseen side effects.

Lübbers: What step functional change (error) in the oxygen measurements is due to the anode being topically attached to the patient?

Rankin: The polarization curve (current vs. polarization voltage), using the in vivo oxygen electrode (gold) and a separate chloridized silver anode, produces a relatively flat curve (current change is small) between 0.600 volts and 0.800 volts. All tests were made at 25°C, 760 mm Hg barometric pressure, air-saturated saline. The current at 0.600 volts was 13 nano amps; at 0.700 volts, it was 13.5 nano amps and at 0.800 volts, it was 14 nano amps. Therefore, a change of ±0.100 volts caused a change of 1 nano amp or approximately ±3.5%. If the tissue oxygen was 60 mm Hg, then the maximum error to be expected for a 100 millivolt shift would be 2 mm Hg.

Numerous tests were made on animals using anodes in several locations: topically applied to the skin, in the mouth (under the

tongue), and subcutaneous on the extremities and the torso.
Switching the anode to any of the four showed no measurable change
in oxygen electrode current (a second oxygen electrode and anode
were used for preference).

The oxygen electrode time response to 90% of an oxygen change
averages 45-60 seconds. Therefore, it does not respond to EEG or
EKG voltages changes. However, it does time average those signals.

Chloride concentration between the silver chloride of the anode
and the skin can change due to sweat gland activity. Measurement
of electrical signals on many patients have been made. Of those
measurements made, the maximum voltage shift found was 50 millivolts
(approximately 1 decade change in chloride concentration).

Summarizing, data collected to date shows minimal error due to
topical application of the anode for in vivo oxygen measurements.

Subsession: OXYGEN MONITORING

Chairmen: Dr. Jose Strauss, Dr. Waldemar Moll and

Dr. Daniel D. Reneau

MEASUREMENT OF OXYGEN IN THE NEWBORN

Jose Strauss, Anthony V. Beran and Rex Baker. Dept.
Ped., U. Miami Med. Sch., Miami, Fla., and Dept. Ped.
Calif. Col. of Med., U.C., Irvine, Calif., U.S.A.

Numerous conditions endanger the life or the health of the
newborn baby apparently through decreased or increased oxygen
supply to various organs. Shortly after birth cardio-respiratory
embarrassment leads to hypoxia and creates the need for increased
oxygen concentration in the inspired gas (F_{IO_2}). Subsequently,
as the baby improves, the need to regulate oxygen intake persists
mainly because of the detrimental effects of high oxygen in the
environment or in the blood and the dependency on external support
of respiration.

Until recently the degree of the baby's oxygenation derange-
ment was evaluated only be means of in vitro determinations of
PaO_2 using a membrane covered O_2 electrode (1). Though this method
has been proven reliable and steady and has become the reference
point for all other determinations, it is obvious that intermittent
sampling of arterial blood may be faulty because of missing events
occuring between samples or prohibitive in terms of amount of blood
needed from a usually small baby (2). Thus, numerous attempts to
obtain continuous information about a baby's oxygenation have been
described in the last few months mainly from meetings of concerned
people like the group attending this Symposium. Our aim here is
to review some of the possible ways to measure oxygen in the
newborn which are currently under evaluation and our experience.

The current clinical approach to management of neonates with
oxygenation problems continues to be based on blood samples. This
trend may continue until systems are manufactured which do not need
in vivo calibration and which supply tissue PO_2 in mm Hg. So far
the systems aim at providing continuous information which can be

converted either to PaO_2 (i.e. from skin current) or directly
supply the equivalent of the in vitro PO_2 determination.

Some of the systems currently available to monitor oxygenation
of the newborn are: a catheter micromodel of the Clark electrode
(A. Huch, D.W. Lübbers and R. Huch)*; a catheter containing a
gold cathode extruded within an umbilical catheter and a micro-
amplifier (E.G. Brown, C.C. Liu, F.E. McDonnell, M.R. Neuman and
A.Y. Sweet)*; a catheter consisting of two silver-plated copper
wires imbedded in teflon with a tip containing dried KCL electrolyte gel which goes into solution after implantation (H.I. Bicher
and J.W. Rubin)*;a catheter with its cathode separated from its
anode which is applied to the skin (T.R. Harris and M. Nugent)*;
two skin electrodes which are Clark type electrodes with heating
elements to induce hyperemia [P. Eberhard, W. Mindt, K. Hammacher
and F. Jann; and R. Huch, D.W. Lübbers and A. Huch (4)]*; a scalp
electrode (W. Erdmann, H. Hünther, H, Schäfer and S. Kunke)*;
and the ear oxymeter (5).

The system we use (6) is based on the polarographic reduction
of O_2 molecules; -0.6 V are applied across the electrode pair.
The cathode (active, negative, 0.005 inch diameter) consists of a
Pt alloy always as a wire; the anode (reference, positive) was in
some designs a wire (0.007 inch diameter) and in one other design
a pellet, all of Ag-AgCl. The cathode wires were three centimeters
in length with their surface exposed and covered with silicone
only in a one mm area inserted in the earlobe; the rest was
insulated with epoxy resin. The anode wires were three centimeters
in length with the whole surface exposed and covered with silicone.
Type I electrode pair had both wires threaded through the earlobe
and making contact with the cable leading to the polarograph-
amplifier system away from the earlobe; Type II pair had the
cathode like in Type I but the anode was a pellet applied to the
surface of the chest; Type III pair, the one currently in use,
has both wires piercing the earlobe as in Type I but contact with
the cable is made at the surface of the earlobe (Fig. 1). Prior
to implantation the wires are treated for at least 24 hs with
-0.6 V applied to them while immersed in a sterile 0.9% NaCl
solution.

Electrode wires were implanted by threading through a 25-gauge
needle piercing the earlobe; for Types I and III the earlobe is
pierced by two needles simultaneously which are separated by the
same distance as that which separates the wires of each electrode.
The needles are then removed leaving the wires implanted. For all
three types contact with the lead cable is produced physically
without soldering or welding to allow the performance of this
procedure in the nursery.

*This Symposium

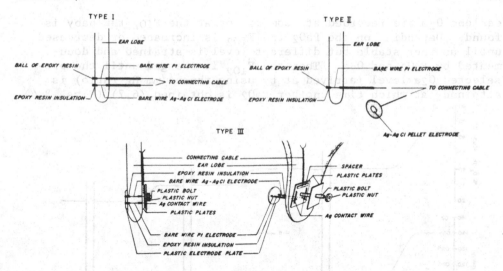

Fig. 1. Illustration of three types of O_2a electrodes
         implanted in the earlobe.

Type II pair has the electrode wires passed through a plastic
plate with two 0.020 inch Ag wires and a protruding bolt. Subsequently, the electrode wires are bent and uninsulated over the
Ag plate wires and another plastic plate and nut are placed on
top and screwed to the bolt. In this manner, the electrode wires
and the lead wire are all secured to the earlobe.

Electrodes are selected by their <u>in vitro</u> response; 40-60
min after implantation they are calibrated <u>in vivo</u> using either
arterial or capillary earlobe blood. F_{IO_2} can be monitored continuously by means of an Automatic Oxygen Controller (IMI, Div. of
Becton, Dickinson, Newport Beach, Calif.) (7) (Fig. 2). PaO_2 and

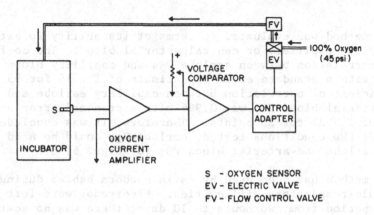

Fig. 2. Schematic diagram of oxygen control system (7).

earlobe O_2a are recorded at room air or at the F_{IO2} the baby is found. Depending on the PaO_2 the F_{IO2} is increased or decreased until another stable but different level is attained and documented by PaO_2 and O_2a. Then the F_{IO2} is changed until the preselected O_2a level (arrived at by using the μA/mm Hg ratio) is attained, at which time another PaO_2 is obtained (6,7) (Figs. 3,4).

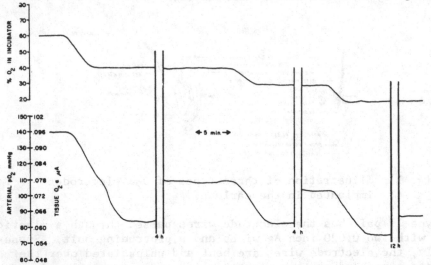

Fig. 3.   "Procedure for "step down acclimatization" from high incubator oxygen to room air applied to a newborn with respiratory distress syndrome requiring oxygen therapy. Preselected incubator oxygen concentrations were decreased step wise while maintaining optimum earlobe tissue O_2a" (7).

The implantation and sampling earlobe area can be warmed up to 41°C. The calibration should be repeated initially every 6-12 hs. (6).

This method was evaluated in terms of its ability to estimate PO_2 of earlobe capillary or central arterial blood. The coefficient of correlation between earlobe O_2a and capillary blood PO_2 was 0.992 with a standard error of estimate of ± 7.15 for 93 points. The coefficient of correlation between capillary earlobe and central arterial blood PO_2 was 0.996 with a standard error of estimate of ± 3.21 for 23 points. Therefore, it was concluded that, under the conditions tested, earlobe O_2a could be used to estimate earlobe and arterial blood PO_2 (6) (Figs. 5, 6).

This method has been used by us in newborn babies during and after cardio-respiratory difficulties. Electrodes were left in place for period from two hours to 10 days; there was no scar or infection at the site of implantation.

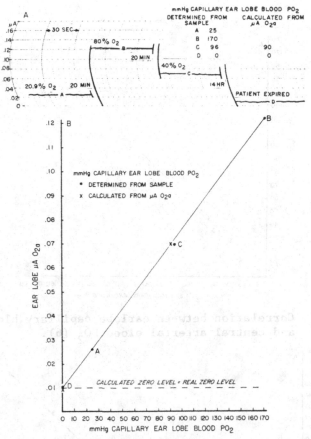

Fig. 4.   "In vivo electrode calibration. A: original tracings
of one study (read from left to right). B: graph
of results obtained in A" (6).

Hyperoxia was frequently detected in babies controlled mainly
by clinical appearance who were in incubators or respirators
recovering from their original illness or under "routeine" nebulizer
treatment. These evaluations have enabled us to conclude that some
form of continuous O_2 measurement is indispensible (a) to detect
clinically unrecognized periods of hypo- or hyperoxia, (b) to
determine the degree of abnormal oxygenation, and (c) to control
respirator, drug and gas mixture therapy, especially stepped down
acclimatization to room air breathing (6-8) (Fig.3). In addition,
the system described here can be used under conditions of extreme
hyperoxia.

Problems encountered include the need for various calibrations,
difficulty in securing the needed equipment, and drift. Drift is
the only problem found in the majority of babies; this should be

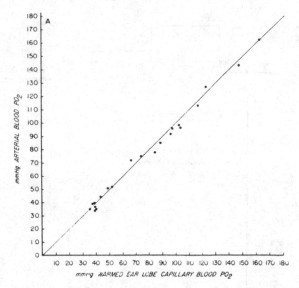

Fig. 5. Correlation between earlobe capillary blood PO_2
and central arterial blood PO_2 (6).

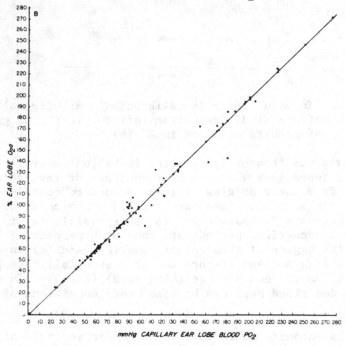

Fig. 6. Correlation between capillary earlobe blood PO_2 and
% earlobe O_2a (6).

minimized by electrode pre-treatment for at least 24 hs. A possibility under evaluation is the application of a sturdy membrane
to cover the sensing area of the cathode and to contain an electrolyte solution; this should decrease drift even further.

Recent use of indwelling arterial catheter electrodes as
described above may provide capability of monitoring hypoxic
babies during the first few days of life and under special conditions. However, afterwards a less invasive monitoring system is
needed, mainly to avoid hyperoxia. The system described here is a
possibility.

REFERENCES

1. Clark, L.C., Jr.: Monitor and control of blood and tissue
   oxygen tension. Trans. Soc. Art. Int. Organs, $\underline{2}$:41, 1956.

2. Approaches to assessing the Oxygenation Status of the High-
   Risk Newborn: Clinical and Experimental Uses of the Bare-
   Wire Oxygen Electrode. Second International Oxygen Electrode
   Conference Proceedings (Vancouver). (J. Strauss, ed.) NICHHD,
   NIH, Washington, D.C. In press.

3. Huch, A.; Lübbers, D.W.; Huch, R.: A catheter electrode for
   the continuous measurement of oxygen pressure in the aorta of
   the newborn infant. In: Oxygen Supply, Theoretical and
   Practical Aspects of Oxygen Supply and Microcirculation of
   Tissue (M. Kessler, D.F. Bruley, L.C. Clark, Jr., D.W. Lübbers,
   I.A. Silver, and J. Strauss, eds.). Urban & Schwarzenberg,
   München, 1973. pp. 110-112.

4. Huch, R.; Lübbers, D.W.; Huch, A.: An electrode for prolonged
   bloodless PO_2 measurements on the skin and a method for in
   situ calibration. In: Oxygen Supply, Theoretical and
   Practical Aspects of Oxygen Supply and Microcirculation of
   Tissue (M. Kessler, D.F. Bruley, L.C. Clark, Jr., D.W. Lübbers,
   I.A. Silver and J. Strauss, eds.). Urban & Schwarzenberg,
   München, 1973. pp. 101-103.

5. Reichert, W.J.: The theory and construction of oximeters.
   In: Oxygen Measurements in Blood and Tissues and their
   Significance (J.P. Payne, and D.W. Hill, eds.). Little,
   Brown and Company, Boston, 1966. pp. 81-102.

6. Strauss, J.; Beran, A.V.; and Baker, R.: Continous O_2 monitoring of newborn and older infants and of children. J. Appl.
   Physiol. $\underline{33}$:238-243, 1972.

7. Beran, A.V.; Taylor, W.F.; Ackerman, B.D.; Sperling, D.R.;
    and Strauss, J.: An automatic oxygen control system for infants.
    Pediatrics 48:315-318, 1971.

8. Strauss, J.; Beran, A.V.; Huxtable, R.F.; Clark, L.C., Jr.; and
    Katurich, N.: Assessment of tissue oxygen availability (O_2a)
    in newborn babies (Abstract and Discussion). Pediat. Res.
    2:424-426, 1968.

Partially aided by National Heart Institute Grants HE-09351 and
HE-14091, by California State Department of Public Health HMD
Study, and by Regional Medical Programs Area VIII Grant 1-G03-
RM-00019.

OXYGEN MONITORING OF NEWBORNS BY SKIN ELECTRODES. CORRELA-

TION BETWEEN ARTERIAL AND CUTANEOUSLY DETERMINED pO_2

Patrick Eberhard, Wolfgang Mindt, Franz Jann,
Konrad Hammacher
Dept. Bioelectronics, Hoffmann-La Roche & Co.,Basle

and Universitätsfrauenklinik, Basle, Switzerland

Measurement of pO_2 in tissues or on the skin surface
has been evaluated recently by several authors as a means of
estimating arterial pO_2 or of qualitatively determining the
state of oxygenation of the newborn (1-6). The present paper
reports results obtained with a skin electrode (Fig. 1) which
is heated to 42°C to create local hyperemization of the tissue
underneath the electrode (7). Contrary to conventional Clark
type electrodes in which the cathode diameter is kept small,
we utilize here a large sized cathode (diameter 3 to 10 mm)
to obtain an average pO_2 value of a sufficiently large skin
area. The permeability of the membrane for oxygen was kept
low to avoid a disturbation of the oxygen profile in the
tissue as a result of oxygen consumption of the sensor. The
response time of the sensor to reach 95% of the steady state
value, $\zeta_{95\%}$, is 60 seconds. The membrane is held by a cap,
the shape of which was chosen such that diffusion of oxygen
from the surrounding air to the cathode is prevented.

As electrolyte, a buffered hygroscopic solution is
used which guarantees a working time of the sensor at 42°C
(without renewing electrolyte and membrane) for about one
week. Inside the sensor are located a heating element consisting of a coil of resistance wire and a thermistor for
the control of the heating temperature.

The sensor is attached to the skin of the newborn by a
double adhesive tape ring, as is commonly used for ECG
electrodes. With this method, the pressure of the sensor
against the skin is extremely low and does not cause

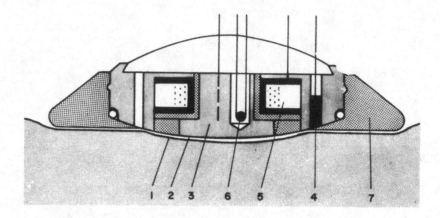

Fig. 1 Schema of sensor
(1) Skin; (2) Membrane; (3) Cathode; (4) Anode; (5)
Heating element; (6) Thermistor; (7) Cap for holding
the membrane.

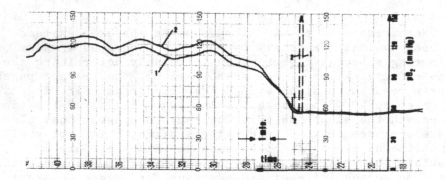

Fig. 2 Response of cutaneous pO_2 to change of oxygen concentration in inspired air: During breathing of room
air, the cutaneous pO_2 varies between 54 and 60 mm
Hg. At the time A, the newborn receives air enriched
with oxygen by means of a mask (3 liter/min.). After
an induction time of 20 secondes, the cutaneous pO_2
increases and within 6 minutes reaches a new "steady
state" with fluctuations between 106 and 128 mm Hg.
The indices 1 and 2 refer to sensor positions on the
right thigh and on the right temple, respectively.
The shift of the time axis of the two channels is
one fifth of one scale unit (corresponding to 12 sec.)

compression of blood vessels underneath the electrode. In
general, movements of the infant do not disturb the measurement. Only extremely strong movements may cause artefacts,
in particular, if the sensor is attached to one of the
extremities. Such artefacts, however, can be identified
immediately in the recording.

After application of the sensor to the skin, the current
reaches a steady-state value within five to seven minutes.
This time is needed for the induction of hyperemization by
heat stimulation. By treating the skin with vasodilating
agents (e.g. histamine) prior to the attachment of the sensor,
this time may be shortened. An example of a recording obtained with two skin electrodes is shown in Figure 2.

Comparative measurements with arterial pO_2 were performed in two ways: First by sampling blood from a centrally
located catheter introduced through the umbilical artery.
The results of 51 measurements in 15 newborns are compiled
in Fig. 3, whereby all cutaneous values were determined on
the subclavicular part of the thorax. For the statistical
analysis of the data, the skin pO_2 has been chosen as independent variable. Actually, the clinical application of
the method is, in general, the estimation of the arterial
pO_2 from the cutaneous measurement. It has been assumed that
within the population of the data a linear relation exists
between the two variables (linear regression). The correlation coefficient is 0.93 with a confidence interval of 0.87
to 0.96. The regression coefficients of the relation
$(pO_2)_{art.} = A \cdot (pO_2)_{cut.} + B$ are : A = 1,30 with a standard
error of 0,08 and B = -14 mm Hg with a standard error of
6 mm Hg. The standard error of estimate is 9 mm Hg. Fig. 6
shows the 95 % confidence interval limits for the ordinate
to the true regression line and the 95 % prediction interval limits for individual values of $(pO_2)_{art}$. For these
investigations, thorax values have been evaluated since,
for long term studies, the thorax was found to be the best
measuring position. The flat region is well suited for the
fixation of the electrode, movement artefacts occur very
rarely and the site is the most practical one for the care
of the child. In addition, the skin of the thorax of newborns is relatively thin and well perfused.

In a second series, comparative measurements are being
performed by continuously determining arterial pO_2 with an
indwelling pO_2 catheter electrode, developed for this purpose in our laboratory. A typical recording of both arterialized skin pO_2 and arterial pO_2 is shown in Fig. 4. The
irregular oscillations of the pO_2 are observed simultane-

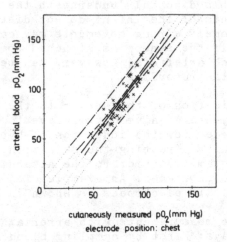

Fig. 3 Relationship between umbilical artery pO$_2$ and cutaneously measured pO$_2$.

    —— true regression line
    --- 95 % confidence interval limits for the ordinate
    to the true regression line
    -.- 95 % prediction interval limits for individual
    values of arterial pO$_2$

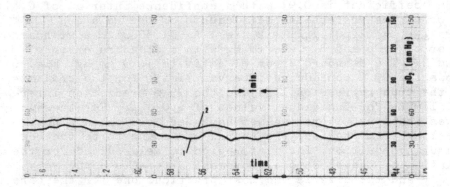

Fig. 4 Simultaneous measurement of arterial and cutaneous
pO$_2$ in a one-day old newborn.
1. Skin electrode. Position of the sensor: Thorax
2. Catheter electrode. Implanted in the aorta des-
cendens through the umbilical artery.

ously on the skin and in the artery. We have observed such
oscillations in most of our cutaneous measurements (e.g.
see Fig. 2). It can be assumed that they are caused by a
central regulatory system which controls the oxygen supply
e.g., via changes of arterial pO_2 (see Fig. 4) or via the
peripheral blood flow. More detailed results of these continuous correlation studies will be published later.

In conclusion, the correlation between arterial pO_2
and skin pO_2 allows a fairly good estimation of the arterial
oxygen partial pressure from the cutaneous measurement. The
accuracy is sufficient to utilize the cutaneous pO_2 measurement for the control of oxygen therapy of high risk newborns.
In addition, information of prognostical value may be obtained from the observation of the trend of the cutaneously
determined pO_2.

References

1. Rodger, J.C., Kerr, M.M., Richards, I.D.G., Hutchison,
   J.H. (1968) Measurements of oxygen tension in subcutaneous tissues of newborn infants under normobaric and
   hyperbaric conditions, The Lancet, 1968:2, 232 - 236.

2. Walker, A.,   Phillips, L., Powe, L., Wood, C. (1968)
   A new instrument for the measurement of tissue pO_2 of
   human fetal scalp. Am. J. Obst. Gynecol. 100, 63 - 71.

3. Huch, A., Huch, R., Lübbers, D.W. (1969) Quantitative
   polarographische Sauerstoffdruckmessung auf der Kopf-
   haut des Neugeborenen, Arch. Gynäk. 207, 443 - 451.

4. Neuman, M.R., Brown, E.G., McDonnel, F.E., Liu, C.C.
   (1971) Application of oxygen cathodes in perinatology,
   Proc. 24th ACEMB, Las Vegas, Nevada, Oct. 31 - Nov. 4,
   1971.

5. Schönjahn, V., Bellé, H., Büchner, M., Franke, R. (1972)
   Einstichelektrode zur kontinuierlichen pO_2-Messung im
   lebenden Hautgewebe, Patentschrift No. $88^2 835$, GDR.

6. Strauss, J., Beran, A.V., Baker, R. (1972) Continuous
   O_2 monitoring of newborn and older infants and of
   children. J. Appl. Physiol. 33, 238 - 243.

7. Eberhard P., Mindt W., Hammacher K. Perkutane Messung
   des Sauerstoffpartialdruckes : Methodik und Anwendungen.
   Presented at the Kongress "Medizin-Technik 1972" in
   Stuttgart. May 14 - 17, 1972.

A UNIQUE ELECTRODE CATHETER FOR CONTINUOUS MONITORING OF ARTERIAL BLOOD OXYGEN TENSION IN NEWBORN INFANTS

Edwin G. Brown, Chung C. Liu*, Francis E. McDonnell,
Michael R. Neuman, and Avron Y. Sweet

Department of Pediatrics and Perinatal Clinical Research
Center, Case Western Reserve University at Cleveland
Metropolitan General Hospital, Cleveland, Ohio, and
*Chemical Engineering Department, University of
Pittsburgh, Pittsburgh, Pennsylvania

Arterial blood oxygen tension (paO$_2$) is frequently determined in the intensive care of newborn infants with lung or heart diseases. Knowledge of paO$_2$ in such patients is imperative because elevation above the normal range of 50-110 mm Hg. may cause damage to the retina of prematurely born infants sufficient to result in permanent blindness. At present the most reliable technique for monitoring paO$_2$ is intermittent and requires periodic removal of arterial blood samples through a catheter which has been passed through an umbilical artery into the aorta. Analysis is done by means of a micro blood gas analyzer which requires 0.4 - 0.6 ml. of blood. A method of continuously monitoring paO$_2$ would permit better control of the oxygen en- riched breathing environment and eliminate the need for frequent sampling which might dangerously deplete the blood volume of a small infant. Several investigators have recently described in- dwelling systems for measuring paO$_2$.[1,2,3,4] The authors have developed a new technique which makes it possible to modify an ordinary umbilical artery catheter for continuously measuring paO$_2$.[5] This is done by means of a polarographic electrode embedded within the wall of the catheter. The lumen of the catheter is, therefore, unobstructed. The small size of the electrode allows the catheter to remain flexible and soft so that the possibility of damage to vessel walls is minimized. The sensing end of the electrode is flush with the catheter tip and covered with a thin layer of material identical to that of the catheter. This acts as an oxygen permeable membrane and protects the electrode from contamination by proteins. The only differ- ence between the catheter containing the electrode and the con- ventional umbilical artery catheter is the negative charge

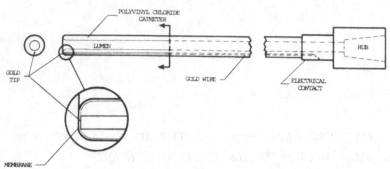

Fig. 1.   Schematic diagram of oxygen catheter.

surrounding the electrode. This may be advantageous since blood
clotting elements are negatively charged and would be repelled
from the cathode.

The oxygen sensor consists of a polarographic cell in which
the cathode is a .003" diameter pure gold wire extruded into the
wall of a 5 French polyvinyl chloride or polyurethane umbilical
vessel catheter. The sensor operates at currents of 10^{-8} – 10^{-9}
Amperes. Insulation is very important, and materials must be used
which adhere to the gold cathode and bond to the catheter. A thin
layer of Isonel®* which is overcoated with polyvinyl chloride
or polyurethane has provided suitable insulation. The anode is
a silver-silver chloride electrode which is coupled to body fluid
by means of an electrolyte solution acting as a salt bridge within
the lumen of the catheter. The actual construction of the polarographic oxygen cell in an umbilical artery catheter is shown in
Fig. 1.

A block diagram of the electronic circuits used with the
electrode is shown in Fig. 2. The electronics must be capable
of carrying out two functions: 1) providing 0.6 V negative bias
to the cathode and 2) detecting currents of the order of 10^{-8} –
10^{-9} Amperes. The preamplifier and bias circuit are battery
operated and isolated from the rest of the electronics and can be
located near the catheter to minimize noise pick up. Shock
hazards are prevented by the isolation. The output of the
electrode is temperature sensitive and, therefore, temperature
compensation is necessary since the body core temperature of the
small premature infant may vary between 34-38° C. Compensation
can be achieved by changing the amplifier gain as the temperature
varies. This can be done by setting a manual control or automatically by using a thermistor which measures the infant's core
temperature. The thermistor has temperature characteristics
which are nearly the inverse of those of the oxygen sensor.

*Schenectady Chemical Co., Schenectady, N. Y.

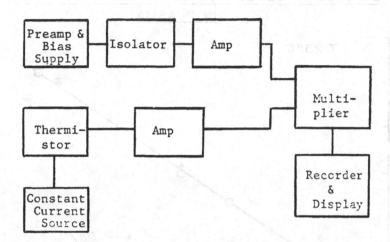

Fig. 2. Block diagram of electronic system

Therefore it is possible to temperature compensate by a thermistor driven by a constant current source to one input of an analog multiplier and connecting the signal from the oxygen sensor to the other input of the multiplier. Data can be presented in analog or digital form and displayed by appropriate recorders or readouts.

The operating characteristics of the instrument have been fully evaluated in the laboratory and in animals prior to clinical use. The following parameters were analyzed: linearity and hysteresis relating paO_2 and current output, effects of motion (flow) on current output, changes in output due to temperature, long term stability, and the effects of blood or plasma on the cathode and membrane.

Calibration of the electrode system has shown a linear relationship between the electrode current output and PO_2 as determined by conventional laboratory methods as shown in Fig. 3. These data were obtained with the sensor immersed into a plasma filled test chamber of a membrane oxygenator at room temperature. They illustrate typical data points obtained by increasing and decreasing oxygen tension during a three day period. No hysteresis is demonstrated and there is stability of the output and calibration constants during the three days.

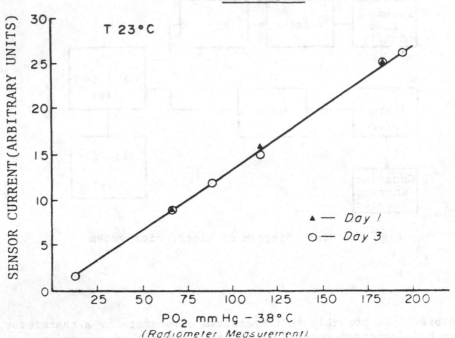

RESPONSE OF SENSOR TO CHANGING OXYGEN TENSIONS
IN PLASMA

Fig. 3

Direct measurement of the effects of flow on the output
current were made under conditions similar to arterial flow.
Saline equilibrated with air was pumped into a test chamber of
constant cross sectional area. The flow rate in the chamber was
varied by changing the infusion rate of the pump. Studies of flow
velocities between 0-6 cm/sec showed the output of the electrode
to increase with flow, but saturation occurred at flow velocities
greater than 3 cm/sec. The total change in output from zero to
maximum flow was no greater than 20% of the overall cell output.

The effects of temperature on the sensor were evaluated by
varying the temperature of a saline solution from 20-40° C. The
saline solution was alternately equilibrated with air and 5%
oxygen - 95% nitrogen. Using several different electrodes, no
difference was found in the slope of the relationship of the log
of the cell output to the reciprocal of the absolute temperatures
as shown in Fig. 4. The slope corresponds to an activation
energy of 3.83×10^{-2} electron volts which represents a temperature
coefficient of about 5% per degree at 37° C.

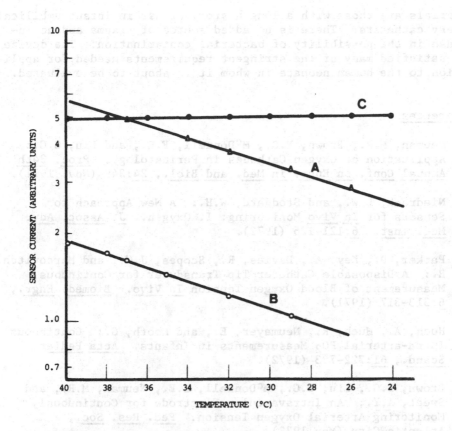

Fig. 4.   Temperature characteristics of high output sensor (A),
           low output sensor (B), and temperature compensated sensor
           A (C) at a PO_2 of 147 mm of Hg.

The sensor showed no significant difference in output over
periods of three days, using albumen, plasma and saline. It,
therefore, appears that protein solutions cause no significant
contamination of the sensor.

Evaluation of the sensor in dogs and one anencephalic human
newborn infant for periods of 8-16 hours has provided data similar
to those obtained in vitro.

It is now possible to produce a catheter tip, extra luminal,
temperature compensated, continuous paO_2 monitor which is un-
effected by flow within the range expected in human neonates. It
is uniquely constructed so that it does not interfere with the
monitoring of other parameters dependent upon systems which require a patent catheter lumen (e.g., blood pressure). The

materials are those with a long history of use in infant umbilical
artery catheters. There is no added source of trauma and no increase in the possibility of bacterial contamination. The device
has satisfied many of the stringent requirements needed for application to the human neonate in whom it is about to be evaluated.

References

1. Neuman, M.R., Brown, E.G., McDonnell, F.E., and Liu, C.C.:
    Application of Oxygen Cathodes in Perinatology. Proc. 24th
    Annual Conf. on Engr. in Med. and Biol., 24:249 (Nov. 1971).

2. Niedrach, L.W., and Stoddard, W.H.: A New Approach to
    Sensors for In Vivo Monitoring: I. Oxygen. J. Assoc. Adv.
    Med. Inst., 6:121-125 (1972).

3. Parker, D., Key, A., Davies, R., Scopes, J.W., and Marcovitch,
    H.: A Disposable Catheter Tip Transducer for Continuous
    Measurement of Blood Oxygen Tension In Vivo. Biomed. Engr.,
    6:313-317 (1971).

4. Huch, A., Huch, R., Neumayer, E., and Rooth, G.: Continuous
    Intra-arterial PO$_2$ Measurements in Infants. Acta Pediat.
    Scand., 61:722-723 (1972).

5. Brown, E.G., Liu, C.C., McDonnell, F.E., Neuman, M.R., and
    Sweet, A.Y.: An Intravascular Electrode for Continuously
    Monitoring Arterial Oxygen Tension. Ped. Res. Soc.,
    Atlantic City (May 1972).

Supported by N.I.H. Grants RR00210, GM-14267 and HD-46250.

LABORATORY AND CLINICAL EVALUATION OF A NEW INDWELLING OXYGEN ELECTRODE FOR CONTINUOUS MONITORING OF PaO$_2$ IN NEONATES

Thomas R. Harris and Michael Nugent

Department of Pediatrics, University of Arizona College

of Medicine, Tucson, Arizona, U.S.A.

Recent development of a relatively inexpensive *in vivo* electrode for monitoring the partial pressure of oxygen (PO$_2$) in adults by International Biophysics Corporation of Irvine, California, prompted us to adapt this system for clinical use in newborn infants. This report describes the final configuration of the monitoring system for neonates, certain *in vitro* characteristics of the oxygen electrode, its *in vivo* accuracy, complications of its use and some clinical applications of the system.

The IBC electrode differs from the standard Clark electrode in that the anode and cathode are separated, with just the cathode placed in the blood stream. Dissolved oxygen in the blood diffuses with water through the hydrophilic (Hydron®) membrane to the gold electrode tip where it is electro-chemically reduced, causing a current to flow by way of the blood and body tissues from the catheter tip to the topically attached anode. The current, which is proportional to the oxygen tension of the blood, is displayed directly by the oxygen analyzer as PO$_2$ in mmHg.

The completed PaO$_2$ monitoring system for neonates includes a strip chart recorder to provide a continuous tracing of PaO$_2$, a safety coupler which electrically isolates the recorder from the analyzer and indwelling cathode entering the Argyle catheter through a side hole, and a topically attached anode. The umbilical catheter may continue to be used for withdrawal of blood samples, measurement of blood pressure, and administration of fluids, medications, blood, etc. The system is recalibrated by comparing laboratory blood gas results with the oxygen analyzer readings, and adjusting the electrical gain of the analyzer accordingly.

In vitro studies demonstrated the electrode to be effected by changes in pH, temperature, and osmolality. An acute change in pH from 6.80 to 8.20 resulted in an average decrease in electrode output of 7% (range 3-11%) in 10 electrodes studied. This effect proved reversible when pH was returned to its original level. A temperature coefficient of plus 3% per degree centigrade was found within the temperature range of 22-42°C. An osmolality change from 300 to 600 miliosmoles caused a statistically significant (P value of 0.01) decrease in electrode output of approximately 7.4% (range 4.3-8.3%). No significant changes in electrode output were noted by alterations in flow of the surrounding medium, or upon exposure to extreme physiologic concentrations of urea, heparin, calcium gluconate and sodium citrate.

The *in vivo* accuracy of the system was assessed while monitoring PaO_2 in 48 acutely ill neonates for a total of 3,538 hours or 147 days. Recalibration was carried out whenever the oxygen electrode was found to be in variance with the lab value by greater than 15-20 mmHg; this proved to be on an average of every 17½ hours. 999 comparisons between laboratory values and the *in vivo* system were made, an average of one check every 3½ hours of operation. The standard error or first standard deviation from the predicted value (the laboratory value is considered absolute) over the entire range of PaO_2 from 10 to 252 mmHg was found to be +18.8 (Table I).

Range PaO_2 mmHg	Standard Error mmHg	Number of Comparisons
>142	38.8	31
132-142	22.8	9
123-132	19.0	11
114-123	28.1	24
105-114	17.7	34
96-105	20.8	49
87-96	18.6	77
78-87	18.8	101
69-78	19.4	155
60-69	13.8	165
51-60	14.8	153
42-51	14.3	115
33-42	12.4	51
24-33	12.0	14
15-24	14.6	10
10-252	18.8	999

Table I. Standard error between bench analyzer and indwelling oxygen electrode PaO_2 values in specific PaO_2 ranges for the 50 cases where the oxygen electrode was consistantly recalibrated.

As can be seen from the table, the error is greatest in the higher PaO$_2$ ranges ($\pm$38.8 for all values above 142), and is equally distributed above and below the predicted values. As regards accuracy and the duration of operation of the electrode, the standard error between bench analyzer and indwelling oxygen electrode PaO$_2$ values for the first 24 hours of operation was $\pm$15.3, while the standard error for operation after the first 24 hours was $\pm$20.2 mmHg.

Partial explanations of the wide standard error found in these studies include the electrode's slow response time in face of rather rapid natural fluctuations in PaO$_2$ which makes it difficult to obtain a truely simultaneous blood sample to compare with the analyzer reading; this also makes accurate recalibration of the system most difficult and lends further to the degree of error. Temperature and pH changes have minor and predictable effects upon the electrode. Glycerol has been used previously by other investigators to test the effects of viscosity on oxygen electrodes of similar design to that used in this study. They interpreted the large decreases in electrode output when exposed to high concentrations of glycerol solution to mean that such electrodes are extremely sensitive to viscosity changes and therefore also to alterations in hematocrit. The *in vitro* results of the present study using a methyl cellulose solution of equal viscosity with the glycerol, and blood of different hematocrits, counter this interpretation. The fact is that large decreases in the output of the electrode can be produced by very high concentrations of substances of small molecular or ionic size such as glucose or sodium chloride as well as glycerol, whereas larger molecules such as methyl cellulose or hemoglobin, irregardless of their viscosity, have no significant influence upon its operation. The effect here may be one of altered degrees of hydration of the hydrophilic membrane which results in altered availability of water and physically dissolved oxygen.

Difficulties and complications of the *in vivo* PaO$_2$ monitoring system proved to be minor. Leakage of infusion fluid at the point of entry of the oxygen electrode into the Argyle catheter was a frequent problem with the earlier catheters. The newer catheters have a layer of Hydron at the marker point of electrode entry which swells to seal the connection once the electrode is advanced to its proper place. An even newer version has the electrode entering the catheter by way of a special 2-way stopcock which had a disc seal at its point of entry making it leak-proof yet interchangable.

Difficulty in withdrawing blood from the catheter with the indwelling oxygen electrode was a recurring problem. Total inability to obtain blood return was eventually encountered in 7 of the 50 catheters; the average operating period of these catheters at the time of blockage was 6 days. Fibrin clots were found protruding from the catheter tip and around the sensing electrode in 2 of the

12 cases coming to autopsy. Four infants demonstrated transient cyanosis in distal portions of one lower extremity, most likely due to air embolism, and another child with omphalitis was heparinized for presumptive partial occlusion of the common iliac arteries, as there was temporary bilateral diminution of femoral pulses and systolic blood pressure readings in both legs.

The tips of all catheters removed were routinely cultured. One of the 42 cultures was positive for E. coli and Strep faecalis, which were considered contaminants, as the infant showed no clinical signs of sepsis.

Continuous monitoring of PaO_2 in acutely ill neonates proves useful in terms of both diagnosis and treatment. In cases where there is marked improvement in PaO_2 following the addition of Positive End Expiratory Pressure (PEEP) to the ventilator or Continuous Positive Airway Pressure (CPAP) via endotracheal (ET) tube or simple mask, it may be assumed that alveolar atelectasis with associated ventilation/perfusion inequalities is contributing largely to the hypoxemia and not hypoventilation or intracardiac right-to-left shunting.

The value of intermittent bag and mask ventilation of infants with Idiopathic Respiratory Distress Syndrome (IRDS) has previously been questioned by other investigators because of the rapid (within one minute) return of PaO_2 to its original pre-bagging level. This is contrasted with the slower return to pre-treatment levels after mask application of CPAP. However, in other cases bagging may be required initially to obtain a rise in PaO_2 which then may be sustained for a more prolonged period of time by back-ground application of CPAP. It is obvious that continuous monitoring of PaO_2 allows one to choose the appropriate mode of therapy for the individual patient.

Real-time monitoring of PaO_2 also allows for a running evaluation of the dangers and/or effectiveness of standard nursing procedures. If it is observed that a procedure such as suctioning of the endotracheal tube of an infant receiving mechanical ventilation with elevated concentrations of oxygen is resulting in a significant drop in PaO_2 as registered on the continuous PaO_2 monitor, the procedure may be expedited or discontinued before an extreme degree of hypoxemia results.

Continuous PaO_2 monitoring is most useful in acute, fast-changing clinical situations. Here one can expeditiously make appropriate and safe changes in ambient oxygen concentration or respirator settings and know immediately what effects these changes are having on PaO_2. In an era when new modes of therapy are capable of achieving rapid and radical changes in PaO_2, the need for equally rapid means of detecting these change becomes obvious.

CONTINUOUS INTRAVASCULAR PO$_2$ MEASUREMENTS WITH CATHETER AND CANNULA ELECTRODES IN NEWBORN INFANTS, ADULTS AND ANIMALS

Albert Huch, Dietrich W. Lübbers, Renate Huch

Department of Obstetrics and Gynaecology,
University of Marburg, Marburg/Germany   and
Max-Planck-Institut für Arbeitsphysiologie,
Dortmund/Germany

In order to perform continuous Po$_2$ measurements in clinical research and in monitoring, we have constructed a catheter electrode (2) with a diameter of 1.5 mm. The 4.5 mm long electrode is fitted on the top of a normal feeding tube no. 5 (Braun, Melsungen).

Using the technical experience we gained with this catheter electrode we then constructed a cannula electrode

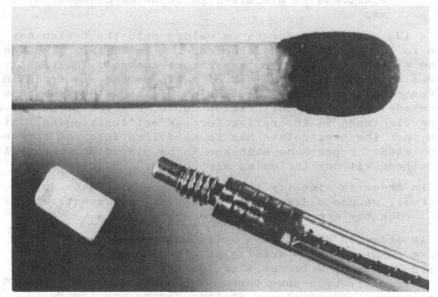

Fig. 1   Po$_2$ catheter electrode compared to the size of a match.

with the same polarographic characteristics. Both electrodes
have 15 µm platinum cathodes electrically welded into glass.
They are surrounded by a silver cylinder, the anode, which has
a groove for screwing on the teflon cap. The diameter of the
teflon cap has to have the same outer diameter of the catheter
or the cannula, namely 1.5 or 1.o mm.

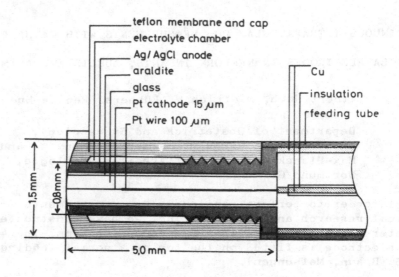

<u>Fig. 2</u>   Scheme of the catheter electrode. The cannula
           electrode is minimized to an outer diameter of 1.o
           mm.

A 12 µm teflon membrane is welded onto the teflon cap
which is screwed on to the electrode. A lo kΩ NTC is glued
in the silver cylinder for temperature control. The tightly
screwed on cap avoids the loss of parts of the electrode in
the vessel. Furthermore, the tension of the welded membrane
guarantees a constant calibration curve. Precise preparation
of the electrode is imperative for perfect functioning. This
means that the electrolyte has to be filled into the cap and
the moistened cuprophane membrane to be placed between teflon
and cathode without including airbubbles.

In order to simplify the application of the cannula
electrode, we use a disposable Braunula(R) and substitute its
needle once the artery is punctured (Fig. 3).

In order to guarantee a well functioning electrode in
vivo, we constructed an artificial circulation system for
in vitro testing. Water at a preset temperature is driven
peristaltically by a pump through an oxygenator into an artificial vessel, containing a Statham-device and the Po_2

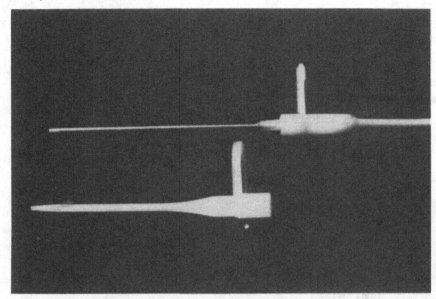

Fig. 3 Cannula electrode fitted in a disposable Braunula(R)

electrode. This model enables us to simulate all changes in
pressure, O_2 tension, flow and temperature.

Tests have shown that even with pressure variations in
the range of 400 mm Hg the Po_2 registration remains constant,

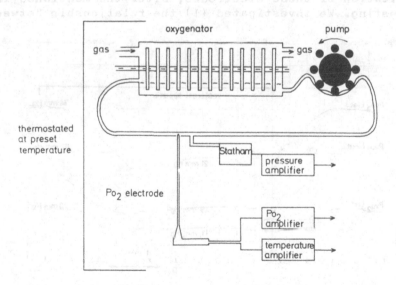

Fig. 4 Circulation system for testing Po_2 electrodes in
       vitro

also considering a 95% response time of the electrode of only
5 sec.

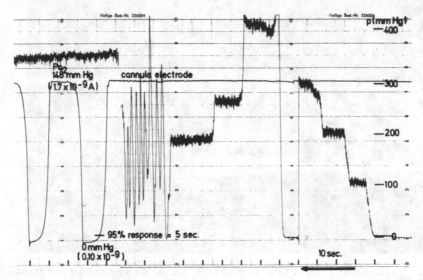

<u>Fig. 5</u> Reproduction of an original chart showing the test
results. Various pressure changes within a range of
4oo mm Hg do not influence the Po_2 recording of a
cannula electrode.

The following examples are meant to show the efficiency
of application of these electrodes, after the mentioned in
vitro testing. We investigated (1) the relationship between

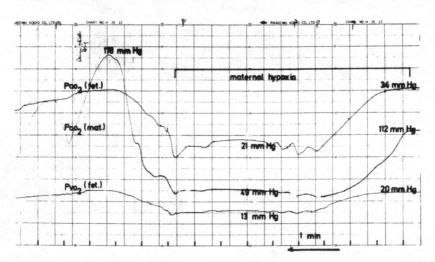

<u>Fig. 6</u> Simultaneous registration of Po_2 with catheter elec-
trodes in the a. carotis of an ewe and in the a.
carotis and v. jugularis of its fetus.

maternal and fetal hypoxaemia by simultaneous and continuous
Po_2 registration in the a. carotis of the ewe and in the a.
carotis and v. jugularis of the fetus. Because response time
of all three electrodes is practically identical, it is possible to measure time factors of gas exchange between mother
and fetus. For example (Fig. 6), a decrease in fetal Po_2 is
observed 2o seconds after the decrease of mother's Po_2. In
fetus near term all Po_2 changes in the mother were promptly
reflected in the fetal curve. As shown in Fig. 6, a decrease
in Po_2 from 112 to 49 mm Hg in the ewe results in a Po_2 decrease from 34 to 21 mm Hg in the fetus.

The application of the catheter electrodes to human infants enabled us to investigate Po_2 changes after birth and
after the onset of respiration (3), see Fig. 7.

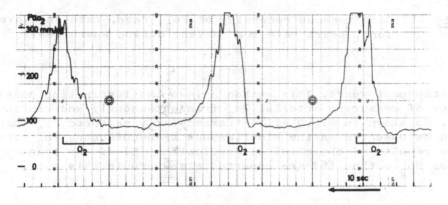

Fig.7   Original chart of Po_2 monitoring in the aorta of a
healthy newborn infant, 12 minutes post partum.

It is evident that continuous and fast information of
Po_2 values, especially in regard to the considerable variations
which occur in the same infant and during short oxygen inhalation, cannot be obtained by blood sampling. Inhalation of supp-
lementary oxygen can increase the Po_2 value up to a few loo
Torr within a few breaths.

The same observations were made also in adults with a
cannula Po_2 electrode in patients during anaesthesia. We observed that arterial Po_2 promptly reflects changes in respiration rate and in O_2 contents of the inhalation gas mixture
(Fig. 8). These examples show that both types of Clark electrodes can be applied in physiological and clinical research, en-
abling the clinician to make observations which could not be
made otherwise.

Finally the intravascular Po_2 method can be used in order to check the reliability of another method we developed

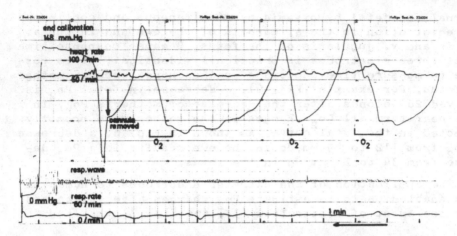

<u>Fig. 8</u>   Simultaneous recording of Po$_2$, resp. rate and wave
form, heart rate in a patient during anaesthesia

in the past years. This method allows the fascinating possibility of measuring arterial Po$_2$ transcutaneously and continuously. As shown in Fig.9 all intravascularly measured Po$_2$ changes
can be observed on the simultaneous transcutaneous Po$_2$ curve.
The results of the clinical application (5) of the transcutaneous Po$_2$ method (6) are reported elsewhere in this book.

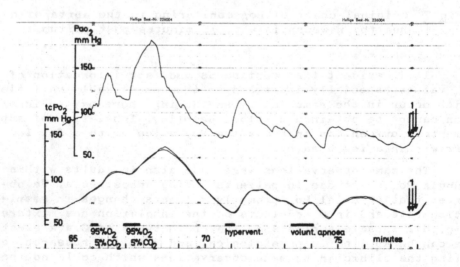

<u>Fig. 9</u>   Simultaneous monitoring of intravascular and trans-
cutaneous Po$_2$ (skin electrode with 2 cathodes). (4)

REFERENCES: (1) BERG, D., SCHULZ, J., WERNICKE, K., HUCH, A.,
HUCH, R.: Perinatal Medicine, Huber, Bern 1973, p. 323; (2)
HUCH, A., HUCH, R., ROOTH, G.: J. Perinat. Med. 1 (1973) 53;(3)
HUCH, A., LÜBBERS, D.W., HUCH, R.: Oxygen Supply, Urban &
Schwarzenberg, München, 1973, 11o; (4) HUCH, A., HUCH, R.,
ARNER,B., ROOTH, G.: Scand. J. Clin. Lab. Invest. - in press;
(5) and (6) LÜBBERS, D.W. and HUCH, A. and R.: in this book.

ROUTINE MONITORING OF THE ARTERIAL PO$_2$ OF NEWBORN INFANTS BY

CONTINUOUS REGISTRATION OF TRANSCUTANEOUS PO$_2$ AND SIMULTANEOUS

CONTROL OF RELATIVE LOCAL PERFUSION

Renate Huch, Dietrich W. Lübbers and Albert Huch

Max-Planck-Institut für Arbeitsphysiologie
Dortmund/Germany   and
Department of Obstetrics and Gynaecology
University of Marburg, Marburg/Germany

In reference to our earlier lecture (9) we would like to report about the results of the transcutaneous Po$_2$ method in the monitoring of newborn babies.

In 1971 in Dortmund an electrode according to Clark (1) was discussed, in which the necessary hyperaemia was induced pharmacologically or by heating the skin surface in the region of the electrode (6, 7, 8). This is now improved to the point that it is possible to produce a regulated and constant long-time hyperaemia by means of a better controlled hyperthermia. The electrode is heated to obtain the necessary perfusion efficiency. This new improvement was the most important requirement for the application of the transcutaneous Po$_2$ method in daily routine (3, 4, 1o) and substitutes the usual blood samplings and gas analysis.

The required temperature is produced in the anode of the electrode by means of a heating coil. Its temperature is regulated by a thermistor fitted in the anode itself close to the outer surface. At the center of the ring-shaped anode there are three distinct 15 µm platinum cathodes which are isolated from each other and which give separate signals to the analyzer. One can test whether the electrode is correctly situated on the skin by seeing whether one gets the same response from all three channels.

Apart from registering Po$_2$ values it is also possible to draw a continuous chart of local perfusion. In fact the heating energy required for keeping the preset temperature at a constant level is used as a relative measurement of the cooling effect of the blood (inverted Gibbs' method (2)).

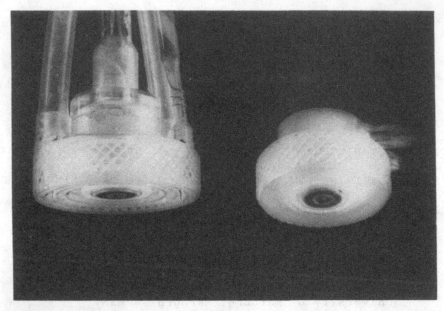

Fig. 1 Electrodes for transcutaneous monitoring of blood
Po_2 (Left: Construction to be applied on skin with vacuum,
with silvertips for simultaneous registration of ECG, e.g.
on fetal scalp. Right: Construction to be applied on skin
with self-adhesive rings.)

After calibration the electrode is applied to the moistened
skin with a self-adhesive ring. In newborn infants the

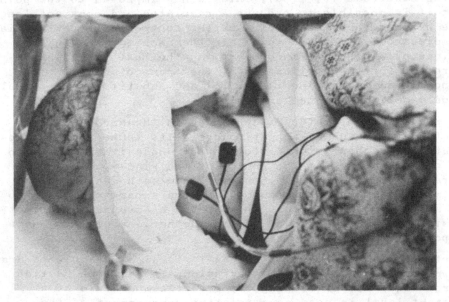

Fig. 2 Electrode applied to the chest of a newborn baby

the electrode is applied to the chest. A temperature of 43° C
on the skin surface produces for hours constant hyperaemia and
is tolerated well by infants.

After ten to fifteen minutes, local skin reaches an optimal
and homogeneous vasodilatation so that from now on all values
which are registered correspond to true arterial Po_2 levels.
It is therefore now possible to monitor even extremely unquiet
babies as long as desired, even for several hours.

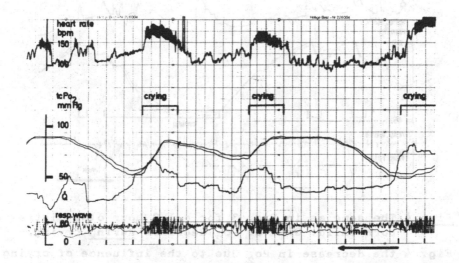

Fig. 3 Part of a continuous transcutaneous Po_2 registration
(tcPo_2 registration). Influence of crying on the Po_2
of a healthy, six hours old baby: Po_2 decrease <u>during</u>
crying.

The above figure shows reproduced original curves to be read
from right to left. Chart speed is 3 cm per minute. From top
to bottom a beat-to-beat registration of the babies' heart
rate can be seen, then the recording of two of the cathodes
of the Po_2 electrode with their parallel behaviour. Then we
have the registration of local perfusion (Q̇) and at the bottom
the recording of two respiratory parameters (resp. wave and
resp. rate). Apart from Po_2 and perfusion all the other parameters on these charts were obtained with instruments manufactured by Hellige (Freiburg/Germany).

These are the values taken from a six hours old, healthy
baby. The Po_2 curves are a further proof of what we have already shown with our intraaortic measurements (5): There is
no such thing as <u>one</u> arterial Po_2 value. As can be seen, as

soon as the baby cries, here indicated by the increase in
heart rate and in the respiratory wave curve, arterial Po_2
starts to go down. The values return to normal levels once the
baby has calmed down.

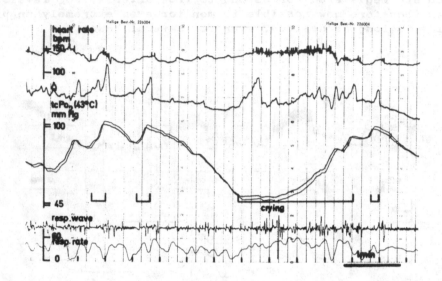

Fig. 4 The same as in Fig. 3 for a 1/2 hour old, distressed
       born child: Greater Po_2 decrease during crying.

In Fig. 4 the decrease in Po_2 due to the influence of crying
is even more evident. This child was born by Caesarian section,
had an Apgar 3 and was 1/2 hour old when we tested it. It is
evident that the usual techniques for monitoring the respiratory rate are not adequate to show insufficient oxygen supply.
In this case the mean respiratory rate curve at low Po_2 values
is normal, and frequent apnoic phases are signalized when the
Po_2 is already back to normal. It is important to point out
here that local perfusion tends to increase during these apnoic phases.

Po_2 normally decreases during crying in babies up to two or
three days old. This does not exclude exceptions such as the
following example (Fig. 5), where the expected dip in Po_2
occurs after crying as a consequence of the apnoea.

These measurements demonstrate clearly the limitations in what
can be learned from a single blood gas analysis. The behaviour
of arterial Po_2 in newborn infants is extremely variable and
this results in a great variability in Po_2 when children cry
during sampling.

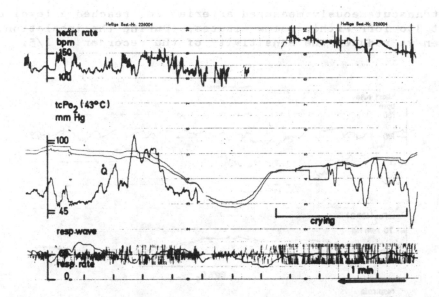

<u>Fig. 5</u>   The same as Fig. 3 and 4 for a 2 days old, healthy
         baby: Po_2 decrease <u>after</u> crying.

When healthy babies breath pure oxygen, Po_2 must increase
rapidly (Fig. 6). In this case the baby was 3 hours old and

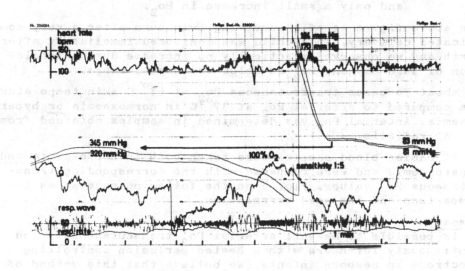

<u>Fig. 6</u>   Part of a continuous $tcPo_2$ registration. Po_2 in-
         crease when a healthy child breathes pure oxygen
         (3 hours old).

the transcutaneously measured arterial Po_2 reached a level of about 33o Torr within a few minutes. (During the inhalation of oxygen we reduced the sensitivity of the recorder to 1/5.

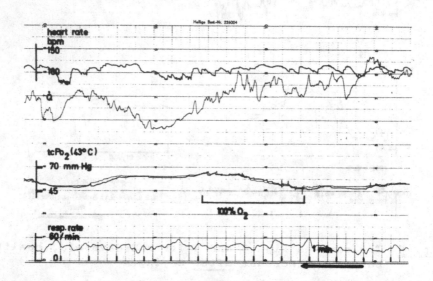

Fig. 7   The same as in Fig. 6 for a surviving twin immediately after a very complicated delivery: Low Po_2 level
and only a small increase in Po_2.

The above figure is of a surviving twin born after a very complicated delivery. This child was monitored immediately after
birth and we observed practically no increase during inhalation of pure oxygen. Po_2 is below the normal range.

In about 7o cases transcutaneous Po_2 at 43° C skin temperature
was compared to arterial Po_2 at 37^{o2} C in normoxaemia <u>or</u> hyperoxaemia. Arterial Po_2 was determined in samples obtained from
the A. radialis.

In 15 babies blood samples were taken both in normoxaemia <u>and</u>
hyperoxaemia and were compared with the corresponding transcutaneous Po_2 values. Fig. 8 on the following page shows the
comparison and the good correlation.

<u>Summary</u>
It is possible to measure arterial Po_2 transcutaneously and
continuously for hours with a heated perfusion controlling
electrode in newborn infants. We believe that this method of
monitoring allows more insight into the state of oxygen supply after birth than the usual methods.

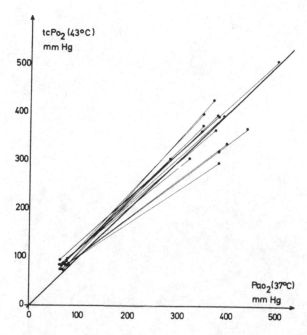

__Fig. 8__ Comparison between tcPo$_2$ and Pao$_2$ in 15 infants

References

1. CLARK, L.C., Jr. : Trans. Amer. Soc. Art. Int. Org. 2, 41,
        1956.
2. GIBBS, F.A.: Proc. Soc. Exper. Biol. a. Med. 31, 141,
        1933.
3. HUCH, A., LÜBBERS, D.W., HUCH, R. in: SALING/DUDENHAUSEN,
        Perinatale Medizin IV, Thieme-Verlag, Stuttgart,
        in press.
4. HUCH, A., HUCH, R., ARNER, B., ROOTH, G.: Scand. J. Clin.
        Lab. Invest. - in press.
5. HUCH, A., HUCH, R., ROOTH, G.: J. Perinat. Med. 1, 53,
        1973.
6.a.7. HUCH, R., LÜBBERS, D.W., HUCH, A.in: Oxygen Supply,
        Urban & Schwarzenberg, München, 1973, lol __and__ 27o
8. HUCH, R., LÜBBERS, D.W., HUCH, R.: Pflügers Arch. 337,
        185, 1972.
9. LÜBBERS, D.W., HUCH, A., HUCH, R. in this book.
lo. LÜBBERS, D.W., HUCH, A., HUCH, R.: KliWo 51, 8, 411,
        1973.

CLINICAL EXPERIENCES WITH CLAMP ELECTRODES IN FETAL SCALP FOR SIMULTANEOUS pO_2 AND ECG - REGISTRATION

W. Erdmann, S. Kunke, J. Heidenreich,
W. Dempsey, H. Günther, and H. Schäfer

Institut für Anaesthesiologie, Physiologisches
Institut und Universitätsfrauenklinik der
Universität Mainz, Germany

The development of a scalp electrode for continuous measurement of fetal mean tissue-PO_2 and simultaneous registration of ECG respectively fetal heart rate has been described previously (1). The electrode mechanics and electronic equipment must be adapted to the various demands of the birth process and the labor room environment. The technical results are not yet refined, when unfavorable conditions prevail, artifacts may be anticipated. A brief review of the proceedings in electrode clamp measurements should be presented.

Methods

The electrode tips (gold wire in glass coating of about 2o-3o ‚u) are inserted into a o,2 mm spring-loaded steel canula. Two canulas are fixed together by glue along their straight segment, a ring is pushed down and the two bent tips are thus brought together and pressed into the scalp tissue. Besides this simple and disposable electrodes (fig.1a) another one is constructed on the basis of a steel spring which presses down a surrounding tube and thus brings the two bent tips of the central steel canulas near to each other after putting them onto the fetal scalp (fig.1b). The indifferent silver clamp electrode is fixed somewhere in the tissue of the vagina.

The electronique equipment for simultaneous registration of PO_2 (gold electrode in the center of the steel canulas) and ECG (steel canulas themselves) has been described before (1).

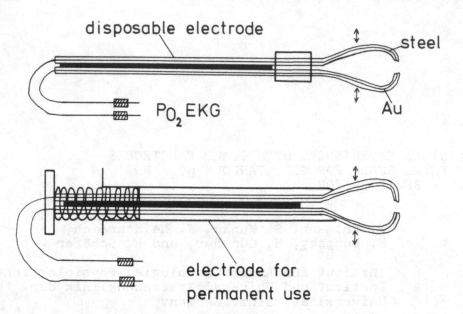

1 a/b: Clamp electrodes for simultaneous PO_2 and ECG-registration.

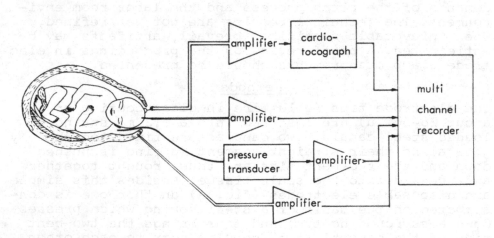

2: Schematic of fetal control during birth.

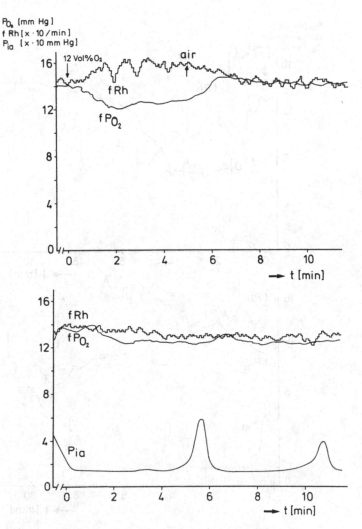

3a: Heart rate and fetal tissue-PO_2 during the hypoxic
    breathing test of the mother: Intact feto-mater-
    nal oxygen exchange

3b: Fetal heart rate and tissue-PO_2 of the fetus with
    intact oxygen supply during uterine contractions
    of the mother

3a/b: f PO_2 = fetal tissue-PO_2, f Rh = fetal heart
      rate, Pia = intraamnial pressure

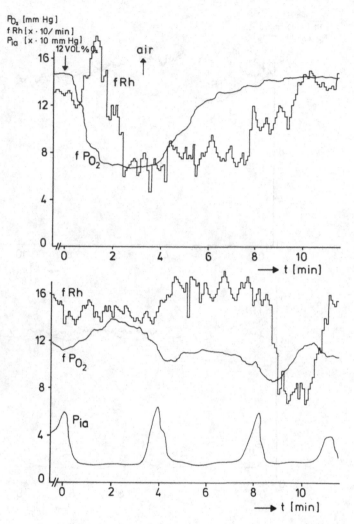

4a: Heart rate and fetal tissue-PO_2 during a hypoxic breathing test of the mother: Disturbed feto-maternal oxygen exchange

4b: Fetal heart rate and fetal tissue-PO_2 during final unterine contractions: Disturbed feto-maternal gas exchange

4a/b: f PO_2 = fetal tissue-PO_2, f Rh = fetal heart rate, Pia = intraamnial pressure

Besides ECG heart rate and fetal tissue PO_2 usually
intramnial pressure is recorded. Before terminal end
of the birth a second electrode can be introduced to
measure PO_2 in uterine tissue (fig.2).

Results

The system has its advantages in the control of the
function of feto-maternal unit before the critical terminal phase of birth is reached. A short period of application to the mother of a 12 Vol% oxygen gas mixture
reveals wether the function of the feto maternal unit
is fully intact or latent failures may produce high
risks during the following birth process.

Normally during inhalation of a 12 Vol% oxygen gas mixture a contrary compensatory effect of oxygen exchange
in the placenta will keep oxygen depression in the fetal scalp below 15% of the initial value. Heart rate
meanwhile increases just a little (fig.3a). During terminal phase of birth in these cases tissue-PO_2 in fetal scalp show relatively constant values as well as
the heart rate. Even during waves no significant
changes are to be perceived (fig.3b).

Latent failures of feto maternal oxygen exchanges are
normally not perceived in the beginning of birth. A
breathing test with 12 Vol% oxygen concentration in
the inspired air reveal this symptomatic. Fetal Tissue-
PO_2 decreases by nearly 5o% and heart rate changes to
bradycardia after a short high frequent phase (fig.4a).

During the same birth tissue-PO_2 jumped down in response to every wave. Heart rate sometimes showed severe tendency to bradycardia and sectio caesarea was
necessary before terminal phase of the birth to hinder
intranatal damage (fig.4b).

References

Kunke, S., Christ, U., Heidenreich, J. and W. Erdmann,
Simultaneous Measurement of PO_2 and ECG in the Fetal
Scalp. In: Oxygen Supply. Eds. Kessler et al.
Urban & Schwarzenberg München - Berlin - Wien (1973)

DISCUSSION OF SESSION V   SUBSESSION: OXYGEN MONITORING

Chairmen: Dr. Jose Strauss, Dr. Waldemar Moll and
           Dr. Daniel D. Reneau

DISCUSSION OF PAPER BY R. HUCH, D. W. LÜBBERS AND A. HUCH

Benzing: In your measurements you use the heat energy to maintain
a constant skin temperature of 43°C. The variation of the heat
energy depends only on the change in local skin temperature. If
you observe no changes in this temperature, you may conclude that
little or no changes in regional blood flow are occurring. However,
if there are changes in temperature, they need not necessarily
reflect changes in blood flow while this skin temperature is
affected by more than the single factor of blood flow.

Bunnell: In view of the reported success of the skin electrodes
for determining arterial pO_2 and their non-invasive character,
what advantages do the indwelling arterial electrodes offer over
these skin electrodes?

Lübbers: As we pointed out in our first paper, arterial pO_2 values
are only measured correctly if the local perfusion efficiency is
high enough. If not, then the pO_2 measured at the skin surface
depends on the blood flow. Since our measuring device allows us to
monitor the local cooling mostly brought about by the local blood
flow, we are able to measure correctly pO_2 changes of the arterial
blood even under these conditions. If blood pressure decreases
very much, then skin blood flow decreases to levels at which the
skin pO_2 does not mirror any more the arterial pO_2. Since there
are big regional differences in skin blood flow, different areas
should be tested. After our experiences, the earlobe skin gave
quick and reliable results.

GENERAL DISCUSSION

Chairmen: Dr. Dietrich W. Lübbers, Dr. Ian S. Longmuir,
Dr. Manfred Kessler and Dr. Ian A. Silver

Bicher: I have reported in our paper that we did not have much problem with clotting. Out of 18 electrodes, there was a problem with only one, and that was probably due to the cannulation, and does not have any clinical significance. The way to prevent clotting is to have some kind of blood compatible material available. Now, there is no such thing. So, if material is put in the blood stream for any length of time, it will induce clotting, it will induce platelet responses, and it will have some interaction with the arterial walls. All we can do is try to prevent this as much as we can, but I would be very leary of saying that we can leave catheters in an artery for a week or so, especially in babies. So the way that we have tried to overcome this is not to heparinize the whole animal during the whole time because it is not clinically feasible, but to improve the flow conditions locally and use the best materials available, like Silastic.

Lübbers: In this respect we have found that some catheter electrodes have reference electrodes on the outside, not on the inside. This type catheter electrode does not work on the Clark principle, so one point is that you have to have a good polarogram, but the other point is that there is a current through the membrane. Does this not facilitate any aggregation of platelets, since you have got some potential? Did you not see anything like this?

Bicher: In our experience we have a semi-pervious membrane. We did not have any problems with inducing clotting. We checked that in a number of situations. The thing that helped the most, is to have a local infusion with a Harvard pump. That infusion keeps the artery open and insures good flow. With the use of Silastic this seems to be a good combination. But I would not want to leave these electrodes <u>in situ</u> for more than a few hours, with a maximum of 36 hours. In accord with Dr. Lübbers, I could not agree with you more, and especially in that it is extremely dangerous to use the electrode interarterially when you have an external reference on the skin.

Harris: I would like to amplify one thing that Dr. Bicher said and would like to contradict another. One point that he made, of the importance of constant infusion around the catheter electrode, is obviously the issue of key importance. If blood is allowed to back up into the catheter, you are going to have a clot forming there, no matter what, unless the patient is fully heparinized, and I agree

1137

with you also that we should not use those doses of heparin. We
are defeating our own purposes. However, to say that a catheter
of this sort whould not be left in the vascular system longer than
a day, I think, is an unreasonable demand, because we found no
correlation between the length of time that the catheters were in
and the clotting. The clotting changed variably by a wide lapse
in nursing procedure, that is, letting blood back up into the
catheter. Using nonthrombogenic coverings, such as Hydron, which
is as good, if not better, than Teflon, but, of course, not better
than Silastic, is more important, and I do not think that time is
the key issue here.

Question: What about the sterilization procedure?

Harris: They come sterilized, and they are disposable. Also, they
are inserted under very careful sterile techniques, and this obviously has to be watched very closely.

Question: What about your calibration curve? Is there a change
from a calibration in vitro to a calibration in vivo?

Harris: We have, due to inconvenience, given up in vitro calibration and have used only the in vivo calibration which requires no
manipulation and has no problems with contamination. The comparison
is made between actual laboratory blood gas analysis and what the
instrument is reading and then the appropriate adjustments in the
gain of the instrument is made.

Kessler: During this meeting we heard many nice things about the
pO_2 microelectrodes but, unfortunately, nobody said anything about
the bad side of the pO_2-sensors. We should point out that all
microelectrodes show a more or less pronounced drift and that they
have a zero current. The amount of both depends on many things,
such as the method of preparing the electrodes, avoidance of
electrical noise, etc.

Bicher: Now, about autoregulation in tissue, we relate that to
oxygen autoregulation. The specific question is, do we have a
sensor in tissue or don't we? The question is open. I think that
we have enough proof now to say that when pO_2 goes down in blood
and remains constant in tissue, the flow rate into the microarea
and the action potentials in brain will not change. It is only
when pO_2 falls in tissue that we have this autoregulatory change.
In our first experiment, it was easy to obtain, as Professor Lübbers'
did, that change because the pO_2 went down and the pO_2 activated
the system. When the pO_2 failed to open the vessels, there was no
increasing flow, and the cells were simultaneously affected. I
think because of the last experiment, when we show that the change

from one side of the brain can bring a response to the other side
of the brain, you have to admit that there is some kind of reflex
mechanism involved. A lot more work is needed, but obviously the
brain has a very delicate autoregulation mechanism.

Lübbers: There are two questions: one, the regulation of the micro
blood flow in the arterioles and the precapillary sphincters; and
the other, the control mechanisms in the mesoderm.

Silver: What is the nature of the oxygen sensor?

Lübbers: The nature of the oxygen sensor is that nobody has seen
one besides the big one, which is the carotid body. We should know
something about this since this is our work. I think one cannot
say anything about this except for some special peculiarities in
the oxygen parallel membrane, which can be destroyed and which is
sensitive to certain drugs. It has an oxidase which behaves dif-
ferently from the other known compound of oxidase. This is completely different from the normal cytochrome oxidase. We could
prove that the critical oxygen tension, or the maximum velocity
of respiration, is possibly at very low oxygen tensions, for the
normal mitochondrial cytochrome oxidase, below 0.05 mm Hg, or something like this. But this special oxidase, which is found in the
carotid body, varies a lot; it starts at zero and goes up to 350
Torr, and it has a linear relationship between oxygen rate and
oxygen tension. This is something very peculiar, and you can
poison it with cyanide and so on and eliminate a section of oxidase
surrounding the tissue.

Kessler: I do not know if any sensor exists in the organ of the
tissue, but I would say that if it exists, we should find it at
the end of longer capillaries which exist in different organs.
From our response with hemorrhage, we know that after each hemorrhage, a short period of local deoxygenation is compensated for,
and therefore the pO_2 sensor must be intravascular. We must ask
how it is measured and how the system is affected by this. I
really do not know the definite answer to this.

Weiss: I would like to say something about what Dr. Bicher said.
Furthermore, I am well aware that I am a borderline case in this
group on the autoregulation of flow. I am a bit unhappy about using
the word 'autoregulation.' They have proposed the word in the literature and defined it as the ability of an isolated organ to keep
the blood flow relatively constant in spite of the variations in
blood pressure. Now, there is no evidence that such an autoregulation mechanism exists. I listened very carefully, but none of the
papers presented here convinced me that an autoregulation of oxygen
exists. I will give you an example. It is well known that in the
kidney, the oxygen consumption and flow diffusion rate are within

a certain limit intimately related. And, it is not the oxygen, or the oxygen tension, which has something to do with the oxygen uptake. It is the situation in which absorption works and is directly related to the main work that is produced. I do very much want to know whether the results, for instance that Dr. Bicher got, cannot be explained by some mechanism analogous to that which exists in the kidney. The answer, to my mind, would not be a classical case of autoregulation of oxygen.

Kessler: How would you define autoregulation of oxygen?

Weiss: We do not have the knowledge that it is the oxygen tension itself which is the regulating parameter. I would make a plea that we use the term'autoregulation,' so far, only for the autoregulation of blood, at least in our kind of society. If we have nothing which commits us to say'autoregulation of oxygen,' we will just try to define it very loosely, and we might always say 'autoregulation of flow' or 'autoregulation of oxygen.' But, so far, I think the term autoregulation' could be referred to according to Bayless' definition.

Bruley: I think that this is a very important point, the matter of terminology. To some degree, I did not realize, myself, when I first started dealing with the term 'autoregulation,' and not until recently, that it had such a restrictive definition. As an engineer, I interpreted 'autoregulation' to be any physiological control mechanism in the body, and I would refer to it that way even in my papers. This is part of the problem here; it is just a matter of definitions. So, if it is defined for oxygen control conditions, I would say that, to define 'autoregulation,' in this case for oxygen, you would have to use the oxygen tension as the critical parameter.

Lubbers: I also refer to what Dr. Weiss said, there has been a tremendous misunderstanding with other people in physiology who are used to the strict sense of the term 'autoregulation.'

Bicher: When I started using the term 'autoregulation,' I was well aware of Bayless' concepts and really, we did not care if the mechanism was parallel or not. It was necessarily Krogh that was doing the regulating, but now, I do not see anything wrong with using the term. I do not understand the problem. I sent a paper to the American Journal of Physiology, and one of the editors completely misunderstood the question and asked what oxygen has to do with autoregulation. But, perhaps, it is a matter of semantics and getting accustomed to the word. It is quite evident that the pO_2 and the pCO_2 are very well regulated and you can call it 'chemoregulation' or whatever, it does not matter what the word is.

Lübbers: Now we can proceed to the other subject we wish to discuss--oxygen transfer.

Panigel: The problem of oxygen transfer across the placenta, or, according to certain people now, oxygen transport across the placenta, should be discussed. I do not know what the American Journal of Physiology says about oxygen transfer or oxygen transport. Indeed, it is very difficult to sum up. I am not able to summarize all of the papers which have been given this morning. I see at least two of the chairmen, Dr. Reneau and Dr. Grote, who could give their opinions. I feel that pediatricians are waiting for the answer we can give, not only in physiology, but in neonatology in general.

Lübbers: Are you sure you measure the blood flow in vitro? In other words, especially in the placenta, it goes so far down that normally we assume it is from the mother. You could only register a very few times the transcutaneous pO_2 during birth. There we have seen, or we can prove, that during uterine contractions there is even a rise in pO_2 or local flow. There are so many questions that, if you have seen this once, could you always assume the blood flow decreases during labor? One assumes this from the calculation, if I understood correctly.

Moll: The uterus control during uterine contractions, expecially the flow calculation, is an untackled problem. You cannot get a good picture of what happens. There are two important elements: the inflow of blood in the interarterial sphincter and the pressure inside the uterus. These are two different problems. Usually, most of the work which has been done has been done on the exterior side of the fetus. During Caesarian sections, people have measured oxygen tension in the uterine artery, and in one of the uterine veins. The inflow is controlled by two important regulators, first the blocking of the uterine vein by uterine contractions, because when the uterus contracts, placental blood inside interarterial space is blocked by the contraction which blocks the normal patient's uterine vein. On the other hand, the inflow to the placenta is controlled, not only by the contraction of the uterine muscle, but by the smooth musculature around the uterine and placental arteries. Blood may be controlled by catecholamines.

Lübbers: Models always have a very good purpose if they give experimental physiologists a new kind of experiment. Also, if the system gets so complicated that you cannot follow the different kinds of changes and different variables, then you need a model to understand the different influences of your different variables. But, if a model is only compared to another model, it is not giving you any new information and only shows that one is a little bit better than another one. Then modeling has no advantage. It can impede physiological progress or throw it in the wrong direction.

Bruley: I am inclined to agree with Dr. Lübbers on mathematical modeling. I do think that the models are not only applicable to designing experiments; they can be used, if they are good enough models, for particular calculations.

Lübbers: Can you have a control-element model with another model? Can you get the true information of what is behind all that from another model? If I build certain equations in my model, and I see how it comes out, then I conclude that my mechanism with which my model was conceived, if correct, corresponds to the mechanism which is in the black box of the organism.

Bruley: I think that it would help your understanding of it and help you to develop the mechanism.

Lübbers: It helps you to make a competent experiment.

Bruley: It depends upon how you define the purpose of your model. In other words, let us say that we are not talking about physiology, but let us talk about chemical engineering.

Lübbers: Do you think one can just forget physiology?

Bruley: No. That is not the point. You see, one field is ahead of the other in the application of mathematical modeling. In chemical engineering twenty years ago, you would never think of doing feedforward control, for instance. Right now, it is common-place in the chemical industries, and these are mathematical models that have been designed and developed from basic considerations of the basic phenomena in nature. They have been developed mathematically all the way and are now used as predictive models that can actually be used in control. This is important in modeling.

Strauss: I would like to ask the leaders of the discussion what way you think the field should go in the area of oxygenation?

Grote: I think it is not possible to say: here is physiology, and here is modeling. I think understanding of physiology and basics of natural sciences is not possible without modeling because description of physiology is always modeling. You cannot describe anything in physiology on the basis of natural sciences without using a model.

Grunewald: In my opinion, the future development of models used for analyzing oxygen supply to tissue should make stronger allowance for the fact that various biologic parameters must be understood rather in a stochastic than a deterministic sense. The models ought to be a combination of deterministic and stochastic portions.

Kessler: As I mentioned, it is certainly always very difficult to
predict things, and mostly, it turns out that the prediction was
wrong. Nevertheless, I will try to do it. I can say, from my
experience, that we began to monitor the pO_2 in the tissue with
one electrode, and now, in the experiments we have done with Dr.
Messmer, we measured 40 values at the same time, not only pO_2
values, but a lot of other ones. This became a big job. I still
have the impression that these 40 values are not enough, and the
only problematic thing is that we cannot do these measurements one
after the other, but we have to do many of these measurements at
the same time in the same animal. I think this is a very big
problem and one of the very big problems is going to be how we can
manage experiments of that size. Maybe some sort of international
cooperation is needed, because Ian Silver has special experiences
and Dr. Longmuir and so on. We need some kind of international
collaboration in this. It is always very difficult to run experiments in several groups at the same time. I was very interested to
hear all these things about the models. My impression is the same
in this field. We have heard presentations of simplified models,
and they are totally useful, but we especially need models which
are built stone by stone through persistence. This is, again, a
problem of managing to produce such big models. At this point too,
I think we need collaboration of groups who are specialized in
different things.

Silver: I entirely agree with Dr. Kessler that mathematical
measurements are things that we think are important. On the
other hand, I think perhaps we have not heard so much about
biochemistry. After all, the end result of oxygen transport
is the production and maintenance of the cell. We have rather
lost sight of that. I think that the combination of physiological
investigation of transport and the biochemical end product is
something we must think about.

Longmuir: I think, in the past the major emphasis has been on how
the tissue gets enough oxygen, but I really think we should start
asking the question: how do the tissues protect themselves against
too much oxygen? After all, oxygen is the major air pollutant, an
excretion product of these horrid green plants!

Lübbers: I would say that we have now accomplished a little bit
towards understanding the tissue oxygen transport, and we will soon
realize that this is only one end of the story. Before coming to
an understanding of the whole story, we must take into account that
the organism as a whole has a certain oxygen-providing system, a
total system for providing the organism with enough oxygen. And
so, if we have solved the question a little as to how the transport
in the organism is brought about, the next problem will be regulation. The regulation will start in the lung; will concern the

chemoreceptors; will have a very narrow connection with the
circulation; and will have a very important end in the tissue.
We will come to the point where we can understand tissue oxygen
transport only in terms of systems analysis, and then we will
hopefully ask the modeling-field people to help us, hoping that
they will have made some progress to deal with complicated systems.
This is a rather difficult but fascinating story before us, and
I hope everybody will join with us to try to look at the many
different parts of these fascinating systems. I think that such
an international meeting gives all of us the opportunity to control
our own work, in view of other work which is done in other laboratories, and to be stimulated with new ideas and to progress further
in science.

INDEX

Pages 1-636 will be found in Volume 37A, pages 637-1144 in Volume 37B.

Printed in the United States
by Baker & Taylor Publisher Services